MW00678396

Student Solutions Manual

to accompany

College Algebra: Visualizing and Determining Solutions

Elaine Hubbard
Ronald Robinson

Laurel Technical Services

Houghton Mifflin Company   Boston   New York

Editor-in-Chief: Charles Hartford
Senior Associate Editor: Maureen Ross
Editorial Associate: Kathy Yoon
Senior Manufacturing Coordinator: Sally Culler
Marketing Manager: Michael Busnach

Copyright © 2000 by Houghton Mifflin Company. All rights reserved.

No part of this work may be reproduced or transmitted in any form or by any means, electronic or mechanical, including photocopying and recording, or by any information storage or retrieval system without the prior written permission of Houghton Mifflin Company unless such copying is expressly permitted by federal copyright law. Address inquiries to College Permissions, Houghton Mifflin Company, 222 Berkeley Street, Boston, MA 02116-3764.

Printed in the U.S.A.

ISBN: 0-395-81857-5

123456789-MA-03 02 01 00 99

Table of Contents

Preamble

1. An algebraic expression contains one or more variables, while a numerical expression does not.

3. (b) mode. The second line of the mode menu allows one to choose a floating decimal or a fixed number of decimal places.

5. (f) quit. The quit command always takes one to the home screen.

7. (a) store. Entering $9 \rightarrow X$ assigns 9 as the value of x.

9. Entering $9 - 5^2$ returns -16. (Use the x^2 key for the 2-exponent.)

11. Entering $\sqrt{(16-12)}+1$ returns 3.

13. Entering abs($4^2 - 17$) returns 1. (Use the x^2 key for the 2-exponent. abs is in the Num menu.)

15. (a) (i) Entering $\sqrt{(27)}/\sqrt{(147)}$ returns .4285714286, or about 0.43.
    (ii) Entering $\pi/3$ returns 1.047197551, or about 1.05.
    (iii) Entering $25\sqrt{(10)}$ returns 7.90569415, or about 7.91.

    (b) Only $\sqrt{27}/\sqrt{(147)}$ converts to a fraction, namely 3/7.

17. Entering 9/20–15/14+5/6 returns .2119047619, which the fraction option converts to 89/420.

19. Entering $1/(2\pi)+\pi/5$ returns .7874734738, or about 0.79.

21. Entering -21/20<-20/21 returns 1. The statement is true.

23. Entering abs(-10)<10 returns 0. The statement is false. The true statement is $|-10| = 10$.

25. −5, the value stored for x.

27. 0, because the statement is false, given that x cannot equal $x + 3$.

29. The expression would need to be entered only once, and can be evaluated for different values of x by using the table function.

31. Set $Y_1 = 3X+7$

Value of X	Value of Y_1
-4	-5
9/4	13.75
8.65	32.95
$\sqrt{(10)}$	16.4868... or about 16.49

33. Set $Y_1 = \pi X^3$

Value of X	Value of Y_1
4	201.0619... or about 201.06
3+2/9	105.1032... or about 105.10
3.7	159.1310... or about 159.13
$2\sqrt{(7)}$	465.4648... or about 465.46

35. Use Table start 0.2 and increase in steps of 0.3 using $\triangle$Table 0.3.

37. Set $Y_1 = 2X - X^2$, use Table start –0.5 and $\triangle$Table 0.5.

    a. −1.25
    b. 0
    c. 0.75
    d. 1

39. Set $Y_1 = X + abs(X)$ and enter the appropriate values in the X-column of the table.

    a. 0
    b. 0
    c. 10
    d. 18.8

© Houghton Mifflin Company. All rights reserved.

Chapter R: Basic Concepts Revisited

R.1 Exercises

1. Because the pattern does not consist of a single block of numbers repeated over and over, the number is not a rational number.

3. a. $\sqrt{64} = 8$

   b. $\sqrt{64} = 8$

   c. $-15, \sqrt{64} = 8$

   d. $\dfrac{\pi}{2}, \sqrt{10} = 3.1622\ldots$(non-repeating), $\sqrt[3]{36} = 3.3019\ldots$(non-repeating)

   e. $-15, 4.\overline{52} = \dfrac{448}{99}$,

   $-\dfrac{9}{5}, 8.2 = \dfrac{-41}{5}$,

   $\sqrt{64} = 8, \ 6\dfrac{3}{7} = \dfrac{45}{7}$

5. $\dfrac{41}{33} = 1.\overline{24}$

7. $6\dfrac{7}{8} = 6.875$

9. Because some numbers, namely the irrational numbers, are members of **R** but not of Q.

11. $0.99999\ldots = 3(0.33333\ldots) = 3\left(\dfrac{1}{3}\right) = 1$

13.

Number	$\dfrac{1319}{500}$	$2.\overline{23}$	$\dfrac{3\pi}{4}$	$\sqrt{5}$	$\dfrac{26}{11}$	$\sqrt{15} - \sqrt{2}$
Decimal name	2.638	2.2323...	2.3561...	2.2360...	2.3636...	2.4587...

In ascending order: $2.\overline{23}, \sqrt{5}, \dfrac{3\pi}{4}, \dfrac{26}{11}, \sqrt{15} - \sqrt{2}, \dfrac{1319}{500}$

15. Either $-3 - 5 = -8$ or $-3 + 5 = 2$.

17. $x - 3$ is either -2 or 2, so x is either $-2 + 3 = 1$ or $2 + 3 = 5$.

19. $|4 - (-7)| = |4 + 7| = |11| = 11$

21. $\left|\dfrac{11}{5} - 1.3\right| = |2.2 - 1.3| = |0.9| = 0.9$

23. $|\pi - 4| \approx |3.14 - 4| = |-0.86| = 0.86$

25. Shifting the number line to the left puts higher numbers on top of points P and Q. The coordinates of points P and Q would be increased by 5 but the distance between P and Q would be unchanged.

27. $|-4| - |-10| = 4 - 10 = -6$

29. $\big|-2 - |-7|\big| = |-2 - 7| = |-9| = 9$

© Houghton Mifflin Company. All rights reserved.

31. If $x > 0$, then $|x| > -x$ (because $|x| > 0$ and $-x < 0$).

33. If $x < 0$, then $-x > 0$.

35. a. If $a > 0$, $b > 0$, then $ab > 0$ and
$$|ab| = ab = a \cdot b = |a| |b|.$$

    b. If $a < 0$, $b < 0$, then $ab > 0$ and
$$|ab| = ab = (-a)(-b) = |a| |b|.$$

    c. If $a > 0$, $b < 0$, then $ab < 0$ and
$$|ab| = -ab = a(-b) = |a| |b|; \text{ if } a < 0, \ b > 0, \text{ then}$$
$$ab < 0 \text{ and } |ab| = -ab = (-a)b = |a| |b|.$$

37. Possible examples: Subtraction is not commutative because $2 - 5 \neq 5 - 2$; subtraction is not associative because $2 - (3 - 7) \neq (2 - 3) - 7$.

39. Distributive Property

41. Associative Property of Multiplication

43. Distributive Property

45. Commutative Property of Addition

47. $3x \cdot \dfrac{1}{3x} = 1$, by the Multiplicative Inverse Property.

49. $-\dfrac{5}{8} \cdot 1 = -\dfrac{5}{8}$, by the Multiplicative Identity Property.

51. $4 - x \ [= 4 + (-x)] = -x + 4$

53. $(x + 5) + 6 = x + (5 + 6)$

55. $5(3a + b) = 5(3a) + 5b = 15a + 5b$

57. $-4(2x - 5y) = (-4)(2x) - (-4)(5y) = -8x + 20y$

59. $15x + 20y = (5)(3x) + (5)(4y) = 5(3x + 4y)$

61. $3ab + a = (a)(3b) + (a)(1) = a(3b + 1)$

R.2 Exercises

1. Algebraic expressions are equivalent if they have the same value for all permissible replacements for the variable or variables in the expressions.

3. $\begin{aligned} -3^2 + 2(-1)^5 + 6 &= -9 + 2(-1) + 6 \\ &= -9 - 2 + 6 \\ &= -11 + 6 \\ &= -5 \end{aligned}$

5. $\begin{aligned} 9 - 3[2 - (5 - 1)] &= 9 - 3[2 - 4] \\ &= 9 - 3(-2) \\ &= 9 - (-6) \\ &= 15 \end{aligned}$

7. $\begin{aligned} \dfrac{-5 + \sqrt{(-5)^2 - 4(2)(-7)}}{2(2)} &= \dfrac{-5 + \sqrt{25 - 8(-7)}}{4} \\ &= \dfrac{-5 + \sqrt{25 - (-56)}}{4} \\ &= \dfrac{-5 + \sqrt{81}}{4} \\ &= \dfrac{-5 + 9}{4} \\ &= \dfrac{4}{4} \\ &= 1 \end{aligned}$

9. $\begin{aligned} 5(-3)^2 + 7(-3) - 6 &= 5(9) + (-21) - 6 \\ &= 45 - 21 - 6 \\ &= 24 - 6 \\ &= 18 \end{aligned}$

11. $\begin{aligned} \dfrac{3}{4} + \dfrac{5}{4 - 2} &= \dfrac{3}{4} + \dfrac{5}{2} \\ &= \dfrac{3}{4} + \dfrac{10}{4} \\ &= \dfrac{13}{4} \end{aligned}$

13. $\begin{aligned} \sqrt{15 - 2(-5)} &= \sqrt{15 - (-10)} \\ &= \sqrt{25} \\ &= 5 \end{aligned}$

15. $\begin{aligned} -(-3)^2 - 5(-2) - 4 &= -9 - (-10) - 4 \\ &= 1 - 4 \\ &= -3 \end{aligned}$

17. $\begin{aligned} (-3)(-2)^4 &= (-3)(16) \\ &= -48 \end{aligned}$

19. (i) $-2^4 = -16$; $(-2)^4 = 16$

    (ii) 0^0 and 0^{-3} both return error messages.

    (iii) $-2^3 = -8$; $(-2)^3 = -8$

    a. $-a^n = (-a)^n$ if n is odd

    b. 0^0 and $0^{-3} = \dfrac{1}{0^3} = \dfrac{1}{0}$ are both undefined.

21. $\begin{aligned} (-2)^3(-2)^4 &= (-2)^{3+4} \\ &= (-2)^7 \\ &= -128 \end{aligned}$

23. $-4^0 = -(4^0) = -1$

25. $6^{-2} = \dfrac{1}{6^2} = \dfrac{1}{36}$

© Houghton Mifflin Company. All rights reserved.

27. $\left(\dfrac{2}{3}\right)^{-1}\left(\dfrac{3}{2}\right)^{3} = \left(\dfrac{3}{2}\right)^{1}\left(\dfrac{3}{2}\right)^{3}$

$\qquad\qquad = \left(\dfrac{3}{2}\right)^{1+3}$

$\qquad\qquad = \left(\dfrac{3}{2}\right)^{4}$

$\qquad\qquad = \dfrac{3^4}{2^4}$

$\qquad\qquad = \dfrac{81}{16}$

29. $\dfrac{-12 \cdot 3^{-2}}{4^{-2}} = \dfrac{-12 \cdot 4^2}{3^2}$

$\qquad\qquad = \dfrac{-12 \cdot 16}{9}$

$\qquad\qquad = \dfrac{-192}{9}$

$\qquad\qquad = -\dfrac{64}{3}$

31. $(-6x^4)(3x^2) = (-6)(3)(x^4)(x^2)$

$\qquad\qquad = -18x^{4+2}$

$\qquad\qquad = -18x^6$

33. $(4y)^3 = 4^3 y^3$

$\qquad\quad = 64y^3$

35. $(-2a^6 b^3)^5 = (-2)^5 (a^6)^5 (b^3)^5$

$\qquad\qquad = -32a^{6\cdot5} b^{3\cdot5}$

$\qquad\qquad = -32a^{30} b^{15}$

37. $\dfrac{z^8}{z^{10}} = z^{8-10}$

$\qquad\quad = z^{-2}$

$\qquad\quad = \dfrac{1}{z^2}$

39. $\dfrac{-15a^6 b}{20ab^6} = \dfrac{-15}{20} a^{6-1} b^{1-6}$

$\qquad\qquad = -\dfrac{3}{4} a^5 b^{-5}$

$\qquad\qquad = -\dfrac{3a^5}{4b^5}$

41. $(3x^4 y)^2 (xy^3)^4 = (3)^2 (x^4)^2 (y)^2 (x)^4 (y^3)^4$

$\qquad\qquad = 9x^{4\cdot2} y^2 x^4 y^{3\cdot4}$

$\qquad\qquad = 9x^8 y^2 x^4 y^{12}$

$\qquad\qquad = 9x^{8+4} y^{2+12}$

$\qquad\qquad = 9x^{12} y^{14}$

43.

x	$2x^{-1} = \dfrac{2}{x}$	$\dfrac{1}{2x}$
1	2	$\dfrac{1}{2}$
2	1	$\dfrac{1}{4}$
3	$\dfrac{2}{3}$	$\dfrac{1}{6}$
4	$\dfrac{1}{2}$	$\dfrac{1}{8}$
5	$\dfrac{2}{5}$	$\dfrac{1}{10}$

For $x > 0$, $2x^{-1} > \dfrac{1}{2x}$.

45. $\dfrac{x^{-2}}{x^3} = x^{-2-3}$

$\qquad\quad = x^{-5}$

$\qquad\quad = \dfrac{1}{x^5}$

47. $\dfrac{3x^{-5}}{x^4} = 3x^{-5-4}$

$\qquad\quad = 3x^{-9}$

$\qquad\quad = \dfrac{3}{x^9}$

49. $\dfrac{1}{5y^{-6}} = \dfrac{y^6}{5}$

51. $\left(\dfrac{2b^{-2}}{c^{-3}}\right)^{-3} = \left(\dfrac{2c^3}{b^2}\right)^{-3}$

$\qquad\qquad = \left(\dfrac{b^2}{2c^3}\right)^{3}$

$\qquad\qquad = \dfrac{(b^2)^3}{(2c^3)^3}$

$\qquad\qquad = \dfrac{b^{2\cdot3}}{2^3 (c^3)^3}$

$\qquad\qquad = \dfrac{b^6}{8c^{3\cdot3}}$

$\qquad\qquad = \dfrac{b^6}{8c^9}$

53. $\dfrac{8x^{-5} y^{-2}}{24x^3 y^{-4}} = \dfrac{8y^4}{24x^3 x^5 y^2}$

$\qquad\qquad = \dfrac{8y^{4-2}}{24x^{3+5}}$

$\qquad\qquad = \dfrac{8y^2}{24x^8}$

$\qquad\qquad = \dfrac{y^2}{3x^8}$

© Houghton Mifflin Company. All rights reserved.

54. $\dfrac{x^{-2}y^3}{x^{-7}y^{-2}} = \dfrac{x^7 y^3 y^2}{x^2}$

$\qquad\qquad = x^{7-2}y^{3+2}$

$\qquad\qquad = x^5 y^5$

55. $\left(\dfrac{3x^2}{y^{-1}}\right)^{-2}\left(\dfrac{y}{x}\right)^{-3} = \left(\dfrac{y^{-1}}{3x^2}\right)^2\left(\dfrac{x}{y}\right)^3$

$\qquad\qquad = \left(\dfrac{1}{3x^2 y}\right)^2\left(\dfrac{x}{y}\right)^3$

$\qquad\qquad = \dfrac{1}{(3x^2 y)^2}\cdot\dfrac{x^3}{y^3}$

$\qquad\qquad = \dfrac{1}{3^2 (x^2)^2 y^2}\cdot\dfrac{x^3}{y^3}$

$\qquad\qquad = \dfrac{1}{9x^4 y^2}\cdot\dfrac{x^3}{y^3}$

$\qquad\qquad = \dfrac{x^3}{9x^4 y^{2+3}}$

$\qquad\qquad = \dfrac{1}{9xy^5}$

57.   **a.** $3000\left(1+\dfrac{0.09}{12}\right)^8 = 3184.796\ldots$

       The interest is $3184.80 - $3000 = $184.80.

     b. $3000\left(1+\dfrac{0.09}{12}\right)^{15} = 3355.807\ldots$

       The interest is $3355.81 - $3000 = $355.81.

59. $15{,}000(1.047)^{2004-1998} = 19{,}759.290\ldots$

    The predicted cost is $19,759.29.

61. $3\cdot10^7 = 30{,}000{,}000$

63. $4.36\cdot10^3 = 4360$

65. $3.34\cdot10^{-6} = 0.00000334$

67. $4.2\cdot10^{-4} = 0.00042$

69. $7{,}030{,}000 = 7.03\cdot10^6$

71. $532{,}000 = 5.32\cdot10^5$

73. $0.000437 = 4.37\cdot10^{-4}$

75. $0.002 = 2\cdot10^{-3}$

77.   Entering 4.6E5*3.1E-2 returns 14260 or $1.426\cdot10^4$.

79.   Entering 4.6E7^5 returns 2.05962976E38 or about $2.06\cdot10^{38}$.

81.   Entering 4.2E-6 / 7E-2 returns 6E-5 or $6\cdot10^{-5}$.

83.   (36 dollars / person / week) $\times$ ($267\cdot10^6$ persons) $\times$ (52 weeks / year)

    $\approx 5.00\times10^{11}$ dollars / year

85. $\dfrac{8.3\cdot10^{10}\ \text{dollars}}{3.6\cdot10^6\ \text{workers}} \approx 23{,}055.56$ dollars / worker

87.   **a.**   Kilo means thousand and mega means million.

     b. $6180\cdot10^6$ watts $= 6.180\cdot10^9$ watts

     c. $(6180\cdot10^6\ \text{watts})\left(\dfrac{0.3\ \text{pound}}{1\cdot10^3\ \text{watts}}\right)$

       $\times\left(\dfrac{1\ \text{ton}}{2000\ \text{pounds}}\right)$

       $= 927$ tons of coal

R.3 Exercises

1.   (i) True. All monomials are also polynomials.

    (ii) False. For example,

    $(4-x^2)+(x^2-x)=4-x$, where the two polynomials on the left are of degree 2 but the sum on the right is of degree 1.

    (iii) False. The middle term equals zero, so this is a binomial.

3.   Yes

5.   No; a polynomial cannot have terms with negative exponents on variables.

7.   Yes

9.   5, which is the greatest exponent on a variable.

11.   0, because $-5^4 = -625x^0$.

13.   5, because that is the greatest sum of the variable exponents in any one term—in this case, the middle term.

15. $(1-3x)+(x^2+2x-4)=1-3x+x^2+2x-4$

$\qquad\qquad = x^2 +(-3x+2x)+(1-4)$

$\qquad\qquad = x^2 - x - 3$

© Houghton Mifflin Company. All rights reserved.

17. $(3x+2y-6)-(x+3y-3) = 3x+2y-6-x-3y+3$
$$= (3x-x)+(2y-3y)+(-6+3)$$
$$= 2x-y-3$$

19. $(4x^3-2x-x^2+3)-(5-2x)+(x^2+2x^3) = 4x^3-2x-x^2+3-5+2x+x^2+2x^3$
$$= (4x^3+2x^3)+(-x^2+x^2)+(-2x+2x)+(3-5)$$
$$= 6x^3-2$$

21. $(2b-3c+7)-(3c-4a)+(a-b-6) = 2b-3c+7-3c+4a+a-b-6$
$$= (4a+a)+(2b-b)+(-3c-3c)+(7-6)$$
$$= 5a+b-6c+1$$

23. (i) False; $(x+y)^2 = (x+y)(x+y) = x^2+2xy+y^2 \neq x^2+y^2$

(ii) True

(iii) False. For example, $(x+2)(x-2) = x^2-4$, which is a binomial, not a trinomial.

25. $-2xy^3(3y^2-5y+6) = -2xy^3(3y^2)-2xy^3(-5y)-2xy^3(6)$
$$= -6xy^5+10xy^4-12xy^3$$

27. $(a-3)(a^2+3a-5) = a(a^2+3a-5)-3(a^2+3a-5)$
$$= a^3+3a^2-5a-3a^2-9a+15$$
$$= a^3-14a+15$$

29. $(2x+1)(x^2-x-4) = 2x(x^2-x-4)+(x^2-x-4)$
$$= 2x^3-2x^2-8x+x^2-x-4$$
$$= 2x^3-x^2-9x-4$$

31. $(x+3y-2)(4x+y-1) = x(4x+y-1)+3y(4x+y-1)-2(4x+y-1)$
$$= 4x^2+xy-x+12xy+3y^2-3y-8x-2y+2$$
$$= 4x^2+13xy-9x+3y^2-5y+2$$

33. $(a+b)^2 = a^2+2ab+b^2$,
so $a^2+b^2 = (a+b)^2-2(ab)$
$$= 7-2(-5)$$
$$= 17$$

35. $(y+7)(y-5) = y^2-5y+7y-35$
$$= y^2+2y-35$$

37. $(3x+2)(2x+3) = 6x^2+9x+4x+6$
$$= 6x^2+13x+6$$

39. $(y-3x)(2y+5x) = 2y^2+5xy-6xy-15x^2$
$$= 2y^2-xy-15x^2$$

41. $(5-3z)(2+3z) = 10+15z-6z-9z^2$
$$= 10+9z-9z^2$$

43.

x	$(x-3)^2$	x^2+9	x^2-9	x^2-6x+9
1	4	10	-8	4

The answer is (iii).

© Houghton Mifflin Company. All rights reserved.

45. Product of a Sum and Difference of Two Terms

$$(7x-2)(7x+2) = (7x)^2 - (2)^2$$
$$= 49x^2 - 4$$

47. Product of a Sum and Difference of Two Terms

$$(y^2+4z)(y^2-4z) = (y^2)^2 - (4z)^2$$
$$= y^4 - 16z^2$$

49. Square of a Binomial

$$(5-y)^2 = (5)^2 - 2(5)(y) + y^2$$
$$= 25 - 10y + y^2$$

51. Square of a Binomial

$$(a+3b)^2 = (a)^2 + 2(a)(3b) + (3b)^2$$
$$= a^2 + 6ab + 9b^2$$

53. Cube of a Binomial

$$(y-2)^3 = (y)^3 - 3(y)^2(2) + 3(y)(2)^2 - (2)^3$$
$$= y^3 - 6y^2 + 12y - 8$$

55. Cube of a Binomial

$$(2y+3z)^3 = (2y)^3 + 3(2y)^2(3z) + 3(2y)(3z)^2 + (3z)^3$$
$$= 8y^3 + 36y^2 + 54yz^2 + 27z^3$$

57. $3x^2(2x-1)(3x+2) = [(3x^2)(2x-1)](3x+2)$
$$= (6x^3 - 3x^2)(3x+2)$$
$$= (6x^3)(3x) + (6x^3)(2) - (3x^2)(3x) - (3x^2)(2)$$
$$= 18x^4 + 12x^3 - 9x^3 - 6x^2$$
$$= 18x^4 + 3x^3 - 6x^2$$

59. $(x+3)(x-2)(x-1) = (x^2 - 2x + 3x - 6)(x-1)$
$$= (x^2 + x - 6)(x-1)$$
$$= x(x^2 + x - 6) - (x^2 + x - 6)$$
$$= x^3 + x^2 - 6x - x^2 - x + 6$$
$$= x^3 - 7x + 6$$

61. $(x+2)(2x-1) - (x-1)(x+3) = [(2x)(x) + (x)(-1) + (2)(2x) + (2)(-1)] - (x^2 + 3x - x - 3)$
$$= (2x^2 - x + 4x - 2) - (x^2 + 2x - 3)$$
$$= 2x^2 + 3x - 2 - x^2 - 2x + 3$$
$$= x^2 + x + 1$$

63. $[(2y+4) - (3-y)] + y(y-4) = (2y + 4 - 3 + y) + y^2 - 4y$
$$= (3y+1) + y^2 - 4y$$
$$= 3y + 1 + y^2 - 4y$$
$$= y^2 - y + 1$$

65. Use $(A+B)(A-B)$ and $(A-B)^2$.

$$(3x+1)(3x-1) - (3x-1)^2 = (3x)^2 - (1)^2 - [(3x)^2 - 2(3x)(1) + (1)^2]$$
$$= 9x^2 - 1 - (9x^2 - 6x + 1)$$
$$= 9x^2 - 1 - 9x^2 + 6x - 1$$
$$= 6x - 2$$

© Houghton Mifflin Company. All rights reserved.

67. Use $(A+B)(A-B)$ twice.

$$(x+6)(x-6)-(x+1)(x-1) = [(x)^2 - (6)^2] - [(x)^2 - (1)^2]$$
$$= (x^2 - 36) - (x^2 - 1)$$
$$= x^2 - 36 - x^2 + 1$$
$$= -35$$

69. First way: $(t+3)(3-t)(t+3) = [(t+3)(t+3)](3-t)$

$$= (t^2 + 6t + 9)(3-t)$$
$$= 3t^2 + 18t + 27 - t^3 - 6t^2 - 9t$$
$$= 27 + 9t - 3t^2 - t^3$$

Second way: $(t+3)(3-t)(t+3) = [(3+t)(3-t)](t+3)$

$$= (9 - t^2)(t+3)$$
$$= 9t - t^3 + 27 - 3t^2$$
$$= 27 + 9t - 3t^2 - t^3$$

71. $(2x+y+5)(2x+y-5) = (2x+y)^2 - (5)^2$

$$= [(2x)^2 + 2(2x)(y) + (y)^2] - 25$$
$$= 4x^2 + 4xy + y^2 - 25$$

73. $(x+2)(x-2)(x^2+4) = [(x)^2 - (2)^2](x^2+4)$

$$= (x^2 - 4)(x^2 + 4)$$
$$= (x^2)^2 - (4)^2$$
$$= x^4 - 16$$

75. $(a+b+3)^2 = (a+b)^2 + 2(a+b)(3) + (3)^2$

$$= (a^2 + 2ab + b^2) + (6a + 6b) + 9$$
$$= a^2 + 2ab + b^2 + 6a + 6b + 9$$

R.4 Exercises

1. The smallest exponent on a is 3 and the smallest exponent on b is 2. So the GCF is $a^3 b^2$.

3. $6a + 9 = 3(2a) + 3(3) = 3(2a+3)$

5. $y^3 - y^2 = y^2(y) - y^2 = y^2(y-1)$

7. $3x(x-2) + y(x-2) = (x-2)(3x+y)$

9. $36x^2 - 1 = (6x)^2 - 1^2 = (6x+1)(6x-1)$

11. $49 - 25c^2 = 7^2 - (5c)^2 = (7+5c)(7-5c)$

13. $x^2 - 6x + 9 = x^2 - 2(x)(3) + 3^2$
$$= (x^2 - 3)^2$$

15. $25 + 30z + 9z^2 = 5^2 + 2(5)(3z) + (3z)^2$
$$= (5+3z)^2$$

17. Produce a table of values for $36x^2 - 1$ and its factored form, $(6x+1)(6x-1)$. The tables should be the same.

19. $64 - w^3 = 4^3 - w^3$
$$= (4-w)(4^2 + 4w + w^2)$$
$$= (4-w)(16 + 4w + w^2)$$

21. $27x^3 + 8 = (3x)^3 + 2^3$
$$= (3x+2)[(3x)^2 - (3x)(2) + 2^2]$$
$$= (3x+2)(9x^2 - 6x + 4)$$

© Houghton Mifflin Company. All rights reserved.

23. (i) $-(x-3)(x+4) = -(x^2+4x-3x-12)$

$$= -x^2 - x + 12$$
$$= 12 - x - x^2$$

(ii) $(3-x)(4+x) = 12 + 3x - 4x - x^2$
$$= 12 - x - x^2$$

Both are correct.

25. $y^2 - y - 6$. Seek m and n such that $mn = -6$ and $m + n = -1$.

Feasible factorizations	Middle term	
$(y+1)(y-6)$	$-5y$	
$(y-1)(y+6)$	$5y$	
$(y+2)(y-3)$	$-y$	Correct middle term

$\vdots$

$$y^2 - y - 6 = (y+2)(y-3)$$

27. $12y^2 + 8y - 15$

Feasible factorizations	Middle term	
$(12y+1)(y-15)$	$179y$	
$(12y-1)(y+15)$	$179y$	
$(12y+5)(y-3)$	$-31y$	
$(12y-5)(y+3)$	$31y$	
$(6y+1)(2y-15)$	$-88y$	
$(6y-1)(2y+15)$	$88y$	
$(6y+5)(2y-3)$	$-8y$	
$(6y-5)(2y+3)$	$8y$	Correct middle term

$\vdots$

$$12y^2 + 8y - 15 = (6y-5)(2y+3)$$

29. $4 + 3x - x^2 = -(x^2 - 3x - 4)$. Seek m and n such that $mn = -4$ and $m + n = -3$.

Feasible factorizations	Middle term	
$(x-1)(x+4)$	$3x$	
$(x+1)(x-4)$	$-3x$	Correct middle term

$\vdots$

$$4 + 3x - x^2 = -1(x+1)(x-4) = (4-x)(1+x)$$

31. $8x^2 - 11x + 3$

Feasible factorizations	Middle term	
$(8x-1)(x-3)$	$-25x$	
$(8x-3)(x-1)$	$-11x$	Correct middle term

$\vdots$

$$8x^2 - 11x + 3 = (8x-3)(x-1)$$

33. $x^4 + 5x^2 + 6 = (x^2)^2 + 5x^2 + 6$. Seek m and n such that $mn = 6$ and $m + n = 5$.

Feasible factorizations	Middle term	
$(x^2+1)(x^2+6)$	$7x^2$	
$(x^2+2)(x^2+3)$	$5x^2$	Correct middle term

$\vdots$

$$x^4 + 5x^2 + 6 = (x^2+2)(x^2+3)$$

© Houghton Mifflin Company. All rights reserved.

35. (i) False. If 16 is written as 4^2, then the middle term needs to be $2(x)(4) = 8x$, not $4x$.

(ii) False. $x^2 - 4$ can be factored into $(x + 2)(x - 2)$.

(iii)True. $x^6 - 64 = (x^3)^2 - 8^2 = (x^2)^3 - 4^3$.

37. $ax + 2bx - ay - 2by = (ax + 2bx) - (ay + 2by)$
$$= x(a + 2b) - y(a + 2b)$$
$$= (a + 2b)(x - y)$$

39. $x^3 - 5x^2 + 2x - 10 = (x^3 - 5x^2) + (2x - 10)$
$$= x^2(x - 5) + 2(x - 5)$$
$$= (x^2 + 2)(x - 5)$$

41. $b^2 + 13b + 42$. Seek m and n such that $mn = 42$ and $m + n = 13$. Use $m = 7$, $n = 6$. $b^2 + 13b + 42 = (b + 7)(b + 6)$.

43. $121m^2 - 4n^2 = (11m)^2 - (2n)^2$
$$= (11m + 2n)(11m - 2n)$$

45. $5x^3 + 25x^2 + x + 5 = (5x^3 + 25x^2) + (x + 5)$
$$= 5x^2(x + 5) + (x + 5)$$
$$= (5x^2 + 1)(x + 5)$$

47. $40y^{40} - 30y^{30} = 10y^{30}(4y^{10}) - 10y^{30}(3)$
$$= 10y^{30}(4y^{10} - 3)$$

49. $20a^4 + a^2b - 12b^2 = 20(a^2)^2 + a^2b - 12b^2$

Feasible factorizations	Middle term	
$(20a^2 + b)(a^2 - 12b)$	$-239a^2b$	
$(5a^2 + b)(4a^2 - 12b)$	$-56a^2b$	
$(5a^2 + 4b)(4a^2 - 3b)$	a^2b	Correct middle term

$\vdots$

$$20a^4 + a^2b - 12b^2 = (5a^2 + 4b)(4a^2 - 3b)$$

51. $4x^2 - 29xy + 42y^2$

Feasible factorizations	Middle term	
$(4x - 6y)(x - 7y)$	$-34xy$	
$(4x - 7y)(x - 6y)$	$-31xy$	
$(4x - 14y)(x - 3y)$	$-26xy$	
$(4x - 21y)(x - 2y)$	$-29xy$	Correct middle term

$\vdots$

$$4x^2 - 29xy + 42y^2 = (4x - 21y)(x - 2y)$$

53. $81 - 36y + 4y^2 = 9^2 - 2(9)(2y) + (2y)^2 = (9 - 2y)^2$

55. $4x^4 + 4x = 4x(x^3 + 1) = 4x(x + 1)(x^2 - x + 1)$

57. $8x - 2x^2 - 3x^3 = -x(3x^2 + 2x - 8)$

Feasible factorizations	Middle term	
$(3x + 2)(x - 4)$	$-10x$	
$(3x + 4)(x - 2)$	$-2x$	
$(3x - 4)(x + 2)$	$2x$	Correct middle term

$\vdots$

$$8x - 2x^2 - 3x^3 = -x(3x - 4)(x + 2)$$

© Houghton Mifflin Company. All rights reserved.

59. $81a^4 - 16 = (9a^2)^2 - 4^2$
$$= (9a^2 + 4)(9a^2 - 4)$$
$$= (9a^2 + 4)[(3a)^2 - 2^2]$$
$$= (9a^2 + 4)(3a + 2)(3a - 2)$$

61. $x^2 - 7xy - 30y^2$. Seek m and n such that
$mn = -30$ and $m + n = -7$. Use $m = -10, n = 3$.
$$x^2 - 7xy - 30y^2 = (x - 10y)(x + 3y).$$

63. $1 + 8x^6 = 1^3 + (2x^2)^3$
$$= (1 + 2x^2)[1^2 - 2x^2 + (2x^2)^2]$$
$$= (1 + 2x^2)(1 - 2x^2 + 4x^4)$$

73. $2t^6 + 11t^3 + 12 = 2(t^3)^2 + 11t^3 + 12$

Feasible factorizations	Middle term
$(2t^3 + 1)(t^3 + 12)$	$25t^3$
$(2t^3 + 2)(t^3 + 6)$	$14t^3$
$(2t^3 + 3)(t^3 + 4)$	$11t^3$ Correct middle term

$\vdots$

$$2t^6 + 11t^3 + 12 = (2t^3 + 3)(t^3 + 4)$$

75. $-3x^3 + 18x^2 y - 27xy^2 = -3x(x^2 - 6xy + 9y^2)$
$$= -3x[x^2 - 2(x)(3y)$$
$$+ (3y)^2]$$
$$= -3x(x - 3y)^2$$

77. $98 - 8y^2 = 2(49 - 4y^2)$
$$= 2[7^2 - (2y)^2]$$
$$= 2(7 + 2y)(7 - 2y)$$

79. $2x^5 y^2 + 5x^4 y^3 - 3x^3 y^4 = x^3 y^2(2x^2 + 5xy - 3y^2)$

Feasible factorizations	Middle term
$(2x + y)(x - 3y)$	$-5xy$
$(2x - y)(x + 3y)$	$5xy$

$\vdots$

$$2x^5 y^2 + 5x^4 y^3 - 3x^3 y^4 = x^3 y^2(2x - y)(x + 3y)$$

81. $16y^2 - 54y^5 = 2y^2(8 - 27y^3)$
$$= 2y^2[2^3 - (3y)^3]$$
$$= 2y^2(2 - 3y)[2^2 + 2(3y) + (3y)^2]$$
$$= 2y^2(2 - 3y)(4 + 6y + 9y^2)$$

83. $9(4 - x^2)^2 + 24(4 - x^2) + 16 = \left[3(4 - x^2)\right]^2 + 2[3(4 - x^2)](4) + 4^2$
$$= \left[3(4 - x^2) + 4\right]^2$$
$$= (12 - 3x^2 + 4)^2$$
$$= (16 - 3x^2)^2$$

85. $(2y + z)^2 - 100 = (2y + z)^2 - 100^2$
$$= [(2y + z) + 10][(2y + z) - 10]$$
$$= (2y + z + 10)(2y + z - 10)$$

65. $4a^2 + 28a + 49 = (2a)^2 + 2(2a)(7) + 7^2$
$$= (2a + 7)^2$$

67. Cannot be factored; prime

69. $x^3 + 3x^2 - 4x - 12 = (x^3 + 3x^2) - (4x + 12)$
$$= x^2(x + 3) - 4(x + 3)$$
$$= (x^2 - 4)(x + 3)$$
$$= (x + 2)(x - 2)(x + 3)$$

71. $y(x - 3) - 4z(3 - x) = y(x - 3) + 4z(x - 3)$
$$= (x - 3)(y + 4z)$$

© Houghton Mifflin Company. All rights reserved.

87. $(3y-2)(y+2)+(4-5y)(2-3y) = (3y-2)(y+2)-(4-5y)(3y-2)$
$= (3y-2)[(y+2-(4-5y)]$
$= (3y-2)(6y-2)$
$= 2(3y-2)(3y-1)$

89. $(x+2)^5-(x+2)^3 = (x+2)^3[(x+2)^2-1]$
$= (x+2)^3[(x+2)^2-1^2]$
$= (x+2)^3[(x+2)+1)][(x+2)-1]$
$= (x+2)^3(x+3)(x+1)$

91. $x^2+6x+9-y^2 = (x+3)^2-y^2$
$= (x+3+y)(x+3-y)$

93. Given that $A^3+B^3 = (A+B)(A^2-AB+B^2)$, substitute $-B$ in for B. Then
$A^3+(-B)^3 = [A+(-B)][A^2-A(-B)+(-B)^2]$, or $A^3-B^3 = (A-B)(A^2+AB+B^2)$.

95. a. $x^2-10 = x^2-(\sqrt{10})^2 = (x+\sqrt{10})(x-\sqrt{10})$

b. $15-y^2 = (\sqrt{15})^2-y^2 = (\sqrt{15}+y)(\sqrt{15}-y)$

c. $6x^2-1 = (x\sqrt{6})^2-1^2 = (x\sqrt{6}+1)(x\sqrt{6}-1)$

R.5 Exercises

1. The tables are the same except for $x=-1$. The table for $x+1$ indicates that the expression has a value of 2 when x is 1. Because $\frac{x^2-1}{x-1}$ is not defined for $x=1$, the table displays an error message when $x=1$.

3. The denominator equals zero for $x=-5$ and $x=4$.

5. The denominator is $x^2+3x-10 = (x+5)(x-2)$, which equals zero for $x=-5$ and $x=2$.

7. The denominator is
$y^3+3y^2-y-3 = (y^2-1)(y+3)$
$= (y+1)(y-1)(y+3)$,
which equals zero for $y=-3$, $y=-1$, and $y=1$.

9. $\frac{4c-8}{c^2-4c+4} = \frac{4(c-2)}{(c-2)^2} = \frac{4}{c-2}$

11. $\frac{x^2-2x-15}{5-x} = \frac{(x-5)(x+3)}{-(x-5)} = \frac{x+3}{-1} = -(x+3)$

13. $\frac{x^3-8}{x^2+2x-8} = \frac{(x-2)(x^2+2x+4)}{(x-2)(x+4)} = \frac{x^2+2x+4}{x+4}$

15. $\frac{m^3-2m^2-8m}{2m^3-m^2-10m} = \frac{m(m+2)(m-4)}{m(2m-5)(m+2)} = \frac{m-4}{2m-5}$

17. $\frac{x^2-4}{x(x+5)+6} = \frac{x^2-4}{x^2+5x+6}$
$= \frac{(x+2)(x-2)}{(x+2)(x+3)}$
$= \frac{x-2}{x+3}$

19. $\frac{a^2-1}{4a-2} \cdot \frac{12}{a^2-a} = \frac{(a+1)(a-1)}{2(2a-1)} \cdot \frac{12}{a(a-1)}$
$= \frac{6(a+1)}{a(2a-1)}$

© Houghton Mifflin Company. All rights reserved.

21. $\dfrac{x^2-x-6}{x^2-9}\cdot\dfrac{x^2+x-6}{x^2+5x+6}=\dfrac{(x-3)(x+2)}{(x+3)(x-3)}\cdot\dfrac{(x+3)(x-2)}{(x+3)(x+2)}$

$\qquad\qquad\qquad\qquad\qquad\qquad =\dfrac{x-2}{x-3}$

23. $\dfrac{x^2+4x}{x+4}\div(x^3-16x)=\dfrac{x(x+4)}{x+4}\cdot\dfrac{1}{x(x+4)(x-4)}=\dfrac{1}{x^2-16}$

25. $\dfrac{4x-8}{-3x}\div\dfrac{12-6x}{15}=\dfrac{4(x-2)}{-3x}\cdot\dfrac{15}{6(2-x)}=\dfrac{10}{3}$

27. $\dfrac{x^2-x-6}{x^2+x-2}\cdot\dfrac{x^2+3x-4}{x^2-2x-3}=\dfrac{(x-3)(x+2)}{(x+2)(x-1)}\cdot\dfrac{(x+4)(x-1)}{(x-3)(x+1)}=\dfrac{x+4}{x+1}$

29. $\dfrac{a^4-1}{a^2-1}\div\dfrac{a^3+a^2+a+1}{4}=\dfrac{(a^2+1)(a^2-1)}{(a+1)(a-1)}\cdot\dfrac{4}{(a^2+1)(a+1)}$

$\qquad\qquad\qquad\qquad\qquad =\dfrac{(a^2+1)(a+1)(a-1)}{(a+1)(a-1)}\cdot\dfrac{4}{(a^2+1)(a+1)}$

$\qquad\qquad\qquad\qquad\qquad =\dfrac{4}{a+1}$

31. $\dfrac{2x-6}{5}\cdot\dfrac{x^2-1}{x^2+2x-3}\div\dfrac{x^2-2x-3}{x+3}=\dfrac{2(x-3)}{5}\cdot\dfrac{(x+1)(x-1)}{(x+3)(x-1)}\cdot\dfrac{x+3}{(x-3)(x+1)}$

$\qquad\qquad\qquad\qquad\qquad\qquad\qquad =\dfrac{2}{5}$

33. $\dfrac{x^2-9}{x+1}\cdot(x^2-1)\div(x-3)=\dfrac{(x+3)(x-3)}{x+1}\cdot\dfrac{x^2-1}{x-3}$

$\qquad\qquad\qquad\qquad\qquad\quad =\dfrac{(x+3)(x-3)}{x+1}\cdot\dfrac{(x+1)(x-1)}{x-3}$

$\qquad\qquad\qquad\qquad\qquad\quad =(x+3)(x-1)$

35. The tables would not be the same because the expressions are not equivalent. The student made a sign error.

$\dfrac{2}{x+1}-\dfrac{x+3}{x+1}=\dfrac{2-(x+3)}{x+1}$

$\qquad\qquad\qquad =\dfrac{2-x-3}{x+1}$

$\qquad\qquad\qquad =\dfrac{-x-1}{x+1}$

$\qquad\qquad\qquad =-1$

37. $\dfrac{2x+y}{x-y}+\dfrac{x-2y}{x-y}=\dfrac{2x+y+x-2y}{x-y}=\dfrac{3x-y}{x-y}$

39. $\dfrac{4x+1}{x}-\dfrac{3x-1}{x}=\dfrac{4x+1-(3x-1)}{x}=\dfrac{x+2}{x}$

41. $\dfrac{2x+9}{x-4}+\dfrac{4x-33}{x-4}=\dfrac{2x+9+4x-33}{x-4}$

$\qquad\qquad\qquad\qquad =\dfrac{6x-24}{x-4}$

$\qquad\qquad\qquad\qquad =\dfrac{6(x-4)}{x-4}$

$\qquad\qquad\qquad\qquad =6$

43. $\dfrac{n^2-4n}{n^2-n-6}-\dfrac{n-6}{n^2-n-6}=\dfrac{n^2-4n-(n-6)}{n^2-n-6}$

$\qquad\qquad\qquad\qquad\qquad\quad =\dfrac{n^2-5n+6}{n^2-n-6}$

$\qquad\qquad\qquad\qquad\qquad\quad =\dfrac{(n-2)(n-3)}{(n+2)(n-3)}$

$\qquad\qquad\qquad\qquad\qquad\quad =\dfrac{n-2}{n+2}$

45. $x^2-9=(x+3)(x-3)$ and
$x^2-x-6=(x+2)(x-3)$,
so LCM $=(x+3)(x-3)(x+2)$.

47. $x^3-1=(x-1)(x^2+x+1)$ and
$x^2-1=(x+1)(x-1)$,
so LCM $=(x+1)(x-1)(x^2+x+1)$.

49. $(x+1)^3=(x+1)^3$,
$x^2+2x+1=(x+1)^2$, and
$x^2-1=(x+1)(x-1)$.
So LCM $=(x+1)^3(x-1)$.

51. $\dfrac{3x}{x-5}+\dfrac{2}{5-x}=\dfrac{3x}{x-5}+\dfrac{-2}{x-5}=\dfrac{3x-2}{x-5}$

© Houghton Mifflin Company. All rights reserved.

53. $\dfrac{1}{a-b} - \dfrac{1-b}{b-a} = \dfrac{1}{a-b} - \dfrac{b-1}{a-b} = \dfrac{1-(b-1)}{a-b} = \dfrac{2-b}{a-b}$

55. $\dfrac{x}{x+3} + \dfrac{2}{x+5} = \dfrac{x(x+5)}{(x+3)(x+5)} + \dfrac{2(x+3)}{(x+3)(x+5)}$

$= \dfrac{x^2+5x}{(x+3)(x+5)} + \dfrac{2x+6}{(x+3)(x+5)}$

$= \dfrac{x^2+5x+2x+6}{(x+3)(x+5)}$

$= \dfrac{x^2+7x+6}{(x+3)(x+5)}$

57. $\dfrac{m+n}{m-n} - \dfrac{m-n}{m+n} = \dfrac{(m+n)(m+n)}{(m-n)(m+n)} - \dfrac{(m-n)(m-n)}{(m-n)(m+n)}$

$= \dfrac{m^2+2mn+n^2}{(m-n)(m+n)} - \dfrac{m^2-2mn+n^2}{(m-n)(m+n)}$

$= \dfrac{4mn}{(m-n)(m+n)}$

59. $\dfrac{x}{x-2} + \dfrac{x-16}{x^2+3x-10} = \dfrac{x}{x-2} + \dfrac{x-16}{(x-2)(x+5)}$

$= \dfrac{x(x+5)}{(x-2)(x+5)} + \dfrac{x-16}{(x-2)(x+5)}$

$= \dfrac{x^2+5x+x-16}{(x-2)(x+5)}$

$= \dfrac{x^2+6x-16}{(x-2)(x+5)}$

$= \dfrac{(x+8)(x-2)}{(x-2)(x+5)}$

$= \dfrac{x+8}{x+5}$

61. $\dfrac{3x}{x^2-7x+10} - \dfrac{2x}{x^2-8x+15} = \dfrac{3x}{(x-2)(x-5)} - \dfrac{2x}{(x-3)(x-5)}$

$= \dfrac{3x(x-3)}{(x-2)(x-3)(x-5)} - \dfrac{2x(x-2)}{(x-2)(x-3)(x-5)}$

$= \dfrac{3x^2-9x-(2x^2-4x)}{(x-2)(x-3)(x-5)}$

$= \dfrac{x^2-5x}{(x-2)(x-3)(x-5)}$

$= \dfrac{x(x-5)}{(x-2)(x-3)(x-5)}$

$= \dfrac{x}{(x-2)(x-3)}$

63. $\dfrac{a^2+2}{a^2-a-2} + \dfrac{1}{a+1} - \dfrac{a}{a-2} = \dfrac{a^2+2}{(a-2)(a+1)} + \dfrac{1}{a+1} - \dfrac{a}{a-2}$

$= \dfrac{a^2+2}{(a-2)(a+1)} + \dfrac{a-2}{(a-2)(a+1)} - \dfrac{a(a+1)}{(a-2)(a+1)}$

$= \dfrac{a^2+2+a-2-(a^2+a)}{(a-2)(a+1)}$

$= \dfrac{0}{(a-2)(a+1)}$

$= 0$

© Houghton Mifflin Company. All rights reserved.

65. $\dfrac{5}{2-n} - \dfrac{1}{2+n} + \dfrac{2n}{n^2-4} = \dfrac{-5}{n-2} - \dfrac{1}{n+2} + \dfrac{2n}{(n+2)(n-2)}$

$$= \dfrac{-5(n+2)}{(n-2)(n+2)} - \dfrac{n-2}{(n-2)(n+2)} + \dfrac{2n}{(n+2)(n-2)}$$

$$= \dfrac{-5n-10-(n-2)+2n}{(n+2)(n-2)}$$

$$= \dfrac{-4n-8}{(n+2)(n-2)}$$

$$= \dfrac{-4(n+2)}{(n+2)(n-2)}$$

$$= \dfrac{-4}{n-2}$$

67. $\dfrac{\frac{x-5}{y}}{\frac{x^2-25}{5y}} = \dfrac{x-5}{y} \cdot \dfrac{5y}{x^2-25}$

$$= \dfrac{x-5}{y} \cdot \dfrac{5y}{(x+5)(x-5)}$$

$$= \dfrac{5}{x+5}$$

69. $\dfrac{\frac{1}{a^3}-a}{\frac{1}{a^2}-1} = \dfrac{\frac{1-a^4}{a^3}}{\frac{1-a^2}{a^2}}$

$$= \dfrac{1-a^4}{a^3} \cdot \dfrac{a^2}{1-a^2}$$

$$= \dfrac{(1+a^2)(1-a^2)}{a^3} \cdot \dfrac{a^2}{1-a^2}$$

$$= \dfrac{1+a^2}{a}$$

71. $\dfrac{\frac{1}{x}-\frac{1}{x+h}}{h} = \dfrac{\frac{x+h}{x(x+h)} - \frac{x}{x(x+h)}}{h}$

$$= \dfrac{h}{x(x+h)} \cdot \dfrac{1}{h}$$

$$= \dfrac{1}{x(x+h)}$$

73. $\dfrac{\frac{x}{1-x}+\frac{1+x}{x}}{\frac{1-x}{x}+\frac{x}{1+x}} = \dfrac{x(1-x)(1+x)\left(\frac{x}{1-x}+\frac{1+x}{x}\right)}{x(1-x)(1+x)\left(\frac{1-x}{x}+\frac{x}{1+x}\right)}$

$$= \dfrac{x^2(1+x)+(1-x)(1+x)^2}{(1+x)(1-x)^2+x^2(1-x)}$$

$$= \dfrac{(1+x)[x^2+(1-x)(1+x)]}{(1-x)[(1-x)(1+x)+x^2]}$$

$$= \dfrac{1+x}{1-x}$$

R.6 Exercises

1. The number 16 has two square roots, –4 and 4, because $(-4)^2 = 4^2 = 16$, but 64 has only one cube root, 4, because $4^3 = 64$ but $(-4)^3 = -64$.

3. $\sqrt{(-7)^2} = \sqrt{49} = 7$

5. $\sqrt{25y^6} = \sqrt{(5y^3)^2} = 5|y^3|$

7. $\sqrt[5]{-32} = \sqrt[5]{(-2)^5} = -2$

9. $\sqrt[3]{-8x^{15}} = \sqrt[3]{(-2x^5)^3} = -2x^5$

11. $\sqrt{(x-5)^2} = |x-5|$

13. $\sqrt{75} = \sqrt{25} \cdot \sqrt{3} = 5\sqrt{3}$

15. $\sqrt[3]{40} = \sqrt[3]{8} \cdot \sqrt[3]{5} = 2\sqrt[3]{5}$

17. $\sqrt{x^5 y^6} = \sqrt{x^4} \cdot \sqrt{y^6} \cdot \sqrt{x} = x^2 y^3 \sqrt{x}$

19. $\sqrt{28x^{11}} = \sqrt{4} \cdot \sqrt{7} \cdot \sqrt{x^{10}} \cdot \sqrt{x} = 2x^5\sqrt{7x}$

21. $\sqrt[3]{-24x^4} = \sqrt[3]{-8x^3} \cdot \sqrt[3]{3x} = -2x\sqrt[3]{3x}$

23. $\sqrt{9x^2+36} = \sqrt{9} \cdot \sqrt{x^2+4} = 3\sqrt{x^2+4}$

25. It must be that $m < n$, because if $m \ge n$, then $\sqrt[n]{x^m}$ can be simplified as

$$\sqrt[n]{x^n} \cdot \sqrt[n]{x^{m-n}} = \begin{cases} |x|\sqrt[n]{x^{m-n}} & \text{for even } n \\ x\sqrt[n]{x^{m-n}} & \text{for odd } n \end{cases}.$$

27. $3\sqrt{6a} \cdot (-2\sqrt{2a}) = -6\sqrt{12a^2}$

$$= -6\sqrt{4a^2} \cdot \sqrt{3}$$

$$= -6(2a)\sqrt{3}$$

$$= -12a\sqrt{3}$$

© Houghton Mifflin Company. All rights reserved.

29. $\left(\sqrt{7ab^3}\right)\left(\sqrt{14a^3b^5}\right) = \sqrt{98a^4b^8}$

$\qquad\qquad\qquad\qquad = \sqrt{49a^4b^8} \cdot \sqrt{2}$

$\qquad\qquad\qquad\qquad = 7a^2b^4\sqrt{2}$

31. Produce a table of values for $3\sqrt{6a} \cdot (-2\sqrt{2a})$ and for $-12a\sqrt{3}$. The tables should be the same for all nonnegative values of a.

33. $\sqrt{\dfrac{24x^{11}y^3}{2xy^9}} = \sqrt{\dfrac{12x^{10}}{y^6}}$

$\qquad\qquad\qquad = \dfrac{\sqrt{12x^{10}}}{\sqrt{y^6}}$

$\qquad\qquad\qquad = \dfrac{\sqrt{4x^{10}} \cdot \sqrt{3}}{\sqrt{y^6}}$

$\qquad\qquad\qquad = \dfrac{2x^5\sqrt{3}}{y^3}$

35. $\dfrac{\sqrt{40x^{15}y^4}}{\sqrt{5x^{10}y^6}} = \sqrt{\dfrac{40x^{15}y^4}{5x^{10}y^6}}$

$\qquad\qquad\qquad = \sqrt{\dfrac{8x^5}{y^2}}$

$\qquad\qquad\qquad = \dfrac{\sqrt{4x^4} \cdot \sqrt{2x}}{\sqrt{y^2}}$

$\qquad\qquad\qquad = \dfrac{2x^2\sqrt{2x}}{y}$

37. $\dfrac{\sqrt{18t^3}}{\sqrt{50t^{13}}} = \dfrac{\sqrt{9t^2} \cdot \sqrt{2t}}{\sqrt{25t^{12}} \cdot \sqrt{2t}} = \dfrac{3t}{5t^6} = \dfrac{3}{5t^5}$

39. $\sqrt[3]{\dfrac{5t^5}{40t^2}} = \sqrt[3]{\dfrac{t^3}{8}} = \dfrac{\sqrt[3]{t^3}}{\sqrt[3]{8}} = \dfrac{t}{2}$

41. $3x\sqrt{7} + x\sqrt{7} = (3x + x)\sqrt{7} = 4x\sqrt{7}$

43. $\sqrt{45} + \sqrt{125} = \sqrt{9} \cdot \sqrt{5} + \sqrt{25} \cdot \sqrt{5}$

$\qquad\qquad\qquad = 3\sqrt{5} + 5\sqrt{5}$

$\qquad\qquad\qquad = 8\sqrt{5}$

45. $2t\sqrt{48} - 4\sqrt{27t^2} = 2t\sqrt{16} \cdot \sqrt{3} - 4\sqrt{9t^2} \cdot \sqrt{3}$

$\qquad\qquad\qquad\qquad = 8t\sqrt{3} - 12t\sqrt{3}$

$\qquad\qquad\qquad\qquad = -4t\sqrt{3}$

47. $3\sqrt{2}(5\sqrt{2} - \sqrt{3}) = (3\sqrt{2})(5\sqrt{2}) - (3\sqrt{2})(\sqrt{3})$

$\qquad\qquad\qquad\qquad = 15 \cdot 2 - 3\sqrt{6}$

$\qquad\qquad\qquad\qquad = 30 - 3\sqrt{6}$

49. $(\sqrt{5a} - 2)(\sqrt{2} + \sqrt{5a}) = (\sqrt{5a} \cdot \sqrt{2}) + (\sqrt{5a})^2 - 2\sqrt{2} - 2\sqrt{5a}$

$\qquad\qquad\qquad\qquad = \sqrt{10a} + 5a - 2\sqrt{2} - 2\sqrt{5a}$

51. $\dfrac{1}{\sqrt{x}+1} = \dfrac{1}{\sqrt{x}+1} \cdot \dfrac{\sqrt{x}-1}{\sqrt{x}-1}$

$\qquad\quad = \dfrac{\sqrt{x}-1}{(\sqrt{x})^2 - 1^2}$

$\qquad\quad = \dfrac{\sqrt{x}-1}{x-1}$

x	$\dfrac{1}{\sqrt{x}+1}$	$\dfrac{\sqrt{x}-1}{x-1}$
0	1	1
1	0.5	undefined
2	0.4142...	0.4142...
$\vdots$	$\vdots$	$\vdots$

The expression $\dfrac{1}{\sqrt{x}+1}$ is defined for all nonnegative values of x. However, the rationalized expression $\dfrac{\sqrt{x}-1}{x-1}$ is not defined for $x = 1$.

53. $\dfrac{12}{\sqrt{3}} = \dfrac{12}{\sqrt{3}} \cdot \dfrac{\sqrt{3}}{\sqrt{3}} = \dfrac{12\sqrt{3}}{3} = 4\sqrt{3}$

55. $\dfrac{5}{2\sqrt{3}} = \dfrac{5}{2\sqrt{3}} \cdot \dfrac{\sqrt{3}}{\sqrt{3}} = \dfrac{5\sqrt{3}}{6}$

57. $\dfrac{7}{\sqrt[3]{x}} = \dfrac{7}{\sqrt[3]{x}} \cdot \dfrac{\sqrt[3]{x^2}}{\sqrt[3]{x^2}} = \dfrac{7\sqrt[3]{x^2}}{x}$

59. $\dfrac{6}{\sqrt{6}+3} = \dfrac{6}{\sqrt{6}+3} \cdot \dfrac{\sqrt{6}-3}{\sqrt{6}-3}$

$\qquad\qquad = \dfrac{6(\sqrt{6}-3)}{(\sqrt{6})^2 - 3^2}$

$\qquad\qquad = \dfrac{6\sqrt{6}-18}{6-9}$

$\qquad\qquad = \dfrac{6\sqrt{6}-18}{-3}$

$\qquad\qquad = 6 - 2\sqrt{6}$

61. $\dfrac{\sqrt{5}+\sqrt{x}}{\sqrt{5}-\sqrt{x}} = \dfrac{\sqrt{5}+\sqrt{x}}{\sqrt{5}-\sqrt{x}} \cdot \dfrac{\sqrt{5}+\sqrt{x}}{\sqrt{5}+\sqrt{x}}$

$\qquad\qquad = \dfrac{(\sqrt{5}+\sqrt{x})^2}{(\sqrt{5})^2 - (\sqrt{x})^2}$

$\qquad\qquad = \dfrac{(\sqrt{5})^2 + 2(\sqrt{5})(\sqrt{x}) + (\sqrt{x})^2}{5-x}$

$\qquad\qquad = \dfrac{5 + 2\sqrt{5x} + x}{5-x}$

© Houghton Mifflin Company. All rights reserved.

63.
$$\frac{7a}{\sqrt{a}+a} = \frac{7a}{\sqrt{a}+a} \cdot \frac{\sqrt{a}-a}{\sqrt{a}-a}$$
$$= \frac{7a(\sqrt{a}-a)}{(\sqrt{a})^2-a^2}$$
$$= \frac{7a(\sqrt{a}-a)}{a-a^2}$$
$$= \frac{7(\sqrt{a}-a)}{1-a}$$

65. (i) $\sqrt[3]{-\sqrt[4]{1}} = \sqrt[3]{-1} = -1$

(ii) $\sqrt[4]{-\sqrt[3]{1}} = \sqrt[4]{-1}$ is not a real number, because it is an even root of a negative number.

(iii) $(-1)^{-\frac{1}{2}} = \frac{1}{\sqrt{-1}}$ is not a real number, because it involves an even root of a negative number.

(iv) $-1^{-\frac{1}{2}} = -\frac{1}{\sqrt{1}} = -1$

67. $25^{-\frac{1}{2}} = \frac{1}{25^{\frac{1}{2}}} = \frac{1}{\sqrt{25}} = \frac{1}{5}$

69. $-27^{-\frac{1}{3}} = -\frac{1}{27^{\frac{1}{3}}} = -\frac{1}{\sqrt[3]{27}} = -\frac{1}{3}$

71.
$$\left(\frac{27}{8}\right)^{-\frac{4}{3}} = \left(\frac{8}{27}\right)^{\frac{4}{3}}$$
$$= \frac{8^{\frac{4}{3}}}{27^{\frac{4}{3}}}$$
$$= \frac{\left(\sqrt[3]{8}\right)^4}{\left(\sqrt[3]{27}\right)^4}$$
$$= \frac{2^4}{3^4}$$
$$= \frac{16}{81}$$

73. $64^{\frac{5}{6}} = \left(\sqrt[6]{64}\right)^5 = 2^5 = 32$

75.

Radical Form	Exponential Form
$\sqrt[3]{y}$	**a.** $y^{\frac{1}{3}}$
b. $\sqrt[4]{2a}$	$(2a)^{\frac{1}{4}}$
c. $2\sqrt[4]{a^3}$	$2a^{\frac{3}{4}}$
$-\sqrt[3]{x^2}$	**d.** $-x^{\frac{2}{3}}$

77. $y^{-\frac{2}{3}} \cdot y^{\frac{5}{3}} = y^{\left(-\frac{2}{3}+\frac{5}{3}\right)} = y^{\frac{3}{3}} = y$

79. $\left(\frac{m^6}{n^{12}}\right)^{-\frac{2}{3}} = \left(\frac{n^{12}}{m^6}\right)^{\frac{2}{3}} = \frac{n^{12\left(\frac{2}{3}\right)}}{m^{6\left(\frac{2}{3}\right)}} = \frac{n^8}{m^4}$

81. $(x^{-6})^{-\frac{1}{3}} = x^{-6\left(-\frac{1}{3}\right)} = x^2$

83. $\frac{x^{\frac{1}{3}}y^{\frac{2}{3}}}{y^{-\frac{1}{3}}} = x^{\frac{1}{3}}y^{\frac{2}{3}-\left(-\frac{1}{3}\right)} = x^{\frac{1}{3}}y$

85. For $0.5 \sqrt[x]{16}$ the calculator returns 256, because $0.5\sqrt[x]{16} = 16^{\frac{1}{0.5}} = 16^2 = 256$.

87. $\sqrt[6]{y^2} = (y^2)^{\frac{1}{6}} = y^{\frac{2}{6}} = y^{\frac{1}{3}} = \sqrt[3]{y}$

89.
$$\left(\sqrt[3]{20}\right)^2 = 20^{\frac{2}{3}}$$
$$= \sqrt[3]{20^2}$$
$$= \sqrt[3]{400}$$
$$= \sqrt[3]{8}\cdot\sqrt{50}$$
$$= 2\sqrt[3]{50}$$

90. $\left(\sqrt[6]{b}\right)^3 = b^{\frac{3}{6}} = b^{\frac{1}{2}} = \sqrt{b}$

91. $\sqrt[3]{\sqrt[4]{6}} = \left(6^{\frac{1}{4}}\right)^{\frac{1}{3}} = 6^{\frac{1}{12}} = \sqrt[12]{6}$

93. $\sqrt[3]{a}\cdot\sqrt{a} = a^{\frac{1}{3}}a^{\frac{1}{2}} = a^{\frac{1}{3}+\frac{1}{2}} = a^{\frac{5}{6}} = \sqrt[6]{a^5}$

95.
$$\sqrt{27\sqrt[3]{9}} = \left(3^3 \cdot 3^{\frac{2}{3}}\right)^{\frac{1}{2}}$$
$$= \left(3^{\frac{11}{3}}\right)^{\frac{1}{2}}$$
$$= 3^{\frac{11}{6}}$$
$$= 3\cdot 3^{\frac{5}{6}}$$
$$= 3\sqrt[6]{3^5}$$
$$= 3\sqrt[6]{243}$$

97. Entering $\sqrt(7)-1)\wedge 4.37$ returns 8.820873554 or about 8.82.

99. Entering $(\sqrt(6)+1)/(3^x\sqrt{20}-5)$ returns -1.509238857 or about -1.51.

101. Entering $4^x\sqrt{100}+5^x\sqrt{-50}$ returns .9755535123 or about 0.98.

103. $\frac{\sqrt{y}+3}{\sqrt{y}} = \frac{\sqrt{y}+3}{\sqrt{y}}\cdot\frac{\sqrt{y}}{\sqrt{y}} = \frac{y+3\sqrt{y}}{y}$

105. $\sqrt{8}-\frac{1}{\sqrt{2}} = 2\sqrt{2}-\frac{1}{\sqrt{2}}\cdot\frac{\sqrt{2}}{\sqrt{2}} = \frac{4\sqrt{2}}{2}-\frac{\sqrt{2}}{2} = \frac{3\sqrt{2}}{2}$

107. $c = \sqrt{a^2+b^2} = \sqrt{5^2+11^2} = \sqrt{146} \approx 12.1$

109. $b = \sqrt{c^2-a^2} = \sqrt{15-5} = \sqrt{10} \approx 3.2$

111. The ship-to-ship distance is
$$\sqrt{8.3^2-4.2^2} = \sqrt{46.36} \approx 6.8 \text{ miles.}$$

© Houghton Mifflin Company. All rights reserved.

113. The radius is half the distance from B to C:

$$\frac{\sqrt{2^2 + \left(\sqrt{21}\right)^2}}{2} = \frac{\sqrt{4+21}}{2} = \frac{\sqrt{25}}{2} = \frac{5}{2} = 2.5$$

Review Exercises

1. a. $\sqrt{9} = 3$

   b. $0, \sqrt{9} = 3$

   c. $0, -6, -\sqrt{25} = -5, \sqrt{9} = 3$

   d. $\frac{2}{3}, 0, -6, -\sqrt{25} = -5, 0.\overline{6} = \frac{2}{3},$

      $-0.29 = \frac{-29}{100}, \sqrt{9} = 3$

   e. $\frac{\pi}{3}$

3.
$$\begin{array}{r} 0.3333... \\ -\ 0.3 \\ \hline 0.0333... \end{array}$$
$$0.\overline{3} - 0.3 = 0.0\overline{3}$$

5. Possible answer: $-\sqrt{3}$ and $\sqrt{3}$, whose sum is 0.

7. a. $|-3| = 3$ and $-|3| = -3$, so $|-3| > -|3|$.

   b. $12.7 < 12.\overline{7} = 12.777...$

   c. $1.6 = 1\frac{6}{10} = \frac{16}{10} = \frac{8}{5}$, so $1.6 = \frac{8}{5}$.

9. a. $\frac{1}{y^2+4} \cdot (y^2+4) = 1$ by the Multiplicative Inverse Property.

   b. $-(3-z) = z-3$, by the Property of the Opposite of a Difference.

11. a. $-2 - (-3)(-6) + 2^5 = -2 - 18 + 32 = 12$

   b. $\frac{-3^2 + (-5)}{\sqrt{36} + 2^3} = \frac{-9-5}{6+8} = \frac{-14}{14} = -1$

13. a. $(2 \cdot 3 - 1)(3 \cdot 3 + 5) = (6-1)(9+5)$
$$= (5)(14)$$
$$= 70$$

   b. $\frac{3^2 - (-1)^2}{3^2 + (-1)^2} = \frac{9-1}{9+1} = \frac{8}{10} = \frac{4}{5}$

15. The expression -5^{-2} means

$$-1 \cdot 5^{-2} = -1 \cdot \frac{1}{5^2} = -\frac{1}{25}, \text{whereas } (-5)^{-2} \text{ means}$$
$$\frac{1}{(-5)^2} = \frac{1}{25}.$$

17. a. $x^{-3} \cdot x^2 = x^{-3+2} = x^{-1} = \frac{1}{x}$

   b. $(3xy^{-4})^{-2} = \left(\frac{3x}{y^4}\right)^{-2} = \left(\frac{y^4}{3x}\right)^2 = \frac{(y^4)^2}{(3x)^2} = \frac{y^8}{9x^2}$

19. $12{,}000{,}000{,}000 = 12 \times 10^9 = 1.2 \times 10^{10}$

21. (i) monomial

   (ii) monomial

   (iii) monomial

   (iv) not a monomial, because of the negative exponent

   (v) monomial

23. $2xy^2 + 3x^2y - 5x^2y + xy^2 = (2xy^2 + xy^2) + (3x^2y - 5x^2y)$
$$= 3xy^2 - 2x^2y$$

25. $a^2 - ab + b^2 - (3ab - 4b^2) = a^2 - ab + b^2 - 3ab + 4b^2$
$$= a^2 - 4ab + 5b^2$$

27. $(x - 2y)(x^2 + 2xy + 3y^2) = x(x^2 + 2xy + 3y^2) - 2y(x^2 + 2xy + 3y^2)$
$$= x^3 + 2x^2y + 3xy^2 - 2x^2y - 4xy^2 - 6y^3$$
$$= x^3 - xy^2 - 6y^3$$

29. a. $(xy + 3)(xy - 3) = (xy)^2 - 3^2 = x^2y^2 - 9$

   b. $(2x - 5)^2 = (2x)^2 - 2(2x)(5) + 5^2 = 4x^2 - 20x + 25$

31. The factors contain a further, common factor. The complete factorization is $2(x+2) \cdot 2(x-2)$
$$= 4(x+2)(x-2).$$

© Houghton Mifflin Company. All rights reserved.

33. $-4x^2 - 2x + 8 = -2(2x^2 + x - 4)$

35. a. $4x^2 - 4xy + y^2 = (2x)^2 - 2(2x)(y) + y^2$
$$= (2x - y)^2$$

    b. $2a^2c + 20ac + 50c = 2c(a^2 + 10a + 25)$
$$= 2c(a^2 + 2 \cdot 5a + 5^2)$$
$$= 2c(a + 5)^2$$

37. a. $a^3 + 1000 = a^3 + 10^3$
$$= (a + 10)(a^2 - 10a + 100)$$

    b. $x^3y^3 + 1 = (xy)^3 + 1^3$
$$= (xy + 1)[(xy)^2 - xy + 1^2]$$
$$= (xy + 1)(x^2y^2 - xy + 1)$$

39. a. $4x^2 - x - 5$

Feasible factorizations	Middle Term
$(2x + 1)(2x - 5)$	$-8x$
$(2x - 1)(2x + 5)$	$8x$
$(4x + 1)(x - 5)$	$-19x$
$(4x - 1)(x + 5)$	$19x$
$(4x + 5)(x - 1)$	x
$(4x - 5)(x + 1)$	$-x$ Correct Middle term

$$4x^2 - x - 5 = (4x - 5)(x + 1)$$

    b. $12x^2 + 7x - 10$

Feasible factorizations	Middle Term
$(12x + 1)(x - 10)$	$-119x$
$(12x + 5)(x - 2)$	$-19x$
$(12x + 10)(x - 1)$	$-2x$
$(6x + 5)(2x - 2)$	$-2x$
$(6x + 10)(2x - 1)$	$14x$
$(3x + 2)(4x - 5)$	$-7x$
$(3x - 2)(4x + 5)$	$7x$ Correct middle term
$\vdots$	

$$12x^2 + 7x - 10 = (3x - 2)(4x + 5)$$

41. a. The denominator is $3x - 3 = 3(x - 1)$, which equals zero for $x = 1$.

    b. The denominator is x^2, which equals zero for $x = 0$.

    c. The denominator is $x^2 + 3x + 2 = (x + 2)(x + 1)$, which equals zero for $x = -2$ and $x = -1$.

43. $\dfrac{2x + 6}{4x^2 + 8x - 12} = \dfrac{2(x + 3)}{4(x^2 + 2x - 3)}$
$$= \dfrac{2(x + 3)}{4(x + 3)(x - 1)}$$
$$= \dfrac{1}{2(x - 1)}$$

© Houghton Mifflin Company. All rights reserved.

45. a. $\dfrac{x+1}{x^2+x-6} \cdot \dfrac{x^2-4}{x^2+3x+2} = \dfrac{x+1}{(x+3)(x-2)} \cdot \dfrac{(x+2)(x-2)}{(x+2)(x+1)}$

$\qquad\qquad\qquad\qquad\qquad = \dfrac{1}{x+3}$

b. $\dfrac{x-3}{x+1} \cdot \dfrac{x^2+2x+1}{6-2x} = \dfrac{x-3}{x+1} \cdot \dfrac{(x+1)^2}{-2(x-3)}$

$\qquad\qquad\qquad\qquad = -\dfrac{x+1}{2}$

47. a. $\dfrac{x}{2x+1} - \dfrac{x-1}{2x+1} = \dfrac{x-(x-1)}{2x+1} = \dfrac{1}{2x+1}$

b. $\dfrac{2x-y}{x+y} + \dfrac{2y-x}{x+y} = \dfrac{2x-y+2y-x}{x+y} = \dfrac{x+y}{x+y} = 1$

49. $\dfrac{\frac{1}{x}+1}{1-\frac{1}{x^2}} = \dfrac{\frac{1}{x}+\frac{x}{x}}{\frac{x^2}{x^2}-\frac{1}{x^2}}$

$\qquad = \dfrac{\frac{x+1}{x}}{\frac{x^2-1}{x^2}}$

$\qquad = \dfrac{x+1}{x} \cdot \dfrac{x^2}{x^2-1}$

$\qquad = \dfrac{x+1}{x} \cdot \dfrac{x^2}{(x+1)(x-1)}$

$\qquad = \dfrac{x}{x-1}$

51. a. The square roots of 49 are –7 and 7.

b. $\sqrt{49} = 7$

53. a. $\sqrt{5x^5 y} \cdot \sqrt{10x^6 y^3} = \sqrt{50x^{11}y^4}$

$\qquad\qquad\qquad\qquad = \sqrt{25x^{10}y^4} \cdot \sqrt{2x}$

$\qquad\qquad\qquad\qquad = 5x^5 y^2 \sqrt{2x}$

b. $\dfrac{\sqrt{28x^{12}y^5}}{\sqrt{7x^3 y^9}} = \sqrt{\dfrac{28x^{12}y^5}{7x^3 y^9}}$

$\qquad\qquad\quad = \sqrt{\dfrac{4x^9}{y^4}}$

$\qquad\qquad\quad = \dfrac{\sqrt{4x^8} \cdot \sqrt{x}}{\sqrt{y^4}}$

$\qquad\qquad\quad = \dfrac{2x^4 \sqrt{x}}{y^2}$

55. $\dfrac{2}{3+\sqrt{x}} = \dfrac{2}{3+\sqrt{x}} \cdot \dfrac{3-\sqrt{x}}{3-\sqrt{x}}$

$\qquad = \dfrac{2(3-\sqrt{x})}{3^2-(\sqrt{x})^2}$

$\qquad = \dfrac{2(3-\sqrt{x})}{9-x}$

57. a. $-8^{\frac{2}{3}} = -(2^3)^{\frac{2}{3}} = -2^{3\left(\frac{2}{3}\right)} = -2^2 = -4$

b. $8^{-\frac{2}{3}} = \dfrac{1}{8^{\frac{2}{3}}} = \dfrac{1}{(2^3)^{\frac{2}{3}}} = \dfrac{1}{2^2} = \dfrac{1}{4}$

c. $(-8)^{-\frac{2}{3}} = \dfrac{1}{(-8)^{\frac{2}{3}}}$

$\qquad\qquad = \dfrac{1}{[(-8)^{\frac{1}{3}}]^2}$

$\qquad\qquad = \dfrac{1}{(-2)^2}$

$\qquad\qquad = \dfrac{1}{4}$

59. Entering $\sqrt[3]{(15)}$ +6∧2.4 returns 76.18242247 or about 76.18.

© Houghton Mifflin Company. All rights reserved.

Chapter 1: Equations and Inequalities

1.1 Exercises

1. Any point $P\,(2,\,y)$ is a point on the vertical line located 2 units to the right of the y-axis.

3. $(4, 2)$, $(3, -5)$, $(0, 2)$, $(-2, 1)$, $(3, 0)$

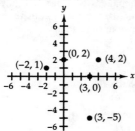

5. Set $y_1 = 3 - \dfrac{1}{2}x$

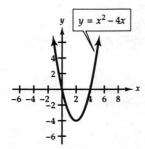

7. Set $y_1 = x^2 - 4x$.

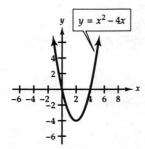

9. Set $y_1 = 3 - |x|$.

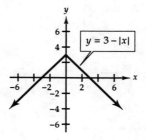

11. Set $y_1 = 2x - 6$

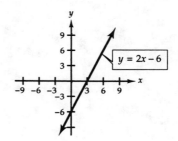

13. Set $y_1 = x^2 + 3x$.

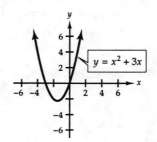

© Houghton Mifflin Company. All rights reserved.

15. Decimal setting is best.

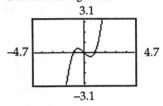

17. Integer setting is best.

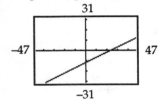

19. Standard setting is best.

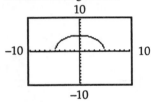

21. We mean that when x is 5, the value of the expression $\sqrt{x^2 - 16}$ is 3.

23. a. $(1, 3)$

    b. $(5, 15)$

    c. $(-6, -6)$

    d. $(-1, -1)$

25. a. $(-2, 0.75)$

    b. $(2, -0.25)$

    c. $(1.6, -0.24)$

    d. $(3.6, -0.09)$

27. The graph of $2x^2 - x^3 + 1$ is (B), using Xmin = -3, Xmax = 5, Xscl = 1, Ymin = -4, Ymax = 6, Yscl = 1.

29. The graph of $\dfrac{2}{x}$ is (A), using Xmin = -6, Xmax = 6, Xscl = 3, Ymin = -6, Ymax = 6, Yscl = 3.

31. The graph of $\sqrt{8 - x}$ is (F), using Xmin = -4, Xmax = 12, Xscl = 4, Ymin = -2, Ymax = 6, Yscl = 2.

33. a. $2x - 7 = 0$ for x = 3.5

    b $(3.5, 0)$ is on the graph.

35. a. $x^2 - 3x - 4 = 0$ for $x = -1$ and $x = 4$.

    b. $(-1, 0)$ and $(4, 0)$ are on the graph.

37. a. $\sqrt{x + 2} = 0$ for $x = -2$.

    b. $(0, -2)$ is on the graph.

39. The first coordinate of the point is the value of the variable for which the two expressions have the same value. (And the second coordinate is the value the expressions have there.)

41. x-intercept: $(2, 0)$
    y-intercept: $(0, -1)$

43. x-intercepts: $(-8, 0)$, $(8, 0)$
    y-intercept: $(0, -16)$

45. x-intercepts: $(-12, 0)$, $(12, 0)$
    y-intercept: $(0, 12)$

47. x-intercepts: $(-2, 0)$, $(0, 0)$, $(2, 0)$
    y-intercept: $(0, 0)$

49. x-intercept: $(15, 0)$
    y-intercept: $(0, 5)$

51. x-intercepts: $(-3.274..., 0)$, $(4.274 ..., 0)$ or about $(-3.3, 0)$, $(4.3, 0)$
    y-intercept: $(0, -14)$

53. The graphs intersect at $(2, 0.\overline{6}) = \left(2, \dfrac{2}{3}\right)$.

55. The graphs intersect at $(-4, 4)$ and $(3, 11)$.

57. The graphs intersect at $(-4, 7)$ and $(24, 21)$.

59. The graphs intersect at $(-0.\overline{6}, \ 10.\overline{3})$ or about $(-0.7, 10.3)$.

61. The graphs intersect at $(1.527 ..., 3.472 ...)$ and $(10.472 ..., -5.472...)$ or about $(1.5, 3.5)$ and $(10.5, -5.5)$.

1.2 Exercises

1. The linear equation $Ax + B = 0$ has one solution, $-\dfrac{B}{A}$.

3. The graphs of $2x - 7$ and $5 - x$ intersect at $x = 4$. So $x = 4$ is the solution of the equation.

5. The graphs of $\dfrac{1}{6}x^2 - \dfrac{1}{2}x$ and 18 (i.e., $y_2 = 18$) intersect at $x = -9$ and $x = 12$. So $x = -9$ and $x = 12$ are the solutions of the equation.

7. The graphs of $x^3 + x^2$ and $-2(x - 2)$ intersect at $x = 1$. So $x = 1$ is the solution of the equation.

9. The graphs of $\sqrt{x - 1}$ and $5 - 2x$ intersect at $x = 2$. So $x = 2$ is the solution of the equation.

11. The graphs of $4x - 3$ and $x + 1$ intersect at $x = 1.\overline{3}$. So $x = 1.3$ is the approximate solution of the equation.

© Houghton Mifflin Company. All rights reserved.

13. The graphs of $\dfrac{1}{x}$ and $x + 1$ intersect at $x = -1.618\ldots$ and $x = .618\ldots$. So $x = -1.6$ and $x = 0.6$ are the approximate solutions of the equation.

15. i. $5x - 7 = 2x + 1 \Rightarrow 3x = 8 \Rightarrow x = \dfrac{8}{3}$

    ii. $5x + 7 = 2x - 1 \Rightarrow 3x = -8 \Rightarrow x = -\dfrac{8}{3}$

    iii. $3x + 8 = 0 \Rightarrow 3x = -8 \Rightarrow x = -\dfrac{8}{3}$

Equations (ii) and (iii) have the same solution and are equivalent. Equation (i) has a different solution and so is not equivalent to the other two.

17. $3 - 2(2x - 3) = -5(x + 1)$

    $3 - 4x + 6 = -5x - 5$ Eliminate parentheses.

    $9 - 4x = -5x - 5$ Combine like terms.

    $x = -14$ Isolate the variable term on the left.

19. $16t + 8(1 - t) = 9t + 7$

    $16t + 8 - 8t = 9t + 7$ Eliminate parentheses.

    $8t + 8 = 9t + 7$ Combine like terms.

    $-t = -1$ Isolate the variable term on the left

    $t = 1$ Isolate the variable on the left.

21. $3x + 5(x - 2) = 2 - 4(3 - 5x)$

    $3x + 5x - 10 = 2 - 12 + 20x$ Eliminate parentheses.

    $8x - 10 = 20x - 10$ Combine like terms.

    $-12x = 0$ Isolate the variable term on the left.

    $x = 0$ Isolate the variable on the left.

23. $\dfrac{2}{3}x + 3 = \dfrac{1}{2}x - 1$

    $6(\dfrac{2}{3}x + 3) = 6(\dfrac{1}{2}x - 1)$ Multiply by LCD of fractions.

    $4x + 18 = 3x - 6$ Eliminate parentheses.

    $x = -24$ Isolate the variable term on the left.

25. $x - 2\left[x - 2(x + 1)\right] = 5x$

    $x - 2x + 4(x + 1) = 5x$ Eliminate parentheses.

    $x - 2x + 4x + 4 = 5x$

    $3x + 4 = 5x$ Combine like terms.

    $-2x = -4$ Isolate the variable term on the left.

    $x = 2$ Isolate the variable on the left.

27. $-\left\{x + 2\left[x - (2x + 1)\right]\right\} = 3(x + 1)$

    $-x - 2\left[x - (2x + 1)\right] = 3(x + 1)$ Eliminate parentheses.

    $-x - 2(-x - 1) = 3(x + 1)$

    $-x + 2x + 2 = 3x + 3$

    $x + 2 = 3x + 3$ Combine like terms.

    $-2x = 1$ Isolate the variable term on the left.

    $x = -\dfrac{1}{2}$ Isolate the variable on the left.

© Houghton Mifflin Company. All rights reserved.

24   *Chapter 1: Equations and Inequalities*

29. $\dfrac{5}{6}\left[\dfrac{2}{3}(x-1)-6(x+1)\right]=\dfrac{1}{6}(x+3)$

$6\bullet\dfrac{5}{6}\left[\dfrac{2}{3}(x-1)-6(x+1)\right]=6\bullet\dfrac{1}{6}(x+3)$ Multiply by LCD of fractions.

$5\left[\dfrac{2}{3}(x-1)-6(x+1)\right]=x+3$ Simplify.

$\dfrac{10}{3}(x-1)-30(x+1)=x+3$ Eliminate parentheses.

$10(x-1)-90(x+1)=3x+9$ Eliminate the remaining fraction.

$10x-10-90x-90=3x+9$ Eliminate the remaining parentheses.

$-80x-100=3x+9$ Combine like terms.

$-83x=109$ Isolate the variable term on the left.

$x=-\dfrac{109}{83}$ Isolate the variable on the left.

31. $3x-(x+1)=2(1-x)+x+2$

$3x-x-1=2-2x+x+2$ Eliminate parentheses.

$2x-1=4-x$ Combine like terms.

$3x=5$ Isolate the variable term on the left.

$x=\dfrac{5}{3}$ Isolate the variable on the left.

33. $\dfrac{4x}{9}-\dfrac{1}{6}=\dfrac{-x}{3}-5$

$18\left(\dfrac{4x}{9}-\dfrac{1}{6}\right)=18\left(-\dfrac{x}{3}-5\right)$ Multiply by LCD of fractions.

$8x-3=-6x-90$ Eliminate parentheses.

$14x=-87$ Isolate the variable term on the left.

$x=-\dfrac{87}{14}$ Isolate the variable on the left.

35. $-2[x-3(x+1)]=-\dfrac{2x}{5}$

$5\bullet(-2)[x-3(x+1)]=5\left(-\dfrac{2x}{5}\right)$ Multiply by LCD of fractions.

$-10[x-3(x+1)]=-2x$ Eliminate parentheses.

$-10[x-3x-3]=-2x$

$-10x+30x+30=-2x$

$20x+30=-2x$ Combine like terms.

$22x=-30$ Isolate the variable term on the left.

$x=-\dfrac{15}{11}$ Isolate the variable on the left.

37.   **a.** The graphs coincide, because a given value for the variable always leads to the same values for each side of the equation.

b. The graphs never intersect, because no value for the variable leads to both sides of the equation having the same value.

© Houghton Mifflin Company. All rights reserved.

39.

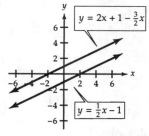

The left and right sides are never equal. The equation is a contradiction.

41.

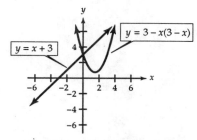

The left and right sides are equal for $x = 0$ and $x = 4$. The equation is conditional.

43.

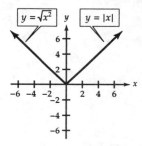

The left and right sides are always equal. The equation is an identity.

45.

$$2(x+3)+3x+7 = 2(x-2)-3(1-x)$$

$\quad 2x+6+3x+7 = 2x-4-3+3x \qquad$ Eliminate parentheses.

$\qquad\qquad 5x+13 = 5x-7 \qquad$ Combine like terms.

$\qquad\qquad\qquad 0 = -20 \qquad$ The equation is a contradiction.

The solution set is $\varnothing$.

47.

$$\frac{1}{2}(x-3)+3 = \frac{1}{2}(1-x)$$

$2\left[\frac{1}{2}(x-3)+3\right] = 2\cdot\frac{1}{2}(1-x) \quad$ Multiply by LCD of fractions.

$\qquad (x-3)+6 = 1-x \qquad$ Eliminate parentheses.

$\qquad x-3+6 = 1-x$

$\qquad\qquad x+3 = 1-x \qquad$ Combine like terms.

$\qquad\qquad\quad 2x = -2 \qquad$ Isolate the variable term on the left.

$\qquad\qquad\quad\; x = -1 \qquad$ Isolate the variable on the left.

The solution is $x = -1$.

© Houghton Mifflin Company. All rights reserved.

49. $7x - 2(3x - 1) = x + 2$

$7x - 6x + 2 = x + 2$ Eliminate parentheses.

$x + 2 = x + 2$ Combine like terms.

$0 = 0$ The equation is an identity.

The solution set is $\mathbb{R}$.

1.3 Exercises

1. i. False, because if c < 0, multiplying each side of a > b by c reverses the inequality symbol: ac < bc.

ii. True.

iii. True. Dividing each side of a > b by c reverses the inequality symbol.

iv True. ac < bc implies –ac > –bc.

3. The solution set of $x < 5$ is $(-\infty, 5)$.

5. $(1, \infty)$ is the solution set of $x > 1$.

7. The solution set of $x \geq -8$ is $[-8, \infty)$.

9. The solution set of $-5 < x < -2$ is $(-5, -2)$

11. $[-5, 3)$ is the solution set of $-5 \leq x < 3$.

13. The solution set of $0 \leq x < 1$ is $[0, 1)$.

15. $9 - 5t \leq -11$

$-5t \leq -20$ Isolate the variable term on the left.

$t \geq 4$ Isolate the variable on the left.

The inequality symbol reverses.

17. $6 - (1 - 3x) < x + 5$

$6 - 1 + 3x < x + 5$ Eliminate parentheses.

$3x + 5 < x + 5$ Combine like terms.

$2x < 0$ Isolate the variable term on the left.

$x < 0$ Isolate the variable on the left.

19. $t - \dfrac{7}{4} \geq \dfrac{5}{4}t - 1$

$$4\left(t - \frac{7}{4}\right) \geq 4\left(\frac{5}{4}t - 1\right)$$

Multiply by LCD of fractions.

$4t - 7 \geq 5t - 4$ Eliminate parentheses.

$-t \geq 3$ Isolate the variable term on the left.

$t \leq -3$ Isolate the variable on the left.

The inequality symbol reverses.

21. $3(2x + 5) > 6(x - 1)$

$6x + 15 > 6x - 6$ Eliminate parentheses.

$0 > -21$ The inequality is always true.
The solution set is $\mathbb{R}$.

© Houghton Mifflin Company. All rights reserved.

23. a. The first solution set is $\varnothing$, because no x can be both less than -5 and greater than 3; the second solution set is $(-\infty, -5) \cup (3, \infty)$.

b. The first solution set is $(-5, 3)$; the second solution set is $\mathbb{R}$, because every x is either greater than -5 or less than 3, or both).

25. $-3x \le 12$ and $-2x \ge -10$

$x \ge -4$ and $x \le 5$

$-4 \le x \le 5$

The solution set is $[-4, 5]$.

26. $\dfrac{1}{4}x - \dfrac{1}{2} \le 0$ and $\dfrac{x}{5} + \dfrac{1}{4} > \dfrac{1}{4}$

$x - 2 \le 0$ and $x + \dfrac{5}{4} > \dfrac{5}{4}$

$x \le 2$ and $x > 0$

$0 < x \le 2$

The solution set is $(0, 2]$.

27. $x - 4 > 1$ or $x + 4 < 3$

$x > 5$ or $x < -1$

The solution set is $(-\infty, -1) \cup (5, \infty)$.

29. $2x - 5 < 3$ and $3x + 1 < -5$

$2x < 8$ and $3x < -6$

$x < 4$ and $x < -2$

The solution set is $(-\infty, -2)$.

31. $3(x+2) \le -3$ or $4x - 2 \ge 0$

$x + 2 \le -1$ or $4x \ge 2$

$x \le -3$ or $x \ge \dfrac{1}{2}$

The solution set is $\left(-\infty, -3\right] \cup \left[\dfrac{1}{2}, \infty\right)$.

33. $-10 < -5t \le 20$

$2 > t \ge -4$

$-4 \le t < 2$

The solution set is $[-4, 2)$.

35. $\dfrac{2}{3} < \dfrac{1}{3} - \dfrac{1}{2}x < \dfrac{10}{3}$

$4 < 2 - 3x < 20$

$2 < -3x < 18$

$-\dfrac{2}{3} > x > -6$

$-6 < x < -\dfrac{2}{3}$

The solution set is $\left(-6, -\dfrac{2}{3}\right)$.

37. i. $|x - 3| = x - 3$

$x - 3 \ge 0$

$x \ge 3$.

The solution is $x \ge 3$.

ii. $|x - 3| = 3 - x$

$x - 3 \le 0$

$x \le 3$

The solution is $x \le 3$.

So the solution set of (i) is $[3, \infty)$ and the solution set of (ii) is $(-\infty, 3]$.

39. $|1 - 2x| = 7$

$1 - 2x = 7$ or $1 - 2x = -7$

$-2x = 6$ or $-2x = -8$

$x = -3$ or $x = 4$

41. $|3x - 11| - 6 = 4$

$|3x - 11| = 10$

$3x - 11 = 10$ or $3x - 11 = -10$

$3x = 21$ or $3x = 1$

$x = 7$ or $x = \dfrac{1}{3}$

43. $|x + 1| + 9 = 2$

$|x + 1| = -7$

The solution set is $\varnothing$.

45. $|3x + 7| - 4 = -4$

$|3x + 7| = 0$

$3x + 7 = 0$

$3x = -7$

$x = -\dfrac{7}{3}$

47. $|2t - 3| > 8$

$2t - 3 > 8$ or $2t - 3 < -8$

$2t > 11$ or $2t < -5$

$t > \dfrac{11}{2}$ or $t < -\dfrac{5}{2}$

The solution set is $\left(-\infty, -\dfrac{5}{2}\right) \cup \left(\dfrac{11}{2}, \infty\right)$.

49. $7 + 3|2x + 1| \ge 22$

$3|2x + 1| \ge 15$

$|2x + 1| \ge 5$

$2x + 1 \ge 5$ or $2x + 1 \le -5$

$2x \ge 4$ or $2x \le -6$

$x \ge 2$ or $x \le -3$

The solution set is $(-\infty, -3] \cup [2, \infty)$.

© Houghton Mifflin Company. All rights reserved.

51. $|4 - 3x| \le 15$

$-15 \le 4 - 3x \le 15$

$-19 \le -3x \le 11$

$\dfrac{19}{3} \ge x \ge -\dfrac{11}{3}$

$-\dfrac{11}{3} \le x \le \dfrac{19}{3}$

The solution set is $\left[-\dfrac{11}{3}, \dfrac{19}{3}\right]$.

53. $|x - 23| + 4 < -1$

$|x - 23| < -5$

The solution set is $\varnothing$.

55. $4x \le 2(x + 3) \le 3x + 13$
$4x \le 2x + 6 \le 3x + 13$
$2x \le 6 \le x + 13$

$2x \le 6$ and $6 \le x + 13$
$x \le 3$ and $-7 \le x$

The solution set is $[-7, 3]$.

57. If $C \le 0$, $|A|$ cannot be less than C. The solution set is $\varnothing$.

59.

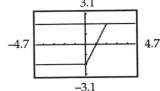

The two sides of the equation are equal for $x \ge 2$. So the solution set is $[2, \infty)$.

61.

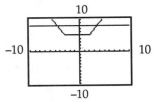

The left side of the inequality is less than the right side for $-4.5 < x < 3.5$. So the solution set is $(-4.5, 3.5)$.

63.

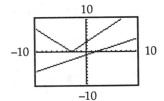

$|x + 3|$ is always greater than $\dfrac{1}{2}x - 1$. So the solution set for (i) is $\mathbb{R}$, and the solution set for (ii) is $\varnothing$.

1.4 Exercises

1. Each method can be used. Graphing, completing the square, and the quadratic formula can always be used, and factoring can be used because

$2x^2 + 5x - 3$ factors into $(2x - 1)(x + 3)$.

3. $w^2 + 4w - 21 = 0$

$(w - 3)(w + 7) = 0$

$w - 3 = 0$ or $w + 7 = 0$
   $w = 3$ or $w = -7$

5. $14x - 2x^2 = 0$

$x^2 - 7x = 0$

$(x - 7)x = 0$

$x - 7 = 0$ or $x = 0$
   $x = 7$ or $x = 0$

7. $x(4x - 11) = 20$

$4x^2 - 11x = 20$

$4x^2 - 11x - 20 = 0$

$(x - 4)(4x + 5) = 0$

$x - 4 = 0$ or $4x + 5 = 0$
   $x = 4$ or $x = -\dfrac{5}{4}$

9. $x^2 + 9 = 6x$

$x^2 - 6x + 9 = 0$

$(x - 3)^2 = 0$

$x - 3 = 0$

$x = 3$

11. $12 - 4x^2 = 0$

$3 - x^2 = 0$

$x^2 = 3$

$x = \pm\sqrt{3} \approx \pm 1.73$

13. $x^6 - 12 = 30$

$x^6 = 42$

$x = \pm\sqrt[6]{42} = \pm 42^{\frac{1}{6}} \approx \pm 1.86$

15. $x^2 + 15 = 3$

$x^2 = -12$

No real solution

17. $(x + 3)^2 = 10$

$x + 3 = \pm\sqrt{10}$

$x = -3 \pm \sqrt{10} \approx -6.16, 0.16$

© Houghton Mifflin Company. All rights reserved.

19. The discriminant $b^2 - 4ac = b^2 + 4$ is never negative, so the equation has real number solutions for any value of b.

21. $x^2 + 6x = -4$ Keep the constant term to the right.

$x^2 + 6x + 9 = -4 + 9$

With $b = 6$ add $\dfrac{b^2}{4}$ to complete the square.

$x^2 + 6x + 9 = 5$

$(x + 3)^2 = 5$ Factor the perfect square.

$x + 3 = \pm\sqrt{5}$

$x = -3 \pm \sqrt{5} \approx -5.24, \, -0.76$

23. $x^2 + 32 = 10x$

$x^2 - 10x + 32 = 0$

$x^2 - 10x = -32$ Move the constant term to the right.

$x^2 - 10x + 25 = -32 + 25$

With $b = -10$ add $\dfrac{b^2}{4}$ to complete the square.

$x^2 - 10x + 25 = -7$

$(x - 5)^2 = -7$ Factor the perfect square.

No real solution.

25. $9x^2 + 6x - 1 = 0$

$x^2 + \dfrac{2x}{3} - \dfrac{1}{9} = 0$ Divide by 9.

$x^2 + \dfrac{2x}{3} = \dfrac{1}{9}$ Move the constant term to the right.

$x^2 + \dfrac{2x}{3} + \dfrac{1}{9} = \dfrac{1}{9} + \dfrac{1}{9}$

With $b = \dfrac{2}{3}$ add $\dfrac{b^2}{4}$ to complete the square.

$x^2 + \dfrac{2x}{3} + \dfrac{1}{9} = \dfrac{2}{9}$

$\left(x + \dfrac{1}{3}\right)^2 = \dfrac{2}{9}$ Factor the perfect square.

$x + \dfrac{1}{3} = \pm\dfrac{\sqrt{2}}{3}$

$x = \dfrac{-1 \pm \sqrt{2}}{3} \approx -0.80, \, -0.14$

© Houghton Mifflin Company. All rights reserved.

27. $x + 5 = 3x^2$

$-3x^2 + x + 5 = 0$

$x^2 - \dfrac{x}{3} - \dfrac{5}{3} = 0$ Divide by –3.

$x^2 - \dfrac{x}{3} = \dfrac{5}{3}$ Move the constant term to the right.

$x^2 - \dfrac{x}{3} + \dfrac{1}{36} = \dfrac{5}{3} + \dfrac{1}{36}$

With $b = -\dfrac{1}{3}$ add $\dfrac{b^2}{4}$ to complete the square.

$x^2 - \dfrac{x}{3} + \dfrac{1}{36} = \dfrac{61}{36}$

$\left(x - \dfrac{1}{6}\right)^2 = \dfrac{61}{36}$ Factor the perfect square.

$x - \dfrac{1}{6} = \pm \dfrac{\sqrt{61}}{6}$

$x = \dfrac{1 \pm \sqrt{61}}{6} \approx -1.14, 1.47|$

29. $9x^2 + 12x - 1 = 0$
     $a = 9, b = 12, c = -1$

$x = \dfrac{-b \pm \sqrt{b^2 - 4ac}}{2a} = \dfrac{-12 \pm \sqrt{144 + 36}}{18}$

$x = \dfrac{-2 \pm \sqrt{5}}{3} \approx -1.41, 0.08$

31. $0 = 6x - 13 + -x^2$

$x^2 - 6x + 13 = 0$
$a = 1, b = -6, c = 13$
$b^2 - 4ac = 36 - 52 = -16 < 0$

No real solution.

33. $3x^2 + 2x = 2$

$3x^2 + 2x - 2 = 0$
$a = 3, b = 2, c\ -2$

$x = \dfrac{-b \pm \sqrt{b^2 - 4ac}}{2a} = \dfrac{-2 \pm \sqrt{4 + 24}}{6}$

$x = \dfrac{-1 \pm \sqrt{7}}{3} \approx -1.22, 0.55$

35. $7 - 3x - 5x^2 = 0$

$-5x^2 - 3x + 7 = 0$

$x = \dfrac{-b \pm \sqrt{b^2 - 4ac}}{2a} = \dfrac{3 \pm \sqrt{9 + 140}}{-10}$

$x = \dfrac{3 \pm \sqrt{149}}{-10} \approx -1.52, 0.92$

37. For a quadratic inequality, either both endpoints are solutions, or neither endpoint is a solution. So either (–3, 5) or [–3, 5] could be a solution, but not $(-3, 5]$.

39. Solve the corresponding equation.

$x^2 + 2x - 15 = 0$
$a = 1, b = 2, c = -15$

$x = \dfrac{-b \pm \sqrt{b^2 - 4ac}}{2a} = \dfrac{-2 \pm \sqrt{4 + 60}}{2}$

$x = -5, 3$

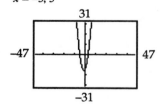

The graph shows that $x^2 + 2x - 15 \geq 0$ for $x \leq -5$ or $x \geq 3$.

The solution set is $(-\infty, -5] \cup [3, \infty)$.

© Houghton Mifflin Company. All rights reserved.

41. Solve the corresponding equation.

$$x(3x+11) = -6$$

$3x^2 + 11x = -6$ Eliminate parentheses.

$3x^2 + 11x + 6 = 0$

$a = 3, b = 11, c = 6$

$x = \dfrac{-b \pm \sqrt{b^2 - 4ac}}{2a} = \dfrac{-11 \pm \sqrt{121 - 72}}{6}$

$x = -3, -\dfrac{2}{3}$

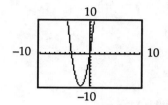

The graph shows that $x(3x+11) + 6 > 0$
for $x < -3$ or $x > -\dfrac{2}{3}$.

The solution set is $\left(-\infty, -3\right) \cup \left(-\dfrac{2}{3}, \infty\right)$.

43. Solve the corresponding equation.

$$8(x+2) = -x^2$$

$8x + 16 = -x^2$ Eliminate parentheses.

$x^2 + 8x + 16 = 0$

$a = 1, b = 8, c = 16$

$x = \dfrac{-b \pm \sqrt{b^2 - 4ac}}{2a} = \dfrac{-8 \pm \sqrt{64 - 64}}{2}$

$x = -4$

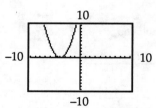

The graph shows that $8(x+2) + x^2 \geq 0$
for all x.

The solution set is $\mathbb{R}$.

45. Solve the corresponding equation.

$$x^2 + 3x = 2$$

$x^2 + 3x - 2 = 0$

$a = 1, b = 3, c = -2$

$x = \dfrac{-b \pm \sqrt{b^2 - 4ac}}{2a} = \dfrac{-3 \pm \sqrt{9 + 8}}{2}$

$x = \dfrac{-3 \pm \sqrt{17}}{2} \approx -3.56, 0.56$

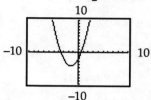

The graph shows that $x^2 + 3x - 2 \leq 0$ for
$-3.56 \leq x \leq 0.56$.

The solution set is $\left[-3.56, 0.56\right]$.

47. $9t^2 + 12t + 4 = 0$

$a = 9, b = 12, c = 4$

$b^2 - 4ac = 144 - 144 = 0$

There is one rational solution.

49. $2 + x^2 = 2x$

$x^2 - 2x + 2 = 0$

$a = 1, b = -2, c = 2$

$b^2 - 4ac = 4 - 8 = -4 < 0$

There is no real solution.

51. $4y^2 + 5y = 0$

$a = 4, b = 5, c = 0$

$b^2 - 4ac = 25 - 0 = 25 > 0$

$25 = 5^2$

There are two rational solutions.

53. $2 + t - 4t^2 = 0$

$-4t^2 + t + 2 = 0$

$a = -4, b = 1, c = 2$

$b^2 - 4ac = 1 + 32 = 33$

33 is not a perfect square.
There are two irrational solutions.

© Houghton Mifflin Company. All rights reserved.

55. $x^4 + 5x^2 - 36 = 0$

$\left(x^2\right)^2 + 5x^2 - 36 = 0$

$\left(x^2 - 4\right)\left(x^2 + 9\right) = 0$

$x^2 - 4 = 0$ or $x^2 + 9 = 0$

$x^2 = 4$ $x^2 = -9$

$x = \pm 2$ Contradiction

The solution is $x = \pm 2$.

57. $x^4 + 20 = 9x^2$

$\left(x^2\right)^2 - 9x^2 + 20 = 0$

$\left(x^2 - 4\right)\left(x^2 - 5\right) = 0$

$x^2 - 4 = 0$ or $x^2 - 5 = 0$

$x^2 = 4$ $x^2 = 5$

$x = \pm 2$ $x = \pm\sqrt{5}$

The solution is $x = \pm 2, \pm\sqrt{5} \approx \pm 2.24$.

59. $x^4 - 25 = 0$

$\left(x^2\right)^2 = 25$

$x^2 = 5$

$x = \pm\sqrt{5}$

The solution is $x = \pm\sqrt{5} \approx \pm 2.24$.

61. Inspect the graph to see the interval(s) for which the graph of $x^3 - 4x$ is above the x-axis.

63. $20 - 5x^3 = 0$

$-5x^3 = -20$

$x^3 = 4$

$x = \sqrt[3]{4} = 4^{\frac{1}{3}} \approx 1.59$

65. $x^3 + 50 = 0$

$x^3 = -50$

$x = \sqrt[3]{-50} = (-50)^{\frac{1}{3}} \approx -3.68$

67. $8x^3 - 50x = 0$

$2x\left(4x^2 - 25\right) = 0$

$2x(2x + 5)(2x - 5) = 0$

$2x = 0$ or $2x + 5 = 0$ or $2x - 5 = 0$

$x = 0$ $x = -\dfrac{5}{2}$ $x = \dfrac{5}{2}$

The solutions are $x = 0, \pm\dfrac{5}{2}$.

69. $x^3 + 3x^2 - 4x - 12 = 0$

$x^2(x + 3) - 4(x + 3) = 0$

$\left(x^2 - 4\right)(x + 3) = 0$

$(x + 2)(x - 2)(x + 3) = 0$

$x + 2 = 0$ or $x - 2 = 0$ or $x + 3 = 0$

$x = -2$ $x = 2$ $x = -3$

The solution is $x = -3, \pm 2$.

71. $y^6 - 64 = 0$

$y^6 = 64$

$y = \sqrt[6]{64} = \pm 2$

The solution is $y = \pm 2$.

73. The equation $8x^3 - 50x = 0$ has the solution

$x = 0, \pm\dfrac{5}{2}$.

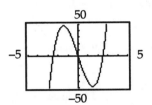

The graph shows that $8x^3 - 50x > 0$ for

$-\dfrac{5}{2} < x < 0$ or $x > \dfrac{5}{2}$. The solution set

is $\left(-\dfrac{5}{2}, 0\right) \cup \left(\dfrac{5}{2}, \infty\right)$.

75. The equation $x^3 + 3x^2 - 4x - 12 = 0$ has the solution $x = -3, \pm 2$.

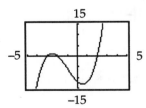

The graph shows that
$x^3 + 3x^2 - 4x - 12 \le 0$ for
$x \le -3$ or $-2 \le x \le 2$.

The solution set is $(-\infty, -3] \cup [-2, 2]$.

© Houghton Mifflin Company. All rights reserved.

77. $x^2 - 2\sqrt{2}x + 1 = 0$

$a = 1, b = -2\sqrt{2}, c = 1$

$x = \dfrac{-b \pm \sqrt{b^2 - 4ac}}{2a} = \dfrac{2\sqrt{2} \pm \sqrt{8 - 4}}{2}$

$x = \sqrt{2} \pm 1 \approx 0.41, 2.41$

79. $x = y^2 + 2y - 1$

$-y^2 - 2y + x + 1 = 0$

$a = -1, b = -2, c = x + 1$

$y = \dfrac{-b \pm \sqrt{b^2 - 4ac}}{2a} = \dfrac{2 \pm \sqrt{4 + 4(x + 1)}}{-2}$

$y = -1 \pm \sqrt{x + 2}$

1.5 Exercises

1. In (i), we write each term as a fraction with the LCD as the denominator. In (ii), we multiply both sides of the equation by the LCD to clear the fractions.

3. $\dfrac{5}{y + 4} = y$

$(y + 4)\left(\dfrac{5}{y + 4}\right) = (y + 4)y$ Multiply by LCD.

$5 = y^2 + 4y$ Eliminate parentheses.

$-y^2 - 4y + 5 = 0$

$y^2 + 4y - 5 = 0$

$(y - 1)(y + 5) = 0$

$y - 1 = 0$ or $y + 5 = 0$

$y = 1$ or $y = -5$

5. $\dfrac{7}{12} + \dfrac{1}{2t} = \dfrac{5}{3t}$

$12t\left(\dfrac{7}{12} + \dfrac{1}{2t}\right) = 12t\left(\dfrac{5}{3t}\right)$ Multiply by LCD.

$7t + 6 = 20$ Eliminate parentheses.

$7t = 14$

$t - 2 = 0$

$t = 2$

7. $x = \dfrac{x + 10}{2x}$

$2x(x) = 2x\left(\dfrac{x + 10}{2x}\right)$ Multiply by LCD.

$2x^2 = x + 10$ Eliminate parentheses.

$2x^2 - x - 10 = 0$

$(x + 2)(2x - 5) = 0$

$x + 2 = 0$ or $2x - 5 = 0$

$x = -2$ or $x = \dfrac{5}{2}$

© Houghton Mifflin Company. All rights reserved.

9. $\dfrac{2}{x} + \dfrac{1}{x+2} = (-1)$

$x(x+2)\left(\dfrac{2}{x} + \dfrac{1}{x+2}\right) = x(x+2)(-1)$ Multiply by LCD.

$x + 2(x+2) = -x^2 - 2x$

$x + 2x + 4 = -x^2 - 2x$

$3x + 4 = -x^2 - 2x$ Combine like terms.

$x^2 + 5x + 4 = 0$

$(x+1)(x+4) = 0$

$x + 1 = 0$ or $x + 4 = 0$

$x = -1$ or $x = -4$

11. $\dfrac{x}{x-4} = 1 + \dfrac{1}{x^2 - 16}$

$(x^2 - 16)\left(\dfrac{x}{x-4}\right) = (x^2 - 16)\left(1 + \dfrac{1}{x^2 - 16}\right)$ Multiply by LCD.

$x(x+4) = x^2 - 16 + 1$ Eliminate parentheses.

$x^2 + 4x = x^2 - 16 + 1$

$x^2 + 4x = x^2 - 15$ Combine like terms.

$4x = -15$

$x = -\dfrac{15}{4}$

13. $\dfrac{t}{t-3} = \dfrac{3}{t-3} - \dfrac{2}{t}$

$t(t-3)\left(\dfrac{t}{t-3}\right) = t(t-3)\left(\dfrac{3}{t-3} - \dfrac{2}{t}\right)$ Multiply by LCD.

$t^2 = 3t - 2(t-3)$ Eliminate parentheses.

$t^2 = 3t - 2t + 6$

$t^2 = t + 6$ Combine like terms.

$t^2 - t - 6 = 0$

$(t-3)(t+2) = 0$

$t - 3 = 0$ or $t + 2 = 0$

$t = 3$ is an extraneous solution, so the solution is $t = -2$.

15. $\dfrac{x}{6} - \dfrac{2x-3}{3x} = \dfrac{1}{x}$

$6x\left(\dfrac{x}{6} - \dfrac{2x-3}{3x}\right) = 6x\left(\dfrac{1}{x}\right)$ Multiply by LCD.

$x^2 - 2(2x-3) = 6$

$x^2 - 4x + 6 = 6$ Eliminate parentheses.

$x^2 - 4x = 0$

$(x-4)x = 0$

$x - 4 = 0$ or $x = 0$

$x = 0$ is an extraneous solution, so the solution is $x = 4$.

17. i $\sqrt{1-x} + 3 = 0$ implies $\sqrt{1-x} = -3$. But the principal square root is never negative.

    ii $\sqrt{1-x}$ is defined only for $x \le 1$, while $\sqrt{x-5}$ is defined only for $x \ge 5$. So the domains of the two radical expressions have no elements in common.

© Houghton Mifflin Company. All rights reserved.

19. $\sqrt{5-3x} = 5$

$5 - 3x = 5^2$

$5 - 3x = 25$

$-3x = 20$

$x = -\dfrac{20}{3}$

21. $\sqrt[3]{1-x} = -2$

$1 - x = (-2)^3$

$1 - x = -8$

$-x = -9$

$x = 9$

23. $\sqrt{4-3x} = 5$

$4 - 3x = 5^2$

$4 - 3x = 25$

$-3x = 21$

$x = -7$

25. $\qquad x - 4 = \sqrt{4-x}$

$(x-4)^2 = 4 - x$

$x^2 - 8x + 16 = 4 - x$

$x^2 - 7x + 12 = 0$

$(x-3)(x-4) = 0$

$x - 3 = 0 \quad \text{or} \quad x - 4 = 0$

$x = 3$ is an extraneous solution, because

$x - 4 = \sqrt{4-x}$ becomes $-1 = \sqrt{1}$.

The solution is $x = 4$.

27. $\qquad x + 2 = \sqrt{12x-11}$

$(x+2)^2 = 12x - 11$

$x^2 + 4x + 4 = 12x - 11$

$x^2 - 8x + 15 = 0$

$(x-3)(x-5) = 0$

$x - 3 = 0 \quad \text{or} \quad x - 5 = 0$

$x = 3 \qquad \text{or} \quad x = 5$

29. $\sqrt{x-7} - \sqrt{x+1} = -2$

$\sqrt{x-7} = \sqrt{x+1} - 2$

$x - 7 = \left(\sqrt{x+1} - 2\right)^2$

$x - 7 = x + 1 - 4\sqrt{x+1} + 4$

$4\sqrt{x+1} = 12$

$\sqrt{x+1} = 3$

$x + 1 = 9$

$x = 8$

31. $\qquad y^{1/4} = \dfrac{3}{2}$

$\left(y^{1/4}\right)^4 = \left(\dfrac{3}{2}\right)^4$

$y = \dfrac{3^4}{2^4}$

$y = \dfrac{81}{16}$

33. $\qquad x^{2/5} = 2$

$\left(x^{2/5}\right)^{5/2} = 2^{5/2}$

$x = \pm 2^2 (2)^{1/2}$ Even Root Property was used.

$ = \pm 4\sqrt{2} \approx \pm 5.66$

35 $\qquad (x+5)^{2/3} = 4$

$\left[(x+5)^{2/3}\right]^{3/2} = 4^{3/2}$

$x + 5 = \pm 8$ Even Root Property was used.

$x = -13, 3$

37. $\quad A = \dfrac{1}{2}h\left(b_1 + b_2\right)$

$\dfrac{2A}{h} = b_1 + b_2$

$b_2 = \dfrac{2A}{h} - b_1$

39. $A = P + Prt$

$A = P(1 + rt)$

$P = \dfrac{A}{1 + rt}$

41. $\qquad A = 2LW + 2LH + 2WH$

$A - 2LW = H(2L + 2W)$

$H = \dfrac{A - 2LW}{2L + 2W}$

43. $x^{2/3} + 2x^{1/3} - 3 = 0$

$\left(x^{1/3}\right)^2 + 2x^{1/3} - 3 = 0$

$\left(x^{1/3} + 3\right)\left(x^{1/3} - 1\right) = 0$

$x^{1/3} + 3 = 0 \qquad \text{or} \quad x^{1/3} - 1 = 0$

$x^{1/3} = -3 \quad \text{or} \qquad x^{1/3} = 1$

$x = -27 \quad \text{or} \qquad\quad x = 1$

© Houghton Mifflin Company. All rights reserved.

Review Exercises

1. Select the Window menu. To see Quadrant II, make Xmin negative and Ymax positive.

3. Set $y_1 = \dfrac{1}{2}x^2 + x$

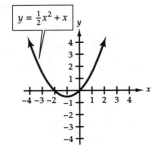

5. Integer setting is best.

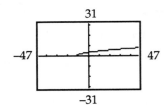

7. $x^2 + 3x - 10 = 0$ for $x = -5$ and $x = 2$. The x-intercepts are $(-5, 0)$ and $(2, 0)$.

9. **a.** x-intercepts: $(-6, 0)$, $(20,0)$
    y-intercept: $(0, 6)$

11. The graphs of $\dfrac{1}{2}x - 1$ and $-2x + \dfrac{3}{2}$ intersect at $x = 1$. So $x = 1$ is the solution of the equation.

13. **a.** A contradiction shows no intersections.

    b. A conditional equation shows one intersection.

    c. An identity shows infinitely many intersections.

15. $3(x - 4) = 2x - (9 - 5x)$

$3x - 12 = 2x - 9 + 5x$	Eliminate parentheses
$3x - 12 = 7x - 9$	Combine like terms.
$-4x = 3$	Isolate the variable term on the left.
$x = -\dfrac{3}{4}$	Isolate the variable on the left.

17. $\dfrac{1}{2}x + \dfrac{1}{3} = 3 - \dfrac{5}{6}x$

$\dfrac{4}{3}x = \dfrac{8}{3}$	Isolate the variable term on the left.
$x = 2$	Isolate the variable on the left.

19. The graphs of $3x - 2$ and $-0.2x + 5$ intersect at $x = 2.1875$, $y = 4.5625$. The solution is $x = 2.1875$.

21. $2x - 5 \le 4(x + 1)$

$2x - 5 \le 4x + 4$	Eliminate parentheses.
$-2x \le 9$	Isolate the variable term on the left.
$x \ge -\dfrac{9}{2}$	Isolate the variable on the left.

The inequality symbol reverses.

The solution set is $\left[-\dfrac{9}{2}, \infty\right)$.

© Houghton Mifflin Company. All rights reserved.

23. $-3 < 2x + 1 < 5$ is called a double inequality. It is equivalent to the compound inequality

$$-3 < 2x + 1 \quad \text{and} \quad 2x + 1 < 5.$$

25. $-8x \le 24$ and $3 - x \ge x - 5$

$\quad\quad x \ge -3 \quad\quad\quad\quad 3 - 2x \ge -5$

$\quad\quad\quad\quad\quad\quad\quad\quad\quad\quad -2x \ge -8$

$\quad\quad\quad\quad\quad\quad\quad\quad\quad\quad\quad\quad x \le 4$

The solution set is [–3, 4].

27. $|2x - 7| - 2 = 3$

$|2x - 7| = 5$

$2x - 7 = 5 \quad \text{or} \quad 2x - 7 = -5$

$2x = 12 \quad\quad\quad\quad 2x = 2$

$x = 6 \quad\quad\quad\quad\quad x = 1$

The solution is $x = 1, 6$.

29. $|x - 4| < 4$

$-4 < x - 4 < 4$

$0 < x < 8$

The solution set is (0, 8).

31. Dividing by zero to turn $2x^2 + 3x = 0$ into $2x + 3 = 0$ discards the solution $x = 0$.

33. $x(x - 6) = -9$

$x^2 - 6x = -9$

$x^2 - 6x + 9 = 0$

$(x - 3)^2 = 0$

$x - 3 = 0$

$x = 3$

35. a. $x^4 - 5 = 11$

$x^4 = 16$

$x = \pm 16^{1/4} = \pm 2$

b. $x^3 + 9 = 0$

$x^3 = -9$

$x = (-9)^{1/3} = -\sqrt[3]{9} \approx -2.08$

37. $\quad 3x^2 + 2x - 6 = 0$

$a = 3, b = 2, c = -6$

$x = \dfrac{-b \pm \sqrt{b^2 - 4ac}}{2a} = \dfrac{-2 \pm \sqrt{4 + 72}}{6}$

$x = \dfrac{-1 \pm \sqrt{19}}{3} \approx -1.79, 1.12$

39. Solve the corresponding equation.

$3 - 2x - x^2 = 0$

$-x^2 - 2x + 3 = 0$

$a = -1, b = -2, c = 3$

$x = \dfrac{-b \pm \sqrt{b^2 - 4ac}}{2a} = \dfrac{2 \pm \sqrt{4 + 12}}{-2}$

$x = -3, 1$

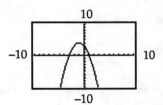

a. The graph shows that $3 - 2x - x^2 < 0$ for $x < -3$ or $x > 1$.

The solution set is $(-\infty, -3) \cup (1, \infty)$.

b. The graph shows that $3 - 2x - x^2 \ge 0$ for $-3 \le x \le 1$. The solution set is [–3, 1].

41. An apparent solution may be extraneous.

© Houghton Mifflin Company. All rights reserved.

43. $\dfrac{1}{t+1} = 3 - \dfrac{5}{t+1}$

$(t+1)\left(\dfrac{1}{t+1}\right) = (t+1)\left(3 - \dfrac{5}{t+1}\right)$

$1 = 3(t+1) - 5$ Eliminate parentheses.

$1 = 3t + 3 - 5$

$1 = 3t - 2$ Combine like terms.

$-3t = -3$

$t = 1$

45. $\dfrac{x+5}{x+1} + \dfrac{4}{x^2+x} = \dfrac{1}{x}$

$x(x+1)\left(\dfrac{x+5}{x+1} + \dfrac{4}{x^2+x}\right) = x(x+1)\left(\dfrac{1}{x}\right)$

$x(x+5) + 4 = x + 1$ Eliminate parentheses.

$x^2 + 5x + 4 = x + 1$

$x^2 + 4x + 3 = 0$

$(x+1)(x+3) = 0$

$x + 1 = 0 \quad \text{or} \quad x + 3 = 0$

$x = -1 \quad \text{or} \quad x = -3$

$x = -1$ is an extraneous solution, because it is a restricted value. The solution is $x = -3$.

47. $x - 1 = \sqrt{x+11}$

$(x-1)^2 = x + 11$

$x^2 - 2x + 1 = x + 11$

$x^2 - 3x - 10 = 0$

$(x+2)(x-5) = 0$

$x + 2 = 0 \quad \text{or} \quad x - 5 = 0$

$x = -2 \qquad x = 5$

$x = -2$ is an extraneous solution, because $x - 1 = \sqrt{x+11}$ becomes $-3 = \sqrt{9} \Rightarrow -3 = 3$. The solution is $x = 5$.

49. $(t-4)^{-2/3} = 25$

$\left[(t-4)^{-2/3}\right]^{-3/2} = 25^{-3/2}$

$t - 4 = \pm 25^{-3/2} = \pm\dfrac{1}{125}$ Even Root Property was used.

$t = 4 \pm \dfrac{1}{125} = 4.008,\ 3.992$

© Houghton Mifflin Company. All rights reserved.

Chapter 2: Basic Functions and Graphs

2.1 Exercises

1. The domain is the set of all first coordinates of the ordered pairs of the relation and the range is the set of all second coordinates.

3. Domain = {−4, 0, 3, 5}.
Range = {−1, 5, 6}
Function, because no domain element is paired with two or more range elements.

5. Domain = {0, 1, 4, 9}
Range = { $\pm\sqrt{0}$, $\pm\sqrt{1}$, $\pm\sqrt{4}$, $\pm\sqrt{9}$ }
= {−3, −2, −1, 0, 1, 2, 3}
Not a function, because elements 1, 4, and 9 of the domain are each paired with two range elements.

7. If a vertical line intersects the graph at more than one point, then two points have the same first coordinate and the graph does not represent a function.

9. $x = y + 5$
$y = x − 5$
Yes y is a function of x.

11. $y^2 = x − 1$
$y = \pm\sqrt{x − 1}$
No, y is not a function of x, because an x-value can correspond to two different y-values.

13. $y = \sqrt{x^2 − 9}$
Yes. y is a function of x.

15. $xy = x^2 y + x + 1$
$y = \dfrac{x + 1}{x − x^2}$
Yes, y is a function of x.

17. Domain: $[−2, \infty)$
Range: $\mathbb{R}$
Not a function; fails the vertical line test.

19. Domain: $\mathbb{R}$
Range: $\mathbb{R}$
A function; passes the vertical line test.

21. Domain: $\mathbb{R}$
Range: $(−3, \infty)$
A function; passes the vertical line test.

23. Domain: $(−\infty, 4]$
Range: $(−\infty, 3]$
A function; passes the vertical line test.

25. No, because an x-coordinate can be paired with more than one y-coordinate. For example, (3, 4) and (3, 5) both satisfy the relation.

27. $f(t) = 4t − 7$

  a $f(3) = 4(3) − 7 = 5$

  b. $f(0) = 4(0) − 7 = −7$

  c. $f(\frac{5}{2}) = 4(\frac{5}{2}) − 7 = 3$

  d. $f(a + 1) = 4(a + 1) − 7 = 4a − 3$

29. $g(y) = −y^2 + 2y + 1$

  a. $g(−1) = −(−1)^2 + 2(−1) + 1 = −2$

  b. $g(2) = −(2)^2 + 2(2) + 1 = 1$

  c. $g(t + 3) = −(t + 3)^2 + 2(t + 3) + 1$
               $= −t^2 − 6t − 9 + 2t + 6 + 1$
               $= −t^2 − 4t − 2$

  d. $g(1.5) = −(1.5)^2 + 2(1.5) + 1 = 1.75$

31. $h(t) = \sqrt{t^2 − 9}$

  a. $h(0) = \sqrt{(0)^2 − 9} = \sqrt{−9}$, which is not a real number

  b. $h(3) = \sqrt{(3)^2 − 9} = 0$

  c. $h(−5) = \sqrt{(−5)^2 − 9} = 4$

  d. $h(3t) = \sqrt{(3t)^2 − 9} = 3\sqrt{t^2 − 1}$

33. The value of the function evaluated at 2 is −6.

35. $f(x) = \dfrac{3}{4}x + 2$

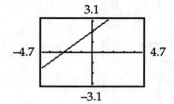

  a. $f(−4.3) = −1.225$

  b. $f(1.4) = 3.05$

  c. $f(−3.7) = −0.775$

© Houghton Mifflin Company. All rights reserved.

37. $s(t) = 1 - \dfrac{1}{t}$

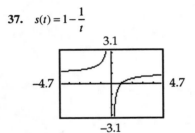

   a. $s(0.4) = -1.5$

   b. $s(1.6) = 0.375$

   c. $s(-0.8) = 2.25$

39. $f(x) = -x^2 + 6x - 5$

   $f(x) = 0$ when $x = 1, 5$.

41. $h(x) = \dfrac{x-3}{x+1}$

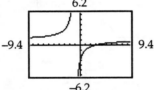

   $h(x) = 0$ when $x = 3$.

43. $f(x) = 2x + x^2$, $g(x) = x - 3$

   a. $f(-4) + g(5) = 2(-4) + (-4)^2 + (5-3)$
$$= 10$$

   b. $g(0) - f(-1) = 0 - 3 - [2(-1) + (-1)^2]$
$$= -2$$

   c. $f(t) + g(t+3) = 2t + t^2 + [(t+3) - 3]$
$$= 3t + t^2$$

45. $s(t) = \sqrt{t+3}$, $r(t) = 3 - |t - 2|$

   a. $s(6) - r(4) = \sqrt{6+3} - (3 - |4-2|) = 2$

   b. $s(-3) + r(7) = \sqrt{-3+3} + 3 - |7-2| = -2$

   c. $s(t^4 - 3) - r(t^2 + 2)$
$$= \sqrt{t^4 - 3 + 3} - (3 - |t^2 + 2 - 2|)$$
$$= 2t^2 - 3$$

47. $\dfrac{f(-2+h) - f(-2)}{h} = \dfrac{1 - (-2+h) - [1 - (-2)]}{h}$
$$= \dfrac{-h}{h} = -1$$

49. $f(x) = 4x + 1$
$$\dfrac{f(x+h) - f(x)}{h} = \dfrac{4(x+h) + 1 - (4x+1)}{h} = 4$$

51. $f(x) = 3x^2 - x$
$$\dfrac{f(x+h) - f(x)}{h}$$
$$= \dfrac{3(x+h)^2 - (x+h) - (3x^2 - x)}{h}$$
$$= \dfrac{3x^2 + 6xh + 3h^2 - x - h - 3x^2 + x}{h}$$
$$= 6x + 3h - 1$$

53. $f(x) = 2 - \dfrac{3}{2}x$

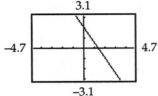

   Domain = $\mathbb{R}$; Range = $\mathbb{R}$

55. $g(x) = 5 - |x + 2|$

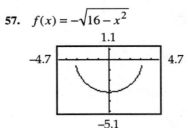

   Domain = $\mathbb{R}$; Range = $(-\infty, 5]$

57. $f(x) = -\sqrt{16 - x^2}$

   Domain = $[-4, 4]$; Range = $[-4, 0]$

59. Domain = $\mathbb{R}$

61. Domain = $[-4, \infty)$, because $\sqrt{x+4}$ is not defined for $x < -4$.

63. Domain = $\{t \mid t \neq 0, 1\}$, because $\dfrac{1}{t}$ is not defined for $t = 0$ and $\dfrac{3}{t-1}$ is not defined for $t = 1$.

© Houghton Mifflin Company. All rights reserved.

65. Two pairs, (5, 8) and (5, 25) belong to the relation. Thus the relation is not a function.

67. **a.** $f(0) = 0^2 = 0$

 b. $f(-1) = -1 + 1 = 0$

 c. $f(3) = 3^2 = 9$

 d. $f(-6) = -6 + 1 = -5$

69. **a.** $g(-2) = -2$

 b. $g(3) = 3 + 1 = 4$

 c. $g(4) = 4 + 2 = 6$

 d. $g(6) = 6 + 2 = 8$

71. **a.** Because the relation contains two pairs with the same first coordinate [for instance, (1, 75) and (1, 80)], the relation is not a function.

 b. Each score corresponds to only one student number. So this set is a function.

73. **a.** $B(2002 - 1976) = \frac{1}{2}(26) + 1 = 14$ gallons.

 b. $B(x) = 8$ when $x = 14$ years after 1976, or in 1990.

75. **a.** Rows = Columns + 1 = $x + 1$, and Panes = Columns × Rows = $x(x + 1)$.

 b. At \$45 per pane, $C(x) = 45x(x + 1)$.

 c. The least number of columns is 1, and the most is 5 (since the maximum for rows is 6). Domain = {1, 2, 3, 4, 5}.

 d. $C(4) = 45(4)(4 + 1) = \$900$

77. **a.** In a given year, \$1.1 billion is spent, plus \$0.2 more for each year since 1992. So the total is $R(n) = 1.1 + 0.2n$ (in billions of dollars).

 b. $R(4) + R(5) + R(6) + R(7) = \$8.8$ billion.

79. **a.** For each year from 1960 to 1972, $f(x)$ is exactly the number shown in the bar graph.

 b. For 1976, $f(16) \approx 45.904\%$ is only off the given value of 46% by just under 0.1%.

2.2 Exercises

1. **a.** The slope is undefined, because the denominator of $m = \frac{y_2 - y_1}{x_2 - x_1}$ is 0.

 b. The slope is 0, because the numerator of $m = \frac{y_2 - y_1}{x_2 - x_1}$ is 0.

3. $m = \frac{2 - (-1)}{-2 - 4} = -\frac{1}{2}$

5. $m = \frac{3 - 3}{-2 - (-5)} = 0$

7. $m = \frac{-1 - (-6)}{3 - 0} = \frac{5}{3}$

9. $y + 9 - 2x = 0$
$y + 9 - 2(0) = 0 \Rightarrow y = -9$
The y-intercept is (0, −9).
$0 + 9 - 2x = 0 \Rightarrow x = \frac{9}{2}$
The x-intercept is $(\frac{9}{2}, 0)$.

11. $4x - y = 12$
$4(0) - y = 12 \Rightarrow y = -12$
The y-intercept is (0, −12).
$4x - 0 = 12 \Rightarrow x = 3$
The x-intercept is (3, 0).

13. $0 = x - y - 5$
$y = f(x) = x - 5$
$m = 1, b = -5$
The slope is 1; the y-intercept is (0, −5).

15. $4x + 3y = 3$
$3y = -4x + 3$
$y = f(x) = -\frac{4}{3}x + 1$
$m = -\frac{4}{3}, b = 1$
The slope is $-\frac{4}{3}$; the y-intercept is (0, 1).

17. $1 - 2y = 0$
$-2y = -1$
$y = f(x) = \frac{1}{2}$
$m = 0, b = \frac{1}{2}$
The slope is 0; the y-intercept is $(0, \frac{1}{2})$.

© Houghton Mifflin Company. All rights reserved.

19.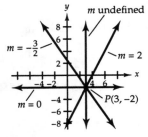

a. $y - (-2) = 2(x - 3)$
$y = 2x - 8$
Two other points are $(0, -8)$ and $(1, -6)$.

b. $y - (-2) = -\dfrac{3}{2}(x - 3)$

$y = -\dfrac{3}{2}x + \dfrac{5}{2}$

Two other points are $(0, \dfrac{5}{2})$ and $(1, 1)$.

c. $y - (-2) = 0(x - 3)$
$y = -2$
Two other points are $(0, -2)$ and $(1, -2)$.

d. $x = 3$
Two other points are $(3, 0)$ and $(3, 1)$.

21. The range of a linear function is $\mathbb{R}$, whereas the range of a constant function $y = b$ is a set containing a single number, $\{b\}$.

23. $y - (-7) = 3(x - 0)$
$y = 3x - 7$

25. $x = -6$

27. $y - 3 = \dfrac{1}{2}(x - 4)$

$y = f(x) = \dfrac{1}{2}x + 1$

29. $y - (-3) = 0(x - 0)$
$y = f(x) = -3$

31. $m = \dfrac{0 - 3}{1 + (-4)} = -\dfrac{3}{5}$

$y - 0 = -\dfrac{3}{5}(x - 1)$

$y = -\dfrac{3}{5}x + \dfrac{3}{5}$

33. The y-coordinate is the same.
$y = 7$

35. $m = \dfrac{1 - (-5)}{-1 - 2} = -2$
$y - (-5) = -2(x - 2)$
$y = -2x - 1$
$f(x) = -2x - 1$

37. $m = \dfrac{-3 - (-3)}{0.5 - 1.5} = 0$
$y - (-3) = 0(x - 1.5)$
$y = -3$
$f(x) = -3$

39. a.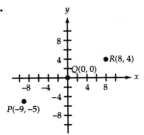

b. The slope of the line containing P and Q is
$m = \dfrac{0 - (-5)}{0 - (-9)} = \dfrac{5}{9}$
The slope of the line containing Q and R is
$m = \dfrac{4 - 0}{8 - 0} = \dfrac{1}{2}$.
The slopes are not equal. Thus, the points are not collinear.

41. $m = \dfrac{c - (-1)}{7 - 3} = \dfrac{c + 1}{4} = 1$
$c + 1 = 4$
$c = 3$

43. If m is undefined, then the x-coordinates are the same.
$c = -4$

45. The slope of the graph of g and the rate of change of g are the same.

47. If b is increased by 6,
$\dfrac{6}{\Delta x} = -\dfrac{3}{5}$
$\Delta x = -10$
a is decreased by 10.

49. If a is decreased by 2,
$\dfrac{\Delta y}{-2} = 2$
$\Delta y = -4$
b is decreased by 4.

51. Given $f(x) = x^2$, we have $f(1) = 1^2 = 1$ and
$f(-2) = (-2)^2 = 4$. Thus, $m = \dfrac{4 - 1}{-2 - 1} = -1$.

© Houghton Mifflin Company. All rights reserved.

53.

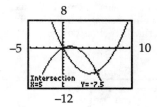

The points of intersection are $(0, 0)$ and $(5, -7.5)$.

$$m = \frac{-7.5 - 0}{5 - 0} = -\frac{7.5}{5} = -\frac{3}{2}$$

55. Given $P(x_1, y_1)$ and $Q(x_2, y_2)$ are points on the

line, the slope is $m = \dfrac{y_2 - y_1}{x_2 - x_1}$

Using the point-slope model with the point
$P(x_1, y_1)$, we have
$$y - y_1 = m(x - x_1)$$
$$\frac{y - y_1}{x - x_1} = m$$
$$\frac{y - y_1}{x - x_1} = \frac{y_2 - y_1}{x_2 - x_1}$$

57. For x equal to the number of years since 1990, we
have the two points $(5, 4)$ and $(40, 8.1)$, where the
population is in millions.
$$m = \frac{8.1 - 4}{40 - 5} = \frac{4.1}{35} = \frac{41}{350}$$

Using the point-slope model with $(5, 4)$, we have
$$y - 4 = \frac{41}{350}(x - 5)$$
$$y = \frac{41}{350}x + \frac{239}{70}$$
Thus, $f(x) = \dfrac{41}{350}x + \dfrac{239}{70}$.

59. For x equal to the number of years since 1970, we
have the two points $(5, 83)$ and $(20, 64)$.
$$m = \frac{64 - 83}{20 - 5} = -\frac{19}{15}$$

Using the point-slope model with $(5, 83)$, we
have
$$y - 83 = -\frac{19}{15}(x - 5)$$
$$y = -\frac{19}{15}x + \frac{268}{3}$$
Thus, $f(x) = -\dfrac{19}{15}x + \dfrac{268}{3}$.

61. a. $f(h) = 300$
This is a constant function.

b. $g(h) = 40h + 100$
The slope is the hourly rate and the y-intercept
corresponds to the annual charge.

63. a. $N(x) = 25 - x$

b. The y-intercept, $(0, 25)$, corresponds to no
workers belonging to the union and the
x-intercept, $(25, 0)$, corresponds to all workers
belonging to the union.

65. a. $C(x) = 5000 + 2x$

b. The y-intercept, $(0, 5000)$, represents the fixed
cost – the cost if no begonias are produced.

c. The slope of the graph of C is 2 and represents
the variable cost.

67. a. $C = 2\pi \cdot 8 = 16\pi$

b. In one minute, the hand will travel $\dfrac{16\pi}{60} = \dfrac{4\pi}{15}$.
Thus, in t minutes the tip of the minute hand
will travel $D(t) = \dfrac{4\pi t}{15}$

c. At 12:35, $t = 35$ and
$$D(35) = \frac{4\pi \cdot 35}{15} = \frac{28\pi}{3} \approx 29.32 \text{ inches.}$$

69. a. For t the number of years since 1985, we have
the two points $(0, 20000)$ and $(10, 28400)$.
$$m = \frac{28,400 - 20,000}{10 - 0} = 840$$

Using the point-slope model with $(0, 20000)$,
we have
$$y - 20,000 = 840(t - 0)$$
$$y = 840t + 20,000$$
Thus, $f(t) = 840t + 20,000$.

b. For the year 2000, $t = 2000 - 1985 = 15$, and
$f(15) = 840(15) + 20,000 = 32,600$.
You should be earning $32,600 to stay even
with inflation. So the purchasing power would
be $2600 less than in 1985.

71. a. We have two points: $(20000, 50000)$ and
$(24000, 60000)$.
$$m = \frac{60,000 - 50,000}{24,000 - 20,000} = \frac{5}{2}$$
Using the point-slope model with $(20000,
50000)$, we have
$$y - 50,000 = \frac{5}{2}(x - 20,000)$$
$$y = \frac{5}{2}x$$
Thus, $P(x) = \dfrac{5}{2}x$.

b. The slope, $\dfrac{5}{2}$, is the profit per subscriber.

c. The y-intercept, $(0, 0)$, indicates that, if the
newspaper has no subscribers, the profit is 0.

© Houghton Mifflin Company. All rights reserved.

73. a. $y = 75x$

b.

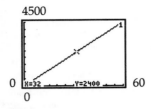

Using the trace feature, $x = 32$. There would be 32 members.

2.3 Exercises

1. Because the three points are not collinear, the linear regression line could contain at most two of the points.

3. Yes

5. No

7. a. $m = \dfrac{-3-5}{6-(-2)} = -1$

Using the point-slope model with the point $P(-2, 5)$, we have $y - 5 = -1(x + 2)$
$$y = -x + 3$$

b. $y = -x + 3$

c. The equations are the same.

d. $r = -1$ because the slope is negative.

9. a. Yes, the function could be linear. Using $(0, 4)$ and $(5, 11)$, $m = \dfrac{11-4}{5-0} = \dfrac{7}{5}$

Using the point-slope model with $(0, 4)$, we

have $y - 4 = \dfrac{7}{5}(x - 0)$
$$y = \dfrac{7}{5}x + 4$$

Check the equation with the remaining points; for example, using $(20, 32)$ we have

$32 \stackrel{?}{=} \dfrac{7}{5}(20) + 4$

$32 \stackrel{?}{=} 7 \cdot 4 + 4$

$32 = 32$

b. $r = 1$ because the points are collinear and the slope is positive.

c. The linear regression model is $y = 1.4x + 4$.

11. a. We use the points $(4, 14.4)$, $(5, 13.0)$, and $(7, 11.2)$ to obtain the linear regression equation $y = -1.04x + 18.43$

b. The coefficient of x, -1.04, is negative which indicates the number of cases is decreasing.

c. For 1996, $x = 6$ and
$y = -1.04(6) + 18.43 = 12.19$
There were approximately 12.2 million cases in 1996.

13. The calculator indicates an error in the data. Because the points belong to a vertical line, the slope is undefined. There is no regression equation.

15.

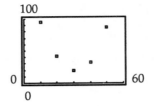

No, a linear model is not appropriate for the data.

17.

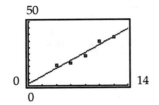

Yes, a linear model is appropriate for the data.

a. Using the data values, we obtain $y = 3.05x + 1.6$. The resulting graph is a good fit to the data.

b. Because $r = 0.97$, a linear model is appropriate.

19. a. Use the data values $(2, 43.33)$, $(6, 42.76)$, $(18, 39.10)$, $(22, 40.33)$, and $(24, 39.25)$ to obtain the linear regression equation $y = -0.19x + 43.65$.

b. For the year 1980, $x = 10$ and $y = -0.19(10) + 43.65 \approx 41.78$ In 1980, the winning time was about 41.78 seconds.

c. Because $r = -0.93$, a linear model is reasonable.

d. The winning times are decreasing.

© Houghton Mifflin Company. All rights reserved.

21. a. (i) $m = \dfrac{121 - 96}{2 - 1} = 25$

Using the point-slope model with (1, 96),
we have $y - 96 = 25(x - 1)$
$$y = 25x + 71$$

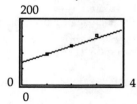

(ii) $m = \dfrac{156 - 96}{3 - 1} = 30$

Using the point-slope model with (1, 96),
we have $y - 96 = 30(x - 1)$
$$y = 30x + 66$$

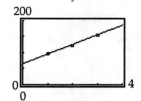

(iii) $m = \dfrac{156 - 121}{3 - 2} = 35$

Using the point-slope model with (2, 121),
we have $y - 121 = 35(x - 2)$
$$y = 35x + 51$$

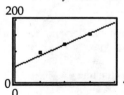

b. (ii) $y = 30x + 66$ appears to model the data best.

c. $y = 30x + 64.33$

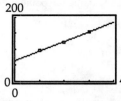

The linear regression equation models the data best.

23. a.

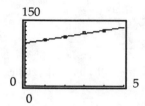

Using the data points (1, 108.0), (2, 115.6),
(3, 126.6), and (4, 129.5), we obtain the linear
regression equation $y = 7.55x + 101.05$.

b. $y = 7.55(0) + 101.05 = 101.05$
According to the model the number of calls in
1992 was about 101 million.

c. Because $r = 0.98$, the prediction is reasonable.

2.4 Exercises

1. Two lines, not in the same plane, may not be parallel even though they do not intersect.

3. $L_1 : 3x - 2y - 2 = 0$
$$2y = 3x - 2$$
$$y = \frac{3}{2}x - 1$$

So $m_1 = \dfrac{3}{2}$

$L_2 : 3y = -2(x - 6)$
$$y = -\frac{2}{3}x + 4$$

So $m_2 = -\dfrac{2}{3}$

Because $m_1 \cdot m_2 = -1$, the lines are perpendicular.

5. $L_1 : m_1 = \dfrac{4 - 0}{5 - (-5)} = \dfrac{2}{5}$

$L_2 : m_2 = \dfrac{-2 - (-4)}{5 - 10} = -\dfrac{2}{5}$

The lines are neither parallel nor perpendicular.

7. $L_1 : m_1 = \dfrac{-5 - 2}{4 - 4} = \dfrac{-7}{0}$ So m_1 undefined

$L_2 : 3 - 2x = 0$ So m_2 undefined
$$x = \frac{3}{2}$$

Because L_1 and L_2 are vertical, the lines are parallel.

9. $f(x) = 2(1 - x) + 1$ $g(x) = 3 - 2(x + 3)$
$\quad\quad = -2x + 3$ $\quad\quad = -2x - 3$
So $m_1 = -2$ So $m_2 = -2$
Because $m_1 = m_2$, the lines are parallel.

11. $f(x) = 2x + 5$ $g(x) = -2x - 3$
So $m_1 = 2$ So $m_2 = -2$
The lines are neither parallel nor perpendicular.

© Houghton Mifflin Company. All rights reserved.

13. f: $m_1 = \dfrac{-4-(-4)}{0-2} = 0$

g: $m_2 = \dfrac{1-1}{-2-5} = 0$

Because $m_1 = m_2$, the lines are parallel.

15. L_1: $x - y = 6$

$\qquad\quad y = x - 6$

So $m_1 = 1$

a. $f \perp L_1$, so $m_2 = -\dfrac{1}{m_1} = -1$

Using the point-slope model with (2, –5), we have $y - (-5) = -1 \cdot (x - 2)$

$\qquad\qquad\qquad y = -x - 3$

Thus, $f(x) = -x - 3$.

b. $f \parallel L_1$ so $m_2 = m_1 = 1$

Using the point-slope model with (2, –5), we have $y - (-5) = 1 \cdot (x - 2)$

$\qquad\qquad\qquad y = x - 7$

Thus, $f(x) = x - 7$.

17. L_1: $m_1 = \dfrac{5-(-7)}{-3-6} = -\dfrac{4}{3}$

The y-intercept of $y = 4 - 3x$ is (0, 4).

a. $f \perp L_1$, so $m_2 = -\dfrac{1}{m_1} = \dfrac{3}{4}$

Using the point-slope model with (0, 4), we have $y - 4 = \dfrac{3}{4}(x - 0)$

$\qquad\qquad\qquad y = \dfrac{3}{4}x + 4$

Thus, $f(x) = \dfrac{3}{4}x + 4$.

b. $f \parallel L_1$, so $m_2 = m_1 = -\dfrac{4}{3}$

Using the point-slope model with (0, 4), we have $y - 4 = -\dfrac{4}{3}(x - 0)$

$\qquad\qquad\qquad y = -\dfrac{4}{3}x + 4$

Thus, $f(x) = -\dfrac{4}{3}x + 4$.

19. L_1: $x + 5 = -1$

$\qquad\qquad x = -6$

So m_1 undefined

a. $f \perp L_1$, so $m_2 = 0$ (L_1 is vertical so the graph of f is horizontal)

Using the point-slope model with (4, 6), we have $y - 6 = 0(x - 4)$

$\qquad\qquad\qquad y = 6$

Thus, $f(x) = 6$.

b. $f \parallel L_1$

Since L_1 is vertical, the graph of f is the vertical line passing through (4, 6). But then f isn't a function.

Thus, $x = 4$.

21. The y-intercept of $y = 2 - 3x$ is (0, 2).

$x - 2y = 4$

$\qquad 2y = x - 4$

$\qquad\quad y = \dfrac{1}{2}x - 2$

So $m_1 = \dfrac{1}{2}$

Thus, $m_2 = -\dfrac{1}{m_1} = -2$. Using the point-slope model, we have $y - 2 = -2(x - 0)$

$\qquad\qquad\qquad y = -2x + 2$

Therefore, L is the line $y = -2x + 2$.

23. First, determine the slope.

$m = \dfrac{-8-4}{3-(-1)} = -3$

Using the point-slope model with $R(5, -2)$, we have $y - (-2) = -3(x - 5)$

$\qquad\qquad\qquad y = -3x + 13$

Thus, L is the line $y = -3x + 13$.

25. For $x = 1$, $y = \dfrac{1}{1} = 1$; for $x = -1$, $y = \dfrac{1}{-1} = -1$.

Therefore, we want the equation of the line containing the points (1, 1) and (–1, –1).

First, $m = \dfrac{-1-1}{-1-1} = 1$. Using the point-slope model with (1, 1), we have $y - 1 = 1(x - 1)$

$\qquad\qquad\qquad\qquad y = x$

Thus, L is the line $y = x$.

© Houghton Mifflin Company. All rights reserved.

27. a. The line $y = -3x + 1$ has slope $m_1 = -3$. A line perpendicular to the line has slope

$m_2 = -\dfrac{1}{m_1} = \dfrac{1}{3}$. Find the slope of our given

line. $2x + cy = 12$

$$cy = -2x + 12$$
$$y = -\frac{2}{c}x + \frac{12}{c}$$

So, $-\dfrac{2}{c} = \dfrac{1}{3}$ or $c = -6$.

b. Find the y-intercept.

$$x + 2y - 16 = 0$$
$$2y = -x + 16$$
$$y = -\frac{1}{2}x + 8$$

The y-intercept is $(0, 8)$. From part (a), the y-intercept of our given line is $\dfrac{12}{c}$. So, $\dfrac{12}{c} = 8$

or $c = \dfrac{3}{2}$.

29. Both (i) and (ii) are possible. Because the endpoints have integer coordinates, the coordinates of the midpoint must be rational numbers so (iii) is not possible.

31. a. $PQ = \sqrt{(-2-3)^2 + (4-(-8))^2}$
$= \sqrt{(-5)^2 + 12^2}$
$= \sqrt{169}$
$= 13$

b. $M(\dfrac{3+(-2)}{2}, \dfrac{-8+4}{2}) = M(\dfrac{1}{2}, -2)$

33. a. $PQ = \sqrt{(5-(-4))^2 + (-8-0)^2}$
$= \sqrt{9^2 + (-8)^2}$
$= \sqrt{145} \approx 12.04$

b. $M(\dfrac{-4+5}{2}, \dfrac{0+(-8)}{2}) = M(\dfrac{1}{2}, -4)$

35.
a. $PQ = \sqrt{[(a-b)-(a+b)]^2 + [(a+b)-(b-a)]^2}$
$= \sqrt{(-2b)^2 + (2a)^2}$
$= \sqrt{2^2(a^2+b^2)}$
$= 2\sqrt{a^2+b^2}$

b. $M(\dfrac{(a+b)+(a-b)}{2}, \dfrac{(b-a)+(a+b)}{2})$
$= (a, b)$

37. $d_1 = \sqrt{5^2 + (-12)^2} = \sqrt{169} = 13$
$d_2 = \sqrt{6^2 + (-11)^2} = \sqrt{157} \approx 12.53$
Since $d_1 > d_2$, $(5, -12)$ is farther from the origin.

39. Let the endpoint B have coordinates (x, y). Then, since $M(\dfrac{0+x}{2}, \dfrac{0+y}{2}) = M(-\dfrac{3}{2}, 2)$, we have

$\dfrac{0+x}{2} = -\dfrac{3}{2}$ and $\dfrac{0+y}{2} = 2$
$x = -3$ $y = 4$

Therefore, B has coordinates $(-3, 4)$.

41. Let the endpoint B have coordinates (x, y). Then, since $M(\dfrac{-3+x}{2}, \dfrac{9+y}{2}) = M(-1, 6)$, we have

$\dfrac{-3+x}{2} = -1$ and $\dfrac{9+y}{2} = 6$
$-3+x = -2$ $9+y = 12$
$x = 1$ $y = 3$

Therefore, B has coordinates $(1, 3)$.

43. The origin has coordinates $(0, 0)$. Using the distance formula to find the distance d from $(0, 0)$ to (a, b), we have

$$d = \sqrt{(a-0)^2 + (b-0)^2} = \sqrt{a^2 + b^2}.$$

45. Label the points $P(-6, 2)$, $Q(0, 5)$, and $R(-1, -8)$

$PQ = \sqrt{[0-(-6)]^2 + (5-2)^2} = 3\sqrt{5}$
$PR = \sqrt{[-1-(-6)]^2 + (-8-2)^2} = 5\sqrt{5}$
$QR = \sqrt{(-1-0)^2 + (-8-5)^2} = \sqrt{170}$

Since $(3\sqrt{5})^2 + (5\sqrt{5})^2 = 45 + 125$,
$= 170 = \sqrt{170}^2$

we have $(PQ)^2 + (PR)^2 = (QR)^2$. The triangle is a right triangle.

47. Label the points $P(-1.5, 2)$, $Q(4.5, 6)$, and $R(6, -9)$.

$PQ = \sqrt{[4.5-(-1.5)]^2 + (6-2)^2} = \sqrt{52}$
$PR = \sqrt{[6-(-1.5)]^2 + (-9-2)^2} = \sqrt{177.25}$
$QR = \sqrt{(6-4.5)^2 + (-9-6)^2} = \sqrt{227.25}$

Since $(PQ)^2 + (PR)^2 \neq (QR)^2$, the triangle is not a right triangle.

49. The figure is a circle with radius $r = 5$ because

$CP = \sqrt{(6-2)^2 + (-2-1)^2} = \sqrt{25} = 5$.

Area: $A = \pi r^2$
$A = \pi(5)^2$
$A = 25\pi \approx 78.54$

Perimeter: $P = 2\pi r$
$P = 2\pi(5)$
$P = 10\pi \approx 31.42$

51. The figure is a rectangle with length $L = 3$ and width $W = 7$.

Area: $A = LW$
$A = 3 \cdot 7$
$A = 21$

Perimeter: $P = 2L + 2W$
$P = 2 \cdot 3 + 2 \cdot 7$
$P = 20$

© Houghton Mifflin Company. All rights reserved.

53.

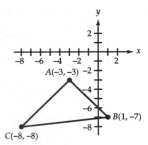

$$AB = \sqrt{[1-(-3)]^2 + [-7-(-3)]^2} = 4\sqrt{2}$$

$$AC = \sqrt{[-8-(-3)]^2 + [-8-(-3)]^2} = 5\sqrt{2}$$

$$BC = \sqrt{(-8-1)^2 + [-8-(-7)]^2} = \sqrt{82}$$

Perimeter $P = 4\sqrt{2} + 5\sqrt{2} + \sqrt{82} \approx 21.78$

Notice

$$m_{AB} = \frac{-7-(-3)}{1-(-3)} = -1$$

$$m_{AC} = \frac{-8-(-3)}{-8-(-3)} = 1$$

Therefore, $AB \perp AC$ and

Area: $A = \dfrac{1}{2}bh = \dfrac{1}{2}AB \cdot AC$

$$= \frac{1}{2}(4\sqrt{2})(5\sqrt{2}) = 20$$

55.

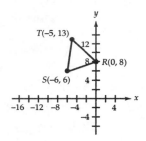

$$RS = \sqrt{(-6-0)^2 + (6-8)^2} = 2\sqrt{10}$$

$$ST = \sqrt{[-5-(-6)]^2 + (13-6)^2} = 5\sqrt{2}$$

$$RT = \sqrt{(-5-0)^2 + (13-8)^2} = 5\sqrt{2}$$

Perimeter: $P = 2\sqrt{10} + 5\sqrt{2} + 5\sqrt{2} \approx 20.47$

We have an isosceles triangle with base b and two equal sides of length a. Using the Pythagorean Theorem, we find the height to be

$$h = \frac{1}{2}\sqrt{4a^2 - b^2} .$$

Area: $A = \dfrac{1}{2}bh$

$$= \frac{1}{2}(2\sqrt{10})(\frac{1}{2}\sqrt{4(5\sqrt{2})^2 - (2\sqrt{10})^2})$$

$$= 20$$

57. First, find the midpoint M of $\overline{PQ}$.

$$M(\frac{-6+2}{2}, \frac{-3+5}{2}) = M(-2, 1)$$

The slope of $\overline{PQ}$ is . $m_{PQ} = \dfrac{5-(-3)}{2-(-6)} = 1$. Thus,

the slope of the perpendicular bisector L is

$$m_L = \frac{-1}{1} = -1.$$

Use the point-slope model for L.

$$y - 1 = -1[x - (-2)]$$
$$y = -x - 1$$

59. First, find the midpoint M of $\overline{PQ}$.

$$M(\frac{2+7}{2}, \frac{-4+8}{2}) = M(\frac{9}{2}, 2)$$

The slope of $\overline{PQ}$ is $m_{PQ} = \dfrac{8-(-4)}{7-2} = \dfrac{12}{5}$. Thus,

the slope of the perpendicular bisector L is

$$m_L = \frac{-1}{\frac{12}{5}} = -\frac{5}{12}.$$

Use the point-slope model for L.

$$y - 2 = -\frac{5}{12}(x - \frac{9}{2})$$

$$y = -\frac{5}{12}x + \frac{31}{8}$$

61. $M_{AD}(\dfrac{0+2b}{2}, \dfrac{0+2c}{2}) = P(b, c)$

$$M_{CD}(\frac{2b+2d}{2}, \frac{2c+2e}{2}) = Q(b+d, c+e)$$

$$M_{BC}(\frac{2a+2d}{2}, \frac{2e+0}{2}) = R(a+d, e)$$

$$M_{AB}(\frac{2a+0}{2}, \frac{0+0}{2}) = S(a, 0)$$

We label the midpoints $P(b, c)$, $Q(b + d, c + e)$, $R(a + d, e)$ and $S(a, 0)$ and calculate the line segments connecting them.

$$PQ = \sqrt{[(b+d)-b]^2 + [(c+e)-c]^2}$$
$$= \sqrt{d^2 + e^2}$$

$$QR = \sqrt{[(a+d)-(b+d)]^2 + [e-(c+e)]^2}$$
$$= \sqrt{(a-b)^2 + c^2}$$

$$RS = \sqrt{[a-(a+d)]^2 + (0-e)^2} = \sqrt{d^2 + e^2}$$

$$SP = \sqrt{(b-a)^2 + (c-0)^2} = \sqrt{(b-a)^2 + c^2}$$

Since $PQ = RS$ and $QR = SP$, the figure is a parallelogram.

© Houghton Mifflin Company. All rights reserved.

63. Assume *a*, *b* and *c* are positive.
First, the coordinates of *P* are

$$M_{AB}\left(\frac{2a+0}{2}, \frac{2b+0}{2}\right) = M_{AB}(a, b)$$

and the coordinates of *Q* are

$$M_{BC}\left(\frac{2a+2c}{2}, \frac{2b+0}{2}\right) = M_{BC}(a+c, b).$$

Now, calculate the slope of $\overline{PQ}$.

$$m_{PQ} = \frac{b-b}{(a+c)-a} = \frac{0}{c} = 0 \text{ and this is equal to the}$$

slope of $\overline{AC}$. Now find the length of $\overline{PQ}$.

$$PQ = \sqrt{[(a+c)-a]^2 + (b-b)^2} = \sqrt{c^2} = c \text{ which}$$

is half the length of $AC = 2c$.

65. For the line $y = -2x + 5$, $m = -2$ and $b = 5$.

$$d = \frac{|mx_1 + b - y_1|}{\sqrt{1+m^2}} = \frac{|-2(-4)+5-(-2)|}{\sqrt{1+(-2)^2}}$$

$$= \frac{15}{\sqrt{5}} = 3\sqrt{5} \approx 6.71$$

67. First, find *m* and *b* for the given line.
$$x + 2y - 2 = 0$$
$$y = -\frac{1}{2}x + 1$$

Therefore, $m = -\frac{1}{2}$ and $b = 1$.

$$d = \frac{|mx_1 + b - y_1|}{\sqrt{1+m^2}} = \frac{\left|\left(-\frac{1}{2}\right) \cdot 7 + 1 - (-3)\right|}{\sqrt{1+\left(-\frac{1}{2}\right)^2}}$$

$$= \frac{\frac{1}{2}}{\sqrt{\frac{5}{4}}} = \frac{\sqrt{5}}{5} \approx 0.45$$

69. We have two points: (1990, 6.5) and (1996, 7.9).
The midpoint is

$$M\left(\frac{1990+1996}{2}, \frac{6.5+7.9}{2}\right) = M(1993, 7.2).$$

The midpoint indicates that in 1993 the median

71.   If points *A* and *C*
belong to the line $y = -\frac{8}{9}x + 400$, the coordinates
of *A* are the
y-intercept, or (0, 400), and the coordinates of
C are the *x*-intercept, or (450, 0).

$$AB = \sqrt{(900-0)^2 + (700-400)^2}$$
$$= 300\sqrt{10} \approx 948.68$$
$$BC = \sqrt{(450-900)^2 + (0-700)^2}$$
$$= 50\sqrt{277} \approx 832.17$$
$$AC = \sqrt{(450-0)^2 + (0-400)^2}$$
$$= 50\sqrt{145} \approx 602.08$$

The perimeter is 948.68 + 832.17 + 602.08, or
about 2383 feet.

2.5 Exercises

1. The *x*-intercepts correspond to the zeros of the
function.

3.

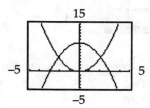

$$f(x) = 9 - x^2$$

The graph of *f* is the graph of $y = x^2$ reflected
across the *x*-axis and shifted upward 9 units.

5.

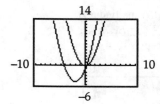

$$f(x) = (x+2)^2 - 5$$

The graph of *f* is the graph of $y = x^2$ shifted to
the left 2 units and downward 5 units.

7.

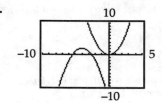

$$f(x) = -(x+4)^2 + 2$$

The graph of *f* is the graph of $y = x^2$ reflected
across the *x*-axis, and then shifted to the left 4
units and upward 2 units.

9. B

11. D

13. C

15. $g(x) = \frac{1}{2}x^2 - 8 = \frac{1}{2}(x-0)^2 - 8$

The vertex is $V(0, -8)$; the axis of symmetry is
$x = 0$.

17. $y(x) = -(x-4)^2 + 1$

The vertex is $V(4, 1)$; the axis of symmetry is
$x = 4$.

19. $p(x) = 2(x+1)^2 + 3$

The vertex is $V(-1, 3)$; the axis of symmetry is
$x = -1$.

© Houghton Mifflin Company. All rights reserved.

21.

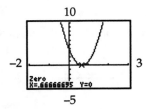

For the x-intercepts, solve the equation $f(x) = 0$.

$$9x^2 - 12x + 4 = 0$$
$$(3x - 2)^2 = 0$$
$$3x - 2 = 0$$
$$x = \frac{2}{3}$$

The x-intercept is $(\frac{2}{3}, 0)$.

For the y-intercept, find $f(0)$.

$$f(0) = 9(0)^2 - 12(0) + 4 = 4$$

The y-intercept is $(0, 4)$.

23.

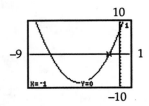

For the x-intercepts, solve the equation $g(x) = 0$.

$$x^2 + 8x + 7 = 0$$
$$(x + 1)(x + 7) = 0$$
$$x + 1 = 0 \quad \text{or} \quad x + 7 = 0$$
$$x = -1 \qquad x = -7$$

The x-intercepts are $(-1, 0)$ and $(-7, 0)$.

For the y-intercept, find $g(0)$.

$$g(0) = (0)^2 + 8(0) + 7 = 7$$

The y-intercept is $(0, 7)$.

25. $h(x) = 3 + 4x - 2x^2$

$$= -2(x^2 - 2x) + 3$$
$$= -2(x^2 - 2x + 1 - 1) + 3$$
$$= -2(x^2 - 2x + 1) + 2(1) + 3$$
$$= -2(x - 1)^2 + 5$$

The vertex is $V(1, 5)$; the axis of symmetry is $x = 1$.

27. $f(x) = 2x^2 - x + 1$

$$= 2(x^2 - \frac{1}{2}x) + 1$$
$$= 2(x^2 - \frac{1}{2}x + \frac{1}{16} - \frac{1}{16}) + 1$$
$$= 2(x^2 - \frac{1}{2}x + \frac{1}{16}) - 2(\frac{1}{16}) + 1$$
$$= 2(x - \frac{1}{4})^2 + \frac{7}{8}$$

The vertex is $V(\frac{1}{4}, \frac{7}{8})$; the axis of symmetry is

$$x = \frac{1}{4}.$$

29. The vertex is the highest (or lowest) point of the graph, so the y-coordinate is the largest (or smallest) value of the elements of the range of the function.

31.

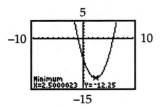

The x-coordinate of the vertex is given by

$$-\frac{b}{2a} = -\frac{3}{2(-1)} = \frac{3}{2}.$$

To determine the y-coordinate, evaluate $f(\frac{3}{2})$.

$$f(\frac{3}{2}) = 6 + 3(\frac{3}{2}) - (\frac{3}{2})^2 = \frac{33}{4}$$

Since $a < 0$, $\frac{33}{4}$ is the maximum value and the

range is $\left(-\infty, \frac{33}{4}\right]$.

33.

The x-coordinate of the vertex is given by

$$-\frac{b}{2a} = -\frac{(-5)}{2(1)} = \frac{5}{2}.$$

To determine the y-coordinate, evaluate $g(\frac{5}{2})$.

$$g(\frac{5}{2}) = (\frac{5}{2})^2 - 5(\frac{5}{2}) - 6 = -\frac{49}{4}$$

Since $a > 0$, $-\frac{49}{4}$ is the minimum value and the

range is $\left[\frac{-49}{4}, \infty\right)$.

© Houghton Mifflin Company. All rights reserved.

35.

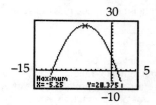

The x-coordinate of the vertex is given by

$$-\frac{b}{2a} = -\frac{(-7)}{2(-\frac{2}{3})} = -\frac{21}{4}.$$

To determine the y-coordinate, evaluate $f(-\frac{21}{4})$.

$$f(-\frac{21}{4}) = -\frac{2}{3}(-\frac{21}{4})^2 - 7(-\frac{21}{4}) + 10$$

$$= \frac{227}{8}$$

Since $a < 0$, $\frac{227}{8}$ is the maximum value and the

range is $\left(-\infty, \frac{227}{8}\right]$.

37. For a quadratic function y with vertex $V(h, k)$, we
have $y = a(x - h)^2 + k$
$$y = a(x + 3)^2 - 2$$
Since the graph contains $(0, 7)$,
$$7 = a(0 + 3)^2 - 2$$
$$7 = 9a - 2$$
$$a = 1$$
Thus, $y = (x + 3)^2 - 2$.

39. For a quadratic function y with vertex $V(h, k)$, we
have $y = a(x - h)^2 + k$
$$y = a(x + 2)^2 - 8$$
Since the graph contains $(0, 0)$,
$$0 = a(0 + 2)^2 - 8$$
$$0 = 4a - 8$$
$$a = 2$$
Thus, $y = 2(x + 2)^2 - 8$.

41. For a quadratic function y with vertex $V(h, k)$, we
have $y = a(x - h)^2 + k$
$$y = a(x + 4)^2 - 5$$
Since the graph contains $(-5, -7)$,
$$-7 = a(-5 + 4)^2 - 5$$
$$-7 = a - 5$$
$$a = -2$$
Thus, $y = -2(x + 4)^2 - 5$.

43. Since $a = 2 > 0$, the parabola opens upward. Since
the vertex is $(3, -3)$, the range is $[-3, \infty)$. The
graph intersects the x-axis in two places, so there
are two zeros.

45. Since $a = -1 < 0$, the parabola opens downward.
Since the vertex is $(-4, -1)$, the range is $(-\infty, -1]$.
The graph does not intersect the x-axis, so there
are no zeros.

47. Since $a = \frac{1}{2} > 0$, the parabola opens upward.
Since the vertex is $(5, 0)$, the range is $[0, \infty)$. The
graph intersects the x-axis in one point, so there is
one zero.

49. Since the vertex is $(2, -1)$ and the graph contains
$(0, -8)$, it must open downward. Therefore, the
graph does not intersect the x-axis so there are no
zeros.

51. If $(7, 3)$ is the maximum value, the graph opens
downward and intersects the x-axis in two places.
There are two zeros.

53. If 4 is a zero of f then $f(4) = 0$.
$$f(x) = kx^2 - 9x + 2k$$
$$f(4) = k(4)^2 - 9(4) + 2k$$
$$0 = 18k - 36$$
$$k = 2$$

55. The minimum value is at $(x, -8)$ where
$$x = -\frac{b}{2a} = -\frac{(-4k)}{2k} = 2 \text{ and } y = f(-\frac{b}{2a})$$
$$f(x) = kx^2 - 4kx + 4$$
$$f(2) = k(2)^2 - 4k(2) + 4$$
$$-8 = -4k + 4$$
$$k = 3$$

57. The axis of symmetry is given by
$$x = -\frac{b}{2a} = -\frac{6k}{2k^2} = -\frac{3}{k}.$$
Thus $-\frac{3}{2} = -\frac{3}{k}$
$$k = 2$$

59. The maximum value occurs at $(-\frac{b}{2a}, 5)$.

$$-\frac{b}{2a} = -\frac{(-3k)}{2(-3)} = -\frac{k}{2}$$
$$f(x) = k - 3kx - 3x^2$$
$$f(-\frac{k}{2}) = k - 3k(-\frac{k}{2}) - 3(-\frac{k}{2})^2$$
$$5 = \frac{3}{4}k^2 + k$$
$$20 = 3k^2 + 4k$$
$$3k^2 + 4k - 20 = 0$$
$$(3k + 10)(k - 2) = 0$$
$$3k + 10 = 0 \quad \text{or} \quad k - 2 = 0$$
$$k = -\frac{10}{3} \qquad k = 2$$
Thus, we have two functions:
$$f(x) = -\frac{10}{3} + 10x - 3x^2 \text{ and } f(x) = 2 - 6x - 3x^2.$$

61. If the discriminant is positive, the function has
2 zeros and the graph has 2 x-intercepts. If the
discriminant is zero, the function has 1 zero and
the graph has 1 x-intercept. If the discriminant is
negative, the function has no zeros and the graph
has no x-intercept.

© Houghton Mifflin Company. All rights reserved.

63.

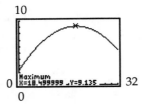

a. The maximum appears to be at (19, 9) (to the nearest integer). The store's highest sales were about $9000 on December 13.

b. The day before Christmas is December 24, 30 days after November 24. So $x = 30$. Using the trace feature, $y = 6$ (to the nearest integer). The store's sales on the day before Christmas were about $6000.

c. The store's sales are predicted to be $5000 when $x = 4$ and when $x = 33$ (to the nearest integer), or on November 28 and December 27.

65. The maximum height will occur at the vertex.

$$x = -\frac{b}{2a} = -\frac{64}{2(-16)} = 2$$

$h(2) = -16(2)^2 + 64(2) = 64$
The maximum height is 64 feet.
The ball will return to the ground when x is equal to one of the zeros of the function.

$$h(t) = 0$$
$$-16t^2 + 64t = 0$$
$$-16t(t - 4) = 0$$
$$t = 0 \text{ or } t - 4 = 0$$
$$t = 4$$

The ball will stay in the air for 4 seconds.

67.

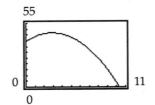

Using the trace feature, the vertex is about (2.91, 46.77). So the maximum number of visitors was about 47 million in 1993.

69. a. $R(x) = 45x - 0.5x^2$

b. $P(x) = R(x) - C(x)$
$P(x) = (45x - 0.5x^2) - (500 + 4.5x)$
$P(x) = -0.5x^2 + 40.5x - 500$

c. P is a quadratic function and its graph is a parabola. Thus, the maximum value of P will occur at the vertex.
Since it is impossible to have 40.5 passengers, the maximum will be produced at the greater of $P(40)$ and $P(41)$.
$P(40) = -0.5(40)^2 + 40.5(40) - 500 = 320$
$P(41) = -0.5(41)^2 + 40.5(41) - 500 = 320$
The maximum profit is $320.

d.   For the maximum profit, either 40 or 41 tickets should be sold.

e.   To break even, the profit is 0. Solve the equation $P(x) = 0$.
$$-0.5x^2 + 40.5x - 500 = 0$$
$$x = \frac{-40.5 \pm \sqrt{(40.5)^2 - 4(-0.5)(-500)}}{2(-0.5)}$$
$$x = \frac{-40.5 \pm \sqrt{640.25}}{-1}$$
$$x = 40.5 \pm 25.3$$
$$x \approx 15.2 \text{ or } x \approx 65.8$$
15 tickets need to be sold to break even.

71. The area is $A(x) = (150 - 3x) \cdot x$
$$A(x) = -3x^2 + 150x$$
The graph is a parabola, opening down, with vertex at
$$x = -\frac{b}{2a} = -\frac{150}{2(-3)} = 25$$
and $A(25) = -3(25)^2 + 150(25) = 1875$.
The maximum area is 1875 square feet.

© Houghton Mifflin Company. All rights reserved.

2.6 Exercises

1.

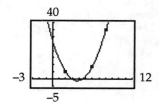

$$y = 2x^2 - 15x + 27$$

3.

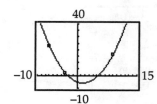

$$y = 0.4x^2 - 1.2x - 5$$

5. A quadratic model would be more appropriate.

7. A linear model would be more appropriate.

9.

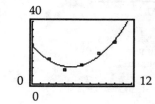

A quadratic model is appropriate.

$$y = 0.63x^2 - 5.89x + 24.71$$

The correlation coefficient is $r^2 \approx .095$, so the model is reasonable.

11.

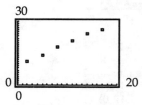

A quadratic model is not appropriate.

13.

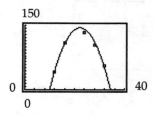

A quadratic model is appropriate.

$$y = -21.88x^2 + 242.43x - 469.65$$

The correlation coefficient is $r^2 \approx 0.97$, so the model is reasonable.

15. a. a must be negative so the sign is minus.

b. (1991, 43)

c.

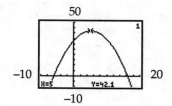

Using the data points (4, 41), (6, 43), (8, 40), (10, 32), and (12, 27), we obtain the quadratic regression equation

$$y = -0.34x^2 + 3.48x + 33.2.$$

The vertex is at (5, 42.1). The coordinates indicate the year, 1985 + 5 = 1990, in which rates were highest and the rate, 42.1, for that year.

17. a

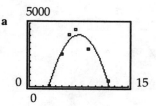

A quadratic regression model is appropriate.

b. We use the data points (3, 90), (5, 2600), (6, 4100), (7, 4500), (9, 3000), and (12, 450). Quadratic regression gives us

$$y = -193x^2 + 2900x - 6800.$$

c. The model most underestimates the data for $x = 6$ and $x = 7$, corresponding to the years 1996 and 1997.

19. a. A quadratic regression model is appropriate. Using the data points (0, 11.2), (5, 11.9), (9, 12.7), (11, 11.3), and (14, 9.4), we obtain

$$y = -0.04x^2 + 0.51x + 11.03.$$

b. Because $r^2 \approx 0.87$, the model is reasonable.

c. The x-intercept is (24.16, 0). This means in the year 2004, the funding will be 0.

21. a. We use the data points (1, 214), (2, 245), (3, 323), (4, 305), and (5, 317). The linear regression equation is $y = 26.6x + 201$. The quadratic regression equation is

$$y = -9.57x^2 + 84.03x + 134.$$

b. For the linear regression model, $r^2 \approx 0.75$.

For the quadratic regression model, $r^2 \approx 0.89$. The larger correlation coefficient for the quadratic model indicates it is the better model.

© Houghton Mifflin Company. All rights reserved.

2.7 Exercises

1. The first coordinate of a *y*-intercept is 0 and the second coordinate tells us where the graph crosses the *y*-axis. The second coordinate of an *x*-intercept is 0 and the first coordinate tells us where the graph crosses the *x*-axis.

3.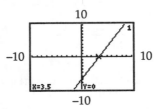

$f(x) = 2x - 7$
y-intercept: (0, –7)
x-intercept: $(\frac{7}{2}, 0)$

5.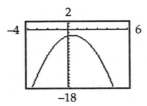

$g(x) = -x^2 + x - 2$
y-intercept: (0, –2)
x-intercept: none

7.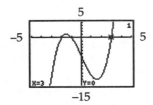

$p(x) = x^3 - 7x - 6$
y-intercept: (0, –6)
x-intercepts: (–2, 0), (–1, 0), (3, 0)

9.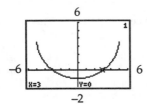

$c(x) = 4 - \sqrt{25 - x^2}$
y-intercept: (0, –1)
x-intercepts: (–3, 0), (3, 0)

11. The function *g* has a relative minimum $x = -1$.

13. Increasing: (–2, 2)
Decreasing: (2, 4)
Constant: (–4, –2)

15. *y*-intercept: (0, 3)
x-intercepts: (1, 0), (5, 0)
Relative minimum: (3, –2)
Increasing: (3, ∞)
Decreasing: (–∞, 3)

17.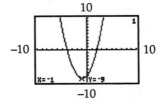

$f(x) = x^2 + 2x - 8$
Relative minimum: (–1, –9)
Increasing: (–1, ∞)
Decreasing: (–∞, –1)

19.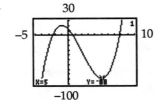

$h(w) = w^3 - 6w^2 - 15w + 12$
Relative maximum: (–1, 20)
Relative minimum: (5, –88)
Increasing: (–∞, –1), (5, ∞)
Decreasing: (–1, 5)

21.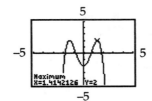

$f(t) = 4t^2 - t^4 - 2$
Relative maxima: (–1.41, 2), (1.41, 2)
Relative minimum: (0, –2)
Increasing (–∞ –1.41), (0, 1.41)
Decreasing: (–1.41, 0), (1.41, ∞)

23.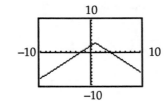

$g(t) = 3 - |t - 1|$
Relative maximum: (1, 3)
Increasing: (–∞, 1)
Decreasing: (1, ∞)

© Houghton Mifflin Company. All rights reserved.

25.

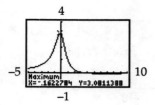

$f(x) = \dfrac{3-x}{x^2+1}$
Relative maximum: (−0.16, 3.08)
Relative minimum: (6.16, −0.08)
Increasing: (−∞, −0.16), (6.16, ∞)
Decreasing: (−0.16, 6.16)

27.

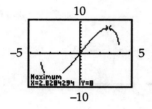

$r(t) = t\sqrt{16-t^2}$
Relative maxima: (−4, 0), (2.83, 8)
Relative minima: (4, 0), (−2.83, −8)
Increasing: (−2.83, 2.83)
Decreasing: (−4, −2.83), (2.83, 4)

29. An even function is symmetric with respect to the *y*-axis. An odd function is symmetric with respect to the origin.

31. a

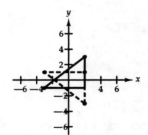

b.

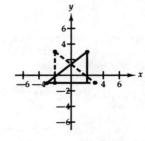

33. a.

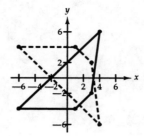

b.

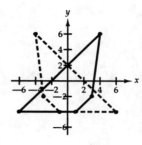

35. a.

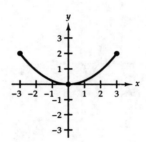

b.

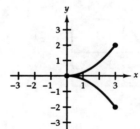

c.

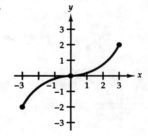

© Houghton Mifflin Company. All rights reserved.

37. a.

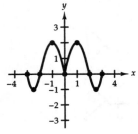

b.

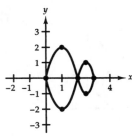

c.

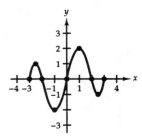

39.

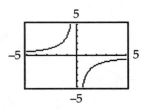

$$f(x) = -\frac{3}{x}$$
$$f(-x) = -\frac{3}{-x} = \frac{3}{x} = -f(x)$$
Because $f(-x) = -f(x)$, f is an odd function.

41.

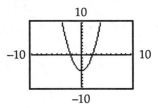

$$h(x) = x^2 - 5$$
$$h(-x) = (-x)^2 - 5 = x^2 - 5 = h(x)$$
Because $h(-x) = h(x)$, h is an even function.

43.

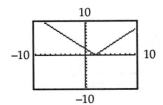

$$f(x) = |x - 2|$$
$$f(-x) = |(-x) - 2| = |-x - 2|$$
$$= |-(x + 2)| = |x + 2|$$
Because $f(-x)$ is equal to neither $f(x)$ nor $-f(x)$, f is neither an even nor odd function.

45.

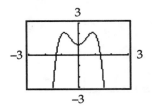

$$f(x) = -2x^4 + 3x^2 + 1$$
$$f(-x) = -2(-x)^4 + 3(-x)^2 + 1$$
$$= -2x^4 + 3x^2 + 1 = f(x)$$
Because $f(-x) = f(x)$, f is an even function.

47.

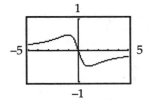

$$g(x) = \frac{-x}{x^2 + 1}$$
$$g(-x) = \frac{-(-x)}{(-x)^2 + 1} = \frac{x}{x^2 + 1} = -g(x)$$
Because $g(-x) = -g(x)$, g is an odd function.

49. $x = y^2 - 5$
$x = (-y)^2 - 5 = y^2 - 5$
Yes, since replacing y with $-y$ results in the same equation, the graph of the equation is symmetric with respect to the x-axis.

51. Since $x = (-y)^2 = y^2$, the graph is symmetric with respect to the x-axis.

53. Since $(-x)(-y) = xy = 1$, the graph is symmetric with respect to the origin.

55. Since $|-x| + |y| = |x| + |y| = 5$, the graph is symmetric with respect to the y-axis.
Since $|x| + |-y| = |x| + |y| = 5$, the graph is symmetric with respect to the x-axis.
Since $|-x| + |-y| = |x| + |y| = 5$, the graph is symmetric with respect to the origin.

© Houghton Mifflin Company. All rights reserved.

57. Both f and h are always increasing. Function g has a relative minimum at $(0, 0)$.

59. $f(x) = \begin{cases} 3, & x \le 0 \\ x, & x > 0 \end{cases}$

61. $f(x) = \begin{cases} |x|, & x \le 0 \\ \sqrt{x}, & x > 0 \end{cases}$

63.

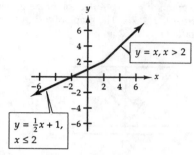

$y = x, x > 2$

$y = \frac{1}{2}x + 1, \; x \le 2$

f is continuous everywhere.

65.

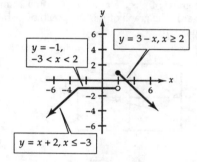

$y = -1, \; -3 < x < 2$

$y = 3 - x, x \ge 2$

$y = x + 2, x \le -3$

h is discontinuous at $x = 2$.

67.

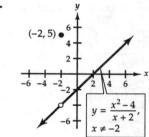

$(-2, 5)$

$y = \dfrac{x^2 - 4}{x + 2}, \; x \ne -2$

f is discontinuous at $x = -2$.

69.

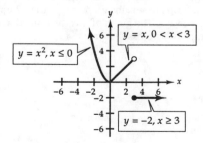

$y = x, 0 < x < 3$

$y = x^2, x \le 0$

$y = -2, x \ge 3$

g is discontinuous at $x = 3$.

71. a.

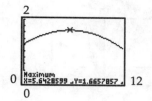

Maximum X=5.6428599 Y=1.6657857

The maximum occurs at $(5.6, 1.7)$. This corresponds to 1.7 million car thefts in 1991 and approximates the data for 1991 on the bar graph.

b. T is increasing on $[0, 4]$ and decreasing on $[6, 10]$.

73. The graph would most resemble $y = x$.

75. The graph would most resemble $y = \dfrac{1}{x}$.

77. a.

$$C(x) \begin{cases} 0.32 & 0 < x \le 1 \\ 0.55 & 1 < x \le 2 \\ 0.78 & 2 < x \le 3 \\ 1.01 & 3 < x \le 4 \\ 1.24 & 4 < x \le 5 \\ 1.47 & 5 < x \le 6 \end{cases}$$

b. (i) $C(0.6) = 0.32$
    (ii) $C(2.9) = 0.78$
    (iii) $C(5.1) = 1.47$

2.8 Exercises

1. A rigid transformation preserves the shape of the graph, whereas a nonrigid transformation distorts the graph.

3. a.

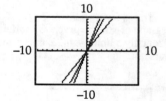

$g(x) = cx$ for $c = 2, 3, 4$
The graph of g is steeper than the graph of f.

b.

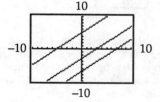

$h(x) = x + c$ for $c = -7, -3, 5$
The graph of h is the graph of f shifted vertically 7 units downward, 3 units downward, and 5 units upward.

c. $y = 3x - 3$

© Houghton Mifflin Company. All rights reserved.

5. a.

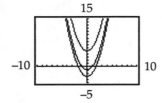

$g(x) = x^2 + c$ for $c = -3, -1, 5$
The graph of g is the graph of f shifted
vertically 3 units downward, 1 unit downward,
and 5 units upward.

b.

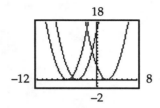

$h(x) = (x + c)^2$ for $c = -2, 4, 6$
The graph of h is the graph of f shifted
horizontally 2 units right, 4 units left and
6 units left.

c. $y = (x + 2)^2 + 3$

7. a.

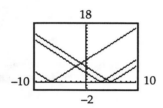

$g(x) = |x + c|$ for $c = -5, -3, 7$
The graph of g is the graph of f shifted
horizontally 5 units right, 3 units right, and
7 units left.

b

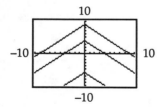

$h(x) = -|x| + c$ for $c = -6, 4, 9$
The graph of h is the reflection of the graph of
f across the x-axis and shifted 6 units
downward, 4 units upward, and 9 units
upward.

c. $y = -|x - 2| - 3$

9.

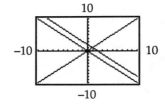

$f(x) = x$, $g(x) = -x$, $h(x) = -x + 2$
The graph of g is the reflection of the graph of
f across the x-axis. The graph of h is the graph of
g shifted upward 2 units.

11.

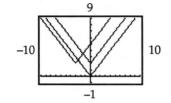

$f(x) = |x|$, $g(x) = |x| + 2$, $h(x) = |x + 3| + 2$
The graph of g is the graph of f shifted 2 units
upward. The graph of h is the graph of g shifted
3 units left.

13.

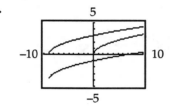

$f(x) = \sqrt{x}$, $g(x) = \sqrt{x + 9}$, $h(x) = \sqrt{x + 9} - 4$
The graph of g is the graph of f shifted 9 units left.
The graph of h is the graph of g shifted 4 units
down.

15.

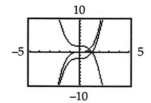

$f(x) = x^3$, $g(x) = x^3 - 2$, $h(x) = 2 - x^3$
The graph of g is the graph of f shifted 2 units
downward. The graph of h is the reflection of the
graph of g across the x-axis.

© Houghton Mifflin Company. All rights reserved.

17.

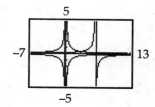

$f(x) = \dfrac{1}{x}$, $g(x) = -\dfrac{1}{x}$, $h(x) = -\dfrac{1}{x-6}$

The graph of g is the reflection of the graph of f across the x-axis. The graph of h is the graph of g shifted 6 units right.

19.

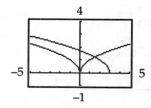

$f(x) = \sqrt{x}$, $g(x) = \sqrt{-x}$, $h(x) = \sqrt{3-x}$

The graph of g is the reflection of the graph of f across the y-axis. The graph of h is the graph of g shifted three units right.

21. Since the graph of $f(x+3)$ is the graph of $f(x)$ shifted 3 units left, the zeros of $f(x+3)$ are $-2-3 = -5$ and $5-3 = 2$.

23.

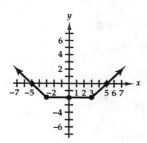

25.

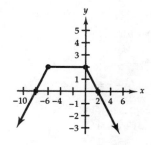

27.

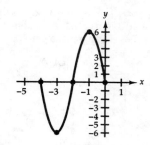

29.

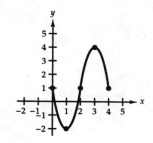

31.

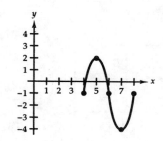

33. As k becomes larger, the graph of f becomes steeper.

35. F

37. A

39. E

41. $f(x) = x^2$ shifted to the left 6 units and upward 5 units is $g(x) = (x+6)^2 + 5$.

43. $f(x) = \dfrac{1}{x}$ reflected across the y-axis is

$g(x) = -\dfrac{1}{x}$.

45. $f(x) = \sqrt{x}$ reflected across the y-axis and shifted to the right 2 units is $g(x) = \sqrt{2-x}$.

47. The basic function is $f(x) = x^2$. The function whose graph is given is $y = (x-2)^2$.

49. The basic function is $f(x) = \sqrt{x}$. The function whose graph is given is $y = \sqrt{x} + 3$.

© Houghton Mifflin Company. All rights reserved.

51. Notice that $g(x) = x - 3 = f(x) - 3$, so g is the graph of f shifted 3 units right. Also notice $g(x) = x - 3 = f(x - 3)$, so g is the graph of f shifted 3 units downward.

53.

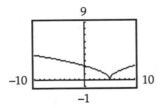

$$y = f(-x) = \sqrt{5 - (-x)} \text{ or } y = \sqrt{5 + x}$$

55.

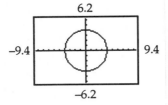

$$y = -c(x) = -\sqrt{16 - x^2}$$

57.

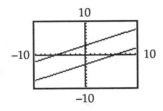

$$y = -g(-x) = -[\tfrac{1}{2}(-x) + 3] \text{ or } y = \tfrac{1}{2}x - 3$$

59. a. If x of the 16 players are boys, then $f(x) = 16 - x$ are girls.

b. The graph of f is the graph of y reflected across the x-axis and shifted 16 units upward.

61. $g(x) = -f(x) + 60$

Review Exercises

1. a. Yes

b. No

3. $\left|x^2 - 3\right| = y - 1$ does define a function because we can write y as a single expression in terms of x.

5. $g(t) = \dfrac{t + 8}{t^2 - 16}$

a. $g(0) = \dfrac{(0) + 8}{(0)^2 - 16} = -\dfrac{1}{2}$

b. $g(4) = \dfrac{(4) + 8}{(4)^2 - 16} = \dfrac{12}{0} = \text{undefined}$

c. $g(2a) = \dfrac{(2a) + 8}{(2a)^2 - 16}$

$= \dfrac{2(a + 4)}{4(a^2 - 4)}$

$= \dfrac{a + 4}{2(a^2 - 4)}$

7.

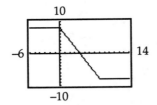

Domain: $\mathbb{R}$; Range: $[-8, 8]$

9. $5x - x^2 \neq 0$

$x(5 - x) \neq 0$

$x \neq 0 \quad \text{and} \quad 5 - x \neq 0$

$x \neq 5$

Domain: $\{ x \mid x \neq 0 \ 5 \}$

11. The x-intercepts correspond to $f(x) = 0$.

13. $m = \dfrac{1 - (-2)}{5 - 7} = -\dfrac{3}{2}$

15. Using the point-slope model:

$y - y_1 = m(x - x_1)$

$y - (-3) = -4 \cdot (x - 1)$

$y + 3 = -4x + 4$

$y = -4x + 1$

17. $f(x) = mx + b$

$f(0) = -2(0) + b$

$0 = b$

Thus, $f(x) = -2x$.

19. x-intercept: $3(0) - x = 15$

$x = -15$

The x-intercept is $(-15, 0)$.

y-intercept: $3y - (0) = 15$

$y = 5$

The y-intercept is $(0, 5)$.

© Houghton Mifflin Company. All rights reserved.

21. a. $1 - 3x = x + 13$
$-4x = 12$
$x = -3$

b. $1 - 3x < x + 3$
$-4x < 2$
$\dfrac{-4x}{-4} > \dfrac{2}{-4}$
$x > -\dfrac{1}{2}$

23. Using the data points (3, 33), (5, 38), and (7, 46), we obtain the linear regression equation $y = 3.25x + 22.75$. To find when the percentage will exceed 60%, we solve the inequality.
$3.25x + 22.75 > 60$
$3.25x > 37.25$
$x > \dfrac{37.25}{3.25}$
$x > 11.46$
The percentage will exceed 60% in $1990 + 12 = 2002$.

25. a. Using the data points (0, 60.2), (2, 61.7), (4, 63.7), (6, 64.3), (8, 66.0), (10, 71.6), and (12, 70.6), we obtain the linear regression equation $y = 0.95x + 59.73$.

b

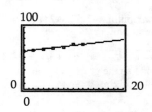

The graph appears to fit the data well.

c. The value of the regression equation is 100 for $x = 42.39$, corresponding to the year $1980 + 43 = 2023$.

27. $L_1: m_1 = \dfrac{-3 - (-4)}{6 - 4} = \dfrac{1}{2}$

$L_2: m_2 = \dfrac{-3 - (-5)}{0 - (-4)} = \dfrac{1}{2}$

Since $m_1 = m_2$, the lines are parallel.

29. $L_1: 4 - x = 7$ $L_2: m_2 = \dfrac{8 - 8}{2 - (-3)} = \dfrac{0}{5} = 0$
$x = -3$
So m_1 is undefined.   So $m_2 = 0$.
Since L_1 is a vertical line and L_2 is a horizontal line, the lines are perpendicular.

31. First, determine the slope.
$m_1 = \dfrac{7 - 5}{0 - (-3)} = \dfrac{2}{3}$
So $m_2 = -\dfrac{1}{m_1} = -\dfrac{1}{\frac{2}{3}} = -\dfrac{3}{2}$
Next, find the y-intercept.

$x - 2y - 8 = 0$
$2y = x - 8$
$y = \dfrac{1}{2}x - 4$
Therefore, the y-intercept is (0, –4). Thus,
$f(x) = -\dfrac{3}{2}x - 4$.

33. Determine the lengths of the sides of the triangle and show that the Pythagorean Theorem is satisfied. Determine the slopes of the sides and show that two of the sides are perpendicular.

35. First find the midpoint of $\overline{AB}$.
$M(\dfrac{-4 + 2}{2}, \dfrac{3 - 7}{2}) = M(-1, -2)$
The slope of $\overline{AB}$ is $m_{AB} = \dfrac{-7 - 3}{2 - (-4)} = -\dfrac{5}{3}$.
Thus, $m_L = \dfrac{-1}{-\frac{5}{3}} = \dfrac{3}{5}$.
Use the point-slope model for L.
$y - y_1 = m(x - x_1)$
$y - (-2) = \dfrac{3}{5}[x - (-1)]$
$y + 2 = \dfrac{3}{5}x + \dfrac{3}{5}$
$y = \dfrac{3}{5}x - \dfrac{7}{5}$

37. The rate of change is
$m = \dfrac{7.1 - 5.3}{1998 - 1996} = \dfrac{1.8}{2} = 0.9$
Or, $0.9 billion per year. Since 2001 is $2001 - 1996 = 5$ years after 1996, the sales in 2001 will be $5.3 + 5 \cdot 0.9 = 9.8$, or $9.8 billion.

39.

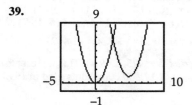

$y = x^2$, $f(x) = (x - 5)^2 + 1$
f is the graph of y shifted 5 units right and 1 unit upward.

© Houghton Mifflin Company. All rights reserved.

41. $g(x) = x^2 - 6x - 10$

y-intercept: $g(0) = -10$

The y-intercept is $(0, -10)$.

x-intercepts:

$x^2 - 6x - 10 = 0$

$$x = \frac{-(-6) \pm \sqrt{(-6)^2 - 4(1)(-10)}}{2(1)}$$

$$x = \frac{6 \pm \sqrt{76}}{2}$$

$$x = \frac{6 - \sqrt{76}}{2} \text{ or } x = \frac{6 + \sqrt{76}}{2}$$

$x \approx -1.36$ or $x \approx 7.36$

The x-intercepts are $(-1.36, 0)$ and $(7.36, 0)$.

$y = x^2 - 6x - 10$

$y = (x^2 - 6x + 9) - 9 - 10$

$y = (x - 3)^2 - 19$

The vertex is $V(3, -19)$.

43. $f(x) = -x^2 + 4x - 5$

$f(x) = -(x^2 - 4x) - 5$

$f(x) = -(x^2 - 4x + 4) + 4 - 5$

$f(x) = -(x - 2)^2 - 1$

Since $a = -1 < 0$, the parabola opens down and the vertex is a maximum. Thus, the range is $(-\infty, -1]$ and the maximum is -1.

45. If the function has a minimum value, the parabola opens up and doesn't intersect the x-axis. Thus, there are no zeros of the function.

47. The single x-intercept is the vertex. So, the vertex lies on the x-axis.

49. We use quadratic regression with the data points $(-3, 26)$, $(6, 8)$, and $(12, 56)$ and obtain

$y = \frac{2}{3}x^2 - 4x + 8$.

51. a. A quadratic regression equation seems most appropriate.

b. We use quadratic regression with the datapoints $(0, 14.7)$, $(1, 17.1)$, $(2, 19.3)$, $(3, 19.9)$, $(4, 19.6)$, $(5, 17.4)$, and $(6, 15.7)$ and obtain $y = -0.53x^2 + 3.30x + 14.62$.

c. The vertex is at $V(3.1, 19.8)$. This represents a maximum of 19,800 complaints filed in 1993.

53.

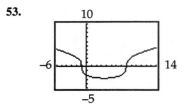

$g(t) = \sqrt[3]{t^2 - 7t - 8}$

y-intercept: $g(0) = \sqrt[3]{0^2 - 7(0) - 8} = \sqrt[3]{-8} = -2$

The y-intercept is $(0, -2)$.

Using the calculator, the x-intercepts are found to be $(-1, 0)$ and $(8, 0)$.

55.

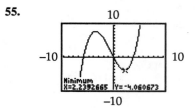

$p(x) = \frac{1}{8}x^3 + \frac{1}{4}x^2 - 3x$

Increasing: $(-\infty, -3.57)$, $(2.24, \infty)$

Decreasing: $(-3.57, 2.24)$

Relative maximum: $(-3.57, 8.21)$

Relative minimum: $(2.24, -4.06)$

57. $f(-x) = \sqrt[3]{-x} - \sqrt[5]{-x} = -\sqrt[3]{x} + \sqrt[5]{x} = -f(x)$

Thus, f is odd so the graph is symmetric with respect to the origin.

59. $y + 3x = x^2$

$y(x) = x^2 - 3x$

$y(-x) = (-x)^2 - 3(-x) = x^2 + 3x$

Thus, y is neither even nor odd so there is no symmetry.

61. No, the pairs (x, y) and $(x, -y)$ belong to the relation.

63.

$y = |x|$, $h(x) = |3 - x|$

The graph of h is the graph of y shifted to the right 3 units.

© Houghton Mifflin Company. All rights reserved.

65.

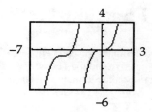

$y = x^3$, $c(x) = -1 + (x+4)^3$
The graph of c is the graph of y shifted to the left
4 units and downward 1 unit.

67.

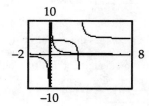

$y = \dfrac{1}{x}$, $r(x) = 5 + \dfrac{1}{x-3}$
The graph of r is the graph of y shifted to the right
3 units and upward 5 units.

69. $f(x) = -\dfrac{1}{x-7}$

71.

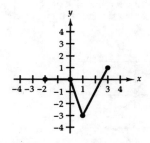

73. $y = |x|$
The graph of f is the graph of y shifted to the left
5 units. Thus, $f(x) = |x + 5|$.

© Houghton Mifflin Company. All rights reserved.

Chapter 3: Polynomial and Rational Functions

3.1 Exercises

1. For even integers n, the graph is symmetric with respect to the y-axis. For odd integers n, the graph is symmetric with respect to the origin.

3. B

5. A

7.

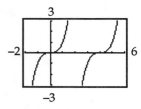

The graph of f is the graph of y shifted to the right 4 units.

9.

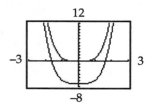

The graph of f is the graph of y shifted downward 7 units.

11.

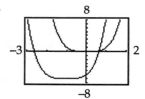

The graph of f is the graph of y shifted to the left 1 unit and downward 7 units.

13. The term $-2x^3$ determines the end behavior. The coefficient is negative and the degree is odd. The graph should rise on the left and fall on the right.

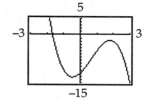

15. The term x^6 determines the end behavior. The coefficient is positive and the degree is even. The graph should rise on the left and rise on the right.

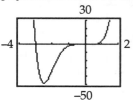

17. The term $-x^4$ determines the end behavior. The coefficient is negative and the degree is even. The graph should fall on the left and fall on the right.

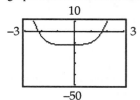

19. A polynomial function with odd degree is continuous and must rise on the left and fall on the right or fall on the left and rise on the right. Thus, it must cross the x-axis and so it must have at least one zero.

21.

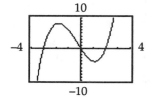

The graph has three x-intercepts. By tracing, we estimate the x-intercepts to be $(-3, 0)$, $(0, 0)$, and $(2, 0)$. To determine the zeros algebraically, we solve $f(x) = 0$.

$$x^3 + x^2 - 6x = 0$$
$$x(x^2 + x - 6) = 0$$
$$x(x + 3)(x - 2) = 0$$
$$x = 0 \quad x + 3 = 0 \quad x - 2 = 0$$
$$x = -3 \qquad x = 2$$

The zeros of the function are 0 (multiplicity 1), -3 (multiplicity 1), and 2 (multiplicity 1).

© Houghton Mifflin Company. All rights reserved.

23.

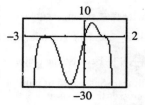

The graph has three x-intercepts. By tracing, we estimate the x-intercepts to be $(-2, 0)$, $(0, 0)$, and $(1, 0)$. To determine the zeros algebraically, we solve $q(x) = 0$.

$$-3x(x+2)^4(x-1)^3 = 0$$
$$x = 0 \quad x+2 = 0 \quad x-1 = 0$$
$$x = -2 \quad x = 1$$

The zeros of the function are 0 (multiplicity 1), -2 (multiplicity 4), and 1 (multiplicity 3).

25.

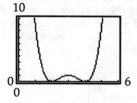

The graph has two x-intercepts. By tracing, we estimate the x-intercepts to be $(2, 0)$ and $(4, 0)$. to determine the zeros algebraically, we solve $p(x) = 0$.

$$\left(x^2 - 6x + 8\right)^2 = 0$$
$$x^2 - 6x + 8 = 0$$
$$(x-2)(x-4) = 0$$
$$x-2 = 0 \quad x-4 = 0$$
$$x = 2 \quad x = 4$$

The zeros of the function are 2 (multiplicity 2) and 4 (multiplicity 2).

27.

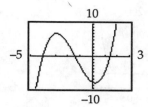

The graph has three x-intercepts. By tracing, we estimate the x-intercepts to be $(-4, 0)$, $(-1.43, 0)$, and $(1.43, 0)$. To determine the zeros algebraically, we solve $h(x) = 0$. We factor by grouping.

$$x^3 + 4x^2 - 2x - 8 = 0$$
$$x^2(x+4) - 2(x+4) = 0$$
$$\left(x^2 - 2\right)(x+4) = 0$$
$$x^2 - 2 = 0 \quad x+4 = 0$$
$$x^2 = 2 \quad x = -4$$
$$x = \pm\sqrt{2}$$

The zeros of the function are -4 (multiplicity 1), $-\sqrt{2} \approx -1.41$ (multiplicity 1), and $\sqrt{2} \approx 1.41$ (multiplicity 1).

29.

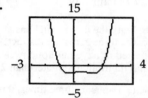

The graph has two x-intercepts. By tracing, we estimate the x-intercepts to be $(-1, 0)$ and $(2, 0)$. To determine the zeros algebraically, we solve $p(x) = 0$. We factor by grouping.

$$x^4 - 2x^3 + x - 2 = 0$$
$$x^3(x-2) + (x-2) = 0$$
$$\left(x^3 + 1\right)(x-2) = 0$$
$$(x+1)\left(x^2 - x + 1\right)(x-2) = 0$$
$$x+1 = 0 \quad x^2 - x + 1 = 0 \quad x-2 = 0$$
$$x = -1 \qquad\qquad x = 2$$
$$x = \frac{-(-1) \pm \sqrt{(-1)^2 - 4(1)(1)}}{2(1)} = \frac{1 + \sqrt{-3}}{2}$$

No real number solution.
The zeros of the function are -1 (multiplicity 1) and 2 (multiplicity 1).

© Houghton Mifflin Company. All rights reserved.

31.

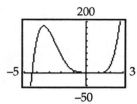

The graph has three x-intercepts. By tracing, we estimate the x-intercepts to be $(-4.37, 0)$, $(0, 0)$, and $(1.37, 0)$. To determine the zeros algebraically, we solve $h(x) = 0$.

$$x^5 + 3x^4 - 6x^3 = 0$$

$$x^3(x^2 + 3x - 6) = 0$$

$$x^3 = 0 \quad x^2 + 3x - 6 = 0$$

$$x = 0 \quad x = \frac{-3 \pm \sqrt{3^2 - 4(1)(-6)}}{2(1)} = \frac{-3 \pm \sqrt{33}}{2}$$

$$x = \frac{-3 - \sqrt{33}}{2} \qquad x = \frac{-3 + \sqrt{33}}{2}$$

The zeros of the function are 0 (multiplicity 3),

$$\frac{-3 - \sqrt{33}}{2} \approx -4.37 \text{ (multiplicity 1), and}$$

$$\frac{-3 + \sqrt{33}}{2} \approx 1.37 \text{ (multiplicity 1).}$$

33. Counting multiplicity the polynomial has at most four zeros. Because three zeros are distinct, only one may be repeated or have multiplicity two. Thus, no zero has multiplicity greater than 2.

35. Because the factor x appears three times, 0 is a zero of multiplicity 3 and the graph crosses the x-axis at $(0, 0)$. Because the factor $(x + 2)$ appears once, -2 is a zero of multiplicity 1 and the graph crosses the x-axis at $(-2, 0)$. Because the factor $(x - 3)$ appears twice, 3 is a zero of multiplicity 2 and the graph is tangent to the x-axis at $(3, 0)$.

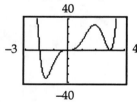

37. Because the factor $(4 - 3x)$ appears twice, $\frac{4}{3}$ is a zero of multiplicity 2 and the graph is tangent to the x-axis at $\left(\frac{4}{3}, 0\right)$. Because the factor $(x + 5)$ appears three times, -5 is a zero of multiplicity 3 and the graph crosses the x-axis at $(-5, 0)$.

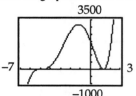

39. Because the factor x appears once, 0 is a zero of multiplicity 1 and the graph crosses the x-axis at $(0, 0)$. Because the factor $(x + 5)$ appears once, -5 is a zero of multiplicity 1 and the graph crosses the x-axis at $(-5, 0)$. Because the factor $(x + 2)$ appears twice, -2 is a zero of multiplicity 2 and the graph is tangent to the x-axis at $(-2, 0)$.

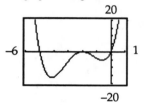

41.

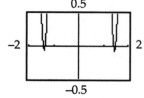

Using the Standard Window, the graph appears to have one intercept. However, if we zoom in on the apparent intercept near 1.5, we see there are actually three intercepts in that vicinity. We estimate the intercepts to be $(1.33, 0)$, $(1.50, 0)$, and $(1.67, 0)$.

43.

Using the Standard Window, the graph appears to have two intercepts. However, if we zoom in on the apparent intercept near -1.5, we see there are actually two intercepts in that vicinity. We estimate the intercepts to be $(-1.41, 0)$ and $(-1.33, 0)$. Similarly, we estimate the intercepts near 1.5 to be $(1.41, 0)$ and $(1.50, 0)$.

Houghton Mifflin Company. All rights reserved.

45. The graph might be close to the point (4, 0) but not contain the point. Zooming in gives a more accurate estimate.

47. To deterine the y-intercept, we evaluate $f(0)$.

$$f(c) = 3(c)^2 - 0^3$$
$$= 0$$

The y-intercept is (0, 0).

To determine the x-intercepts, we solve $f(x) = 0$.

$$3x^2 - x^3 = 0$$
$$x^2(3 - x) = 0$$
$$x^2 = 0 \quad 3 - x = 0$$
$$x = 0 \quad\quad x = 3$$

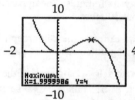

The x-intercepts are (0, 0) and (3, 0). Using the calculator's *minimum* feature, we estimate the relative minimum to be (0, 0). Using the calculator's *maximum* feature, we estimate the relative maximum to be (2, 4).

49. To determine the y-intercept, we evaluate $g(0)$.

$$g(0) = (0)^4 - 5(0)^2 + 4$$
$$= 4$$

The y-intercept is (0, 4).

To determine the x-intercepts, we solve $g(x) = 0$.

$$x^4 - 5x^2 + 4 = 0$$
$$(x^2 - 4)(x^2 - 1) = 0$$
$$(x + 2)(x - 2)(x + 1)(x - 1) = 0$$
$$x + 2 = 0 \quad x - 2 = 0 \quad x + 1 = 0 \quad x - 1 = 0$$
$$x = -2 \quad\quad x = 2 \quad\quad x = -1 \quad\quad x = 1$$

The x-intercepts are (-2, 0), (-1, 0), (1, 0), and (2, 0).

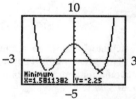

Using the calculator's *minimum* feature, we estimate the relative minima to be (-1.58, -2.25) and (1.58, -2.25). Similarly, we estimate the relative maximum to be (0, 4).

51. To determine the y-intercept, we evaluate $g(0)$.

$$g(0) = 0(4 - 0)(0 + 2) = 0$$

The y-intercept is (0, 0).

To determine the x-intercepts, we solve $g(x) = 0$.

$$x(4 - x)(x + 2) = 0$$
$$x = 0 \quad 4 - x = 0 \quad x + 2 = 0$$
$$x = 4 \quad\quad x = -2$$

The x-intercepts are (-2, 0), (0, 0), and (4, 0).

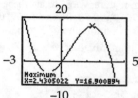

Using the calculator's *minimum* feature, we estimate the relative minimum to be (-1.10, -5.05). Similarly, we estimate the relative maximum to be (2.43, 16.90).

53.

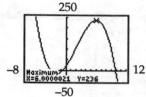

Using the calculator's *minimum* feature, we estimate the relative minimum to be (-2, -20). Using the calculator's *maximum* feature, we estimate the relative maximum to be (6, 236).

55.

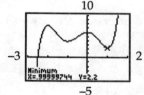

Using the calculator's *maximum* feature, we estimate the relative maxima to be (-2, 7.6) and (0, 6). Using the calculator's *minimum* feature, we estimate the relative minima to be (-1, 3.8) and (1, 2.2).

57.

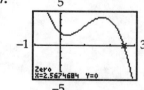

The graph has one x-intercept. Using the calculater's zero feature, we estimate it to be (2.57, 0). Because the degree of the function f is odd, the range is $\mathbb{R}$.

© Houghton Mifflin Company. All rights reserved.

59.

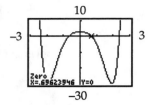

The graph has four x-intercepts. Using the calculater's zero feature, we estimate these to be $(-2.37, 0)$, $(-0.70, 0)$, $(0.70, 0)$, and $(2.37, 0)$. We estimate the relative minimum of F to be -29.60.

Thus the range is approximately $[-29.60, \infty)$.

61.

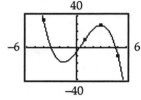

Using the cubic regression option, we obtain the third-degree regression equation

$y = -x^3 + 2x^2 + 14x - 5$ which appears to model the data well.

63.

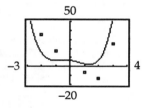

Using the quartic regression option, we obtain the fourth-degree regression equation

$y = x^4 - 3x^2 - 12x + 7$ which appears to model the data well.

65. If $f(-2) = 0$, then $x + 2 = -2$ or $x = -4$

If $f(0) = 0$, then $x + 2 = 0$ or $x = -2$.

If $f(5) = 0$, then $x + 2 = 5$ or $x = 3$.

67.

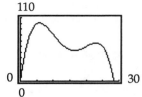

a. We use the calculator's *minimum* feature to estimate the relative minimum. After its first ascent, the kite dips to about 53 feet.

b. We use the calculator's *maximum* feature to estimate the relative maximum. The kite achieves its maximum height after approximately six minutes. The maximum height is about 100 feet.

c. By tracing, we estimate the x-intercept. The kite's total flying time is about 28 minutes.

69. a. The base has area $w \cdot w$ and each of the four sides has area $w \cdot h$. Thus, the total surface area is $w^2 + 4wh = 100$.

b. $w^2 + 4wh = 100$

$$4wh = 100 - w^2$$

$$h = \frac{100 - w^2}{4w}$$

c. The volume of the box is given by

$$V = w \cdot w \cdot h$$

$$V(w) = w^2 \left(\frac{100 - w^2}{4w} \right)$$

$$V(w) = 25w - \frac{1}{4}w^3$$

d.

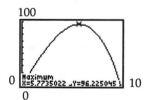

We use the calculator's *maximum* feature to estimate the relative maximum. The width of the base that would maximize the volume of the bin is about 5.77 feet.

Houghton Mifflin Company. All rights reserved.

71. a.

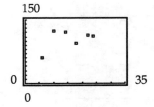

b. Using the cubic regression option, we obtain a third-degree regression equation

$$y = 0.096x^3 - 4.67x^2 + 70.36x - 211.88$$

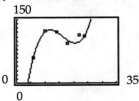

c. Using the quartic regression option, we obtain a fourth-degree regression equation

$$y = 0.0083x^4 + 0.595x^3 - 15.165x^2 \\ + 160.746x - 475.348$$

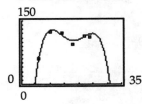

d. Answers may vary.

e. The cubic model indicates a trend of increasing prices whereas the quartic model indicates a trend of decreasing prices.

73. a. For the year 1980, $t = 1980 - 1980 = 0$.

$$B(t) = 0.61t^3 - 12.7t^2 + 66.4t + 259.5$$
$$B(0) = 0.61(0)^3 - 12.7(0)^2 + 66.4(0) + 259.5$$
$$B(0) = 259.5$$

The production in 1980 was about 259.5 million barrels.
For the year 1993, $t = 1993 - 1980 = 13$.

$$B(13) = 0.61(13)^3 - 12.7(13)^2 + 66.4(13) + 259.5$$
$$B(13) = 316.57$$

The production in 1993 was about 316.57 million barrels.

b. Using the calculator's *maximum* feature, we estimate the relative maximum to be (3.49, 362.48). Production increased to about 362 million barrels in 1983 and then decreased.

c. Using the calculator's *minimum* feature, we estimate the relative minimum to be (10.39, 262.59). Production decreased to about 263 million barrels in 1990 and then increased.

© Houghton Mifflin Company. All rights reserved.

3.2 Exercises

1. You can determine R without performing the division by evaluating $P(7)$.

3.
$$x-2\overline{)x^3-x^2+2x+4}$$
quotient x^2+x+4
$$\underline{x^3-2x^2}$$
$$x^2+2x$$
$$\underline{x^2-2x}$$
$$4x+4$$
$$\underline{4x-8}$$
$$12$$

$$\left(x^2+x+4\right)(x-2)+12$$

5.
$$2x+1\overline{)2x^3-11x^2+4x+5}$$
quotient x^2-6x+5
$$\underline{2x^3+x^2}$$
$$-12x^2+4x$$
$$\underline{-12x^2-6x}$$
$$10x+5$$
$$\underline{10x+5}$$
$$0$$

$$\left(x^2-6x+5\right)(2x+1)+0$$

7.
$$x+1\overline{)x^5\quad+5x^3\quad\quad-10}$$
quotient $x^4-x^3+6x^2-6x+6$
$$\underline{x^5+x^4}$$
$$-x^4+5x^3$$
$$\underline{-x^4-\ x^3}$$
$$6x^3$$
$$\underline{6x^3+6x^2}$$
$$-6x^2$$
$$\underline{-6x^2-6x}$$
$$6x-10$$
$$\underline{6x+\ 6}$$
$$-16$$

$$\left(x^4-x^3+6x^2-6x+6\right)(x+1)-16$$

9.
$$x^2-5\overline{)x^4+x^3\quad\quad-x-8}$$
quotient x^2+x+5
$$\underline{x^4\quad-5x^2}$$
$$x^3+5x^2\ -x$$
$$\underline{x^3\quad\quad-5x}$$
$$5x^2+4x-8$$
$$\underline{5x^2\quad-25}$$
$$4x+17$$

$$\left(x^2+x+5\right)\left(x^2-5\right)+4x+17$$

11.
$$\begin{array}{r|rrrr} -2 & 4 & -3 & -8 & 4 \\ & & -8 & 22 & -28 \\ \hline & 4 & -11 & 14 & -24 \end{array}$$
$Q(x)\colon 4x^2-11x+14;\ R(x)\colon -24$

13.
$$\begin{array}{r|rrrr} -\frac{2}{3} & 3 & -10 & 1 & 6 \\ & & -2 & 8 & -6 & -10 \\ \hline & 3 & -12 & 9 & 0 \end{array}$$
$Q(x)\colon 3x^2-12x+9;\ R(x)\colon 0$

15.
$$\begin{array}{r|rrrrr} 3 & 2 & -5 & 0 & -1 & -15 \\ & & 6 & 3 & 9 & 24 \\ \hline & 2 & 1 & 3 & 8 & 9 \end{array}$$
$Q(x)\colon 2x^3+x^2+3x+8;\ R(x)\colon 9$

17. The value of $P(-2)$ is the remainder when is divided synthetically by -2.
$$\begin{array}{r|rrrrr} -2 & 2 & -1 & 5 & -3 & 1 \\ & & -4 & 10 & -30 & 66 \\ \hline & 2 & -5 & 15 & -33 & 67 \end{array}$$
$P(-2)=67$

19. To determine the remainder if $P(x)$ is divided by $x-1$, we evaluate $P(1)$.
$$P(1)=(1)^{12}-3(1)^7+5=3$$
Because $P(1)=3$, the remainder is 3.

Houghton Mifflin Company. All rights reserved.

21. We use synthetic division to evaluate $P(x)$.

a.

$$
\begin{array}{r|rrrrr}
1 & 2 & 3 & -12 & -7 & 6 \\
  &   & 2 & 2 & -7 & -14 \\
\hline
  & 2 & 5 & -7 & -14 & -8
\end{array}
$$

Because $P(1) = -8 \neq 0$, 1 is not a zero of $P(x)$.

b.

$$
\begin{array}{r|rrrrr}
\frac{1}{2} & 2 & 3 & -12 & -7 & 6 \\
  &   & 1 & 2 & -5 & -6 \\
\hline
  & 2 & 4 & -10 & -12 & 0
\end{array}
$$

Because $P\left(\dfrac{1}{2}\right) = 0$, $\dfrac{1}{2}$ is a zero of $P(x)$.

c.

$$
\begin{array}{r|rrrrr}
2 & 2 & 3 & -12 & -7 & -6 \\
  &   & 4 & 14 & 4 & -6 \\
\hline
  & 2 & 7 & 2 & -3 & 0
\end{array}
$$

Because $P(2) = 0$, 2 is a zero of $P(x)$.

23. To determine the remainder when $P(x)$. is divided by $x - c$, we divide synthetically by c.

a.

$$
\begin{array}{r|rrrr}
1 & 3 & -7 & -1 & 8 \\
  &   & 3 & -4 & -6 \\
\hline
  & 3 & -4 & -6 & 2
\end{array}
$$

Because the remainder $\neq 0$, $x - 1$. is not a factor of $P(x)$.

b. $3x - 4 = 3\left(x - \dfrac{4}{3}\right)$, thus $c = \dfrac{4}{3}$.

$$
\begin{array}{r|rrrr}
\frac{4}{3} & 3 & -7 & -2 & 8 \\
  &   & 4 & -4 & -8 \\
\hline
  & 3 & -3 & -6 & 0
\end{array}
$$

Because the remainder $= 0$, $3x - 4$ is a factor of $P(x)$.

c.

$$
\begin{array}{r|rrrr}
2 & 3 & -7 & -2 & 8 \\
  &   & 6 & -2 & -8 \\
\hline
  & 3 & -1 & -4 & 0
\end{array}
$$

Because the remainder $= 0$, $x - 2$ is a factor of $P(x)$.

25. $\dfrac{\text{Factors of } -8}{\text{Factors of } 1} = \dfrac{\pm 1, \pm 2, \pm 4, \pm 8}{\pm 1}$

$$= \pm 1, \pm 2, \pm 4, \pm 8$$

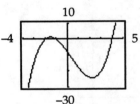

The graph suggests that $(-2, 0)$, $(-1, 0)$, and $(4, 0)$ are possible x-intercepts. We use synthetic division to evaluate the function and verify that -2, -1, and 4 are the zeros of the function.

27. $\dfrac{\text{Factors of } 10}{\text{Factors of } 2} = \dfrac{\pm 1, \pm 2, \pm 5, \pm 10}{\pm 1, \pm 2}$

$$= \pm 1, \pm 2, \pm 5, \pm 10, \pm \frac{1}{2}, \pm \frac{5}{2}$$

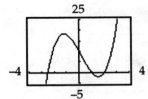

The graph suggests that $\left(-\dfrac{5}{2}, 0\right)$, $(1, 0)$, and $(2, 0)$ are possible x-intercepts. We use synthetic division to evaluate the function and verify that $-\dfrac{5}{2}$, 1, and 2 are the zeros of the function.

29. $\dfrac{\text{Factors of } 15}{\text{Factors of } 16} = \dfrac{\pm 1, \pm 3, \pm 5, \pm 15}{\pm 1, \pm 2, \pm 4, \pm 8, \pm 16}$

$$= \pm 1, \pm 3, \pm 5, \pm 15, \pm \frac{1}{2}, \pm \frac{3}{2} \pm \frac{5}{2}, \pm \frac{15}{2}, \pm \frac{1}{4}, \pm \frac{3}{4}, \pm \frac{5}{4}, \pm \frac{15}{4}, \pm \frac{1}{8}, \pm \frac{3}{8}, \pm \frac{5}{8}, \pm \frac{15}{8}, \pm \frac{1}{16}, \pm \frac{3}{16}, \pm \frac{5}{16}, \pm \frac{15}{16}$$

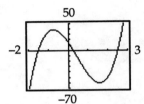

The graph suggests that $\left(-\dfrac{3}{2}, 0\right)$, $\left(\dfrac{1}{4}, 0\right)$ and $\left(\dfrac{5}{2}, 0\right)$ are possible x-intercepts. We use synthetic division to evaluate the function and verify that $-\dfrac{3}{2}, \dfrac{1}{4}$, and $\dfrac{5}{2}$ are the zeros of the function.

© Houghton Mifflin Company. All rights reserved.

31. If c is a solution of the equation $P(x) = 0$, then c is a zero of P and $(c, 0)$ is an x-intercept of the graph of P.

33. Possible rational number zeros of f are $\pm 1, \pm 2, \pm 3, \pm 6$.

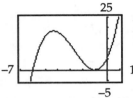

The graph suggests possible x-intercepts at $(-6, 0)$ and $(-1, 0)$. Synthetic division verifies that -6 and -1 are zeros of f. Now we can factor the polynomial.

$x^3 + 8x^2 + 13x + 6 = (x+6)(x+1)(x+1)$

Thus, the zeros are -6 (multiplicity 1) and -1 (multiplicity 2).

35. Possible rational zeros of P are $\pm 1, \pm 2, \pm 4$.

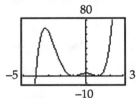

The graph suggests possible x-intercepts at $(-4, 0)$, $(-1, 0)$ and $(1, 0)$. Synthetic division verifies that $-4, -1$, and 1 are zeros of P. Now we can factor the polynomial.

$x^5 + 4x^4 - 2x^3 - 8x^2 + x + 4$

$= (x+4)(x+1)(x-1)(x^2 - 1)$

$= (x+4)(x+1)(x-1)(x+1)(x-1)$

Thus, the zeros are -4 (multiplicity 1), -1 (multiplicity 2), and 1 (multiplicity 2).

37. Possible rational zeros of h are $\pm 1, \pm 2, \pm 4, \pm 11, \pm 22, \pm 44$.

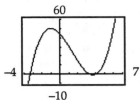

The graph indicates a possible x-intercept at $(4, 0)$. We check the feasible zero with synthetic division.

$$
\begin{array}{r|rrrr}
4 & 1 & -4 & -11 & 44 \\
  &   & 4 & 0   & -44 \\
\hline
  & 1 & 0 & -11 & 0
\end{array}
$$

Thus, 4 is a zero and the function can now be written in factored form.

$h(x) = (x-4)(x^2 - 11)$

The other two zeros are found using the Zero Factor Property.

$x^2 - 11 = 0$

$x^2 = 11$

$x = \pm\sqrt{11}$

Thus, the real number zeros are 4, $\sqrt{11} \approx 3.32$, and $-\sqrt{11} \approx -3.32$.

Houghton Mifflin Company. All rights reserved.

39. Possible rational zeros of Q are ± 1, ± 3, ± 7,

± 21, $\pm\dfrac{1}{2}$, $\pm\dfrac{3}{2}$, $\pm\dfrac{7}{2}$, $\pm\dfrac{21}{2}$, $\pm\dfrac{1}{4}$, $\pm\dfrac{3}{4}$,

$\pm\dfrac{7}{4}$, $\pm\dfrac{21}{4}$, $\pm\dfrac{1}{8}$, $\pm\dfrac{3}{8}$, $\pm\dfrac{7}{8}$, $\pm\dfrac{21}{8}$.

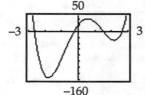

The graph indicates x-intercepts between -3 and -2, -1 and 0, 1 and 2, and 2 and 3. Possible zeros in the interval $(1, 2)$ are $\dfrac{3}{2}$ and $\dfrac{7}{4}$. We check the feasible zeros with synthetic division.

$$\begin{array}{r|rrrrr} \frac{3}{2} & 8 & -10 & -59 & 7\text{-} & 21 \\ & & 12 & 3 & -84 & -21 \\ \hline & 8 & 21 & -56 & -14 & 0 \end{array}$$

Possible zeros in the interval $(-1, 0)$ are $-\dfrac{1}{2}$, $-\dfrac{1}{4}$, $-\dfrac{3}{4}$, $-\dfrac{1}{8}$, and $-\dfrac{3}{8}$. We check the feasible zeros with synthetic division.

$$\begin{array}{r|rrrr} -\frac{1}{2} & 8 & 2 & -56 & -14 \\ & & -4 & 1 & \frac{55}{2} \\ \hline & 8 & -2 & -55 & \frac{27}{2} \end{array}$$

$$\begin{array}{r|rrrr} 3 & 8 & 2 & -56 & -14 \\ & & -2 & 0 & 14 \\ \hline & 8 & 0 & -56 & 0 \end{array}$$

Thus, $\dfrac{3}{2}$ and $-\dfrac{1}{4}$ are zeros and the function can be written in factored form.

$$Q(x) = \left(x - \dfrac{3}{2}\right)\left(x + \dfrac{1}{4}\right)\left(8x^2 - 56\right)$$

The other two zeros are found using the Zero Factor Property.

$$8x^2 - 56 = 0$$
$$8x^2 = 56$$
$$x^2 = 7$$
$$x = \pm\sqrt{7}$$

Thus, the real number zeros are

$\dfrac{3}{2}$, $-\dfrac{1}{4}$, $\sqrt{7} \approx 2.65$ and $-\sqrt{7} \approx -2.65$.

41. The list of possible rational zeros of C is ± 1, ± 7.

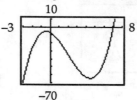

The graph indicates an x-intercept at $(7, 0)$. We check the feasible zero with synthetic division.

$$\begin{array}{r|rrrr} 7 & 1 & -6 & -6 & -7 \\ & & 7 & 7 & 7 \\ \hline & 1 & 1 & 1 & 0 \end{array}$$

Thus, 7 is a zero and the function can be written in factored form.

$$C(x) = (x - 7)\left(x^2 + x + 1\right)$$

The other two zeros are found using the Quadratic Formula.

$$x^2 + x + 1 = 0$$

$$x = \dfrac{-1 \pm \sqrt{1^2 - 4(1)(1)}}{2(1)} = \dfrac{-1 \pm \sqrt{-3}}{2}$$

This produces no real number solutions. Thus, the only real zero is 7.

43.

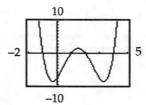

The graph indicates possible x-intercepts at $(-1, 0)$, $(1, 0)$, $(2, 0)$, and $(4,0)$. We check the feasible zeros with synthetic division.

$$\begin{array}{r|rrrr} -1 & -6 & 7 & 6 & -8 \\ & & -1 & 7 & -14 & 8 \\ \hline 1 & -7 & 14 & -8 & 0 \\ & & 1 & -6 & 8 \\ \hline 2 & -6 & 8 & 0 \\ & & 2 & -8 \\ \hline 4 & -4 & 0 \\ & & 4 \\ \hline & 0 \end{array}$$

Thus, the real number zeros are -1, 1, 2, and 4. And the polynomial can be factored as follows.

$$G(x) = (x + 1)(x - 1)(x - 2)(x - 4)$$

© Houghton Mifflin Company. All rights reserved.

45.

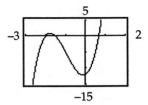

The graph indicates an *x*-intercept at (–2, 0). We check this feasible zero with synthetic division.

$$
\begin{array}{r|rrrr}
-2 & 6 & 17 & 4 & -12 \\
   &   & -12 & -10 & 12 \\
\hline
   & 6 & 5 & -6 & 0
\end{array}
$$

Thus, –2 is a zero and the function can be factored as follows.

$$f(x) = (x+2)\left(6x^2 + 5x - 6\right)$$

$$f(x) = (x+2)(2x+3)(3x-2)$$

47.

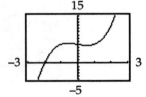

The graph indicates an *x*-intercept at (–2, 0). We check this feasible zero with synthetic division.

$$
\begin{array}{r|rrrr}
-2 & 1 & 0 & -1 & 6 \\
   &   & -2 & 4 & -6 \\
\hline
   & 1 & -2 & 3 & 0
\end{array}
$$

Thus, –2 is a zero and the function can be factored as follows.

$$R(x) = (x+2)\left(x^2 - 2x + 3\right)$$

49.

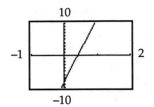

The graph indicates an *x*-intercept between 0 and 1. The list of possible real zeros is ± 1, ± 2, ± 4, ± 8, $\pm\frac{1}{5}$, $\pm\frac{2}{5}$, $\pm\frac{4}{5}$, $\pm\frac{8}{5}$. Of these $\frac{1}{5}$, $\frac{2}{5}$ and $\frac{4}{5}$ lie between 0 and 1. We check the feasible zeros with synthetic division.

$$
\begin{array}{r|rrrr}
\dfrac{1}{5} & 5 & -7 & 22 & -8 \\
   &   & 1 & -\dfrac{6}{5} & \dfrac{104}{25} \\
\hline
   & 5 & -6 & \dfrac{104}{5} & -\dfrac{96}{25}
\end{array}
$$

$$
\begin{array}{r|rrrr}
\dfrac{2}{5} & 5 & -7 & 22 & -8 \\
   &   & 2 & -2 & 8 \\
\hline
   & 5 & -5 & 20 & 0
\end{array}
$$

Thus, $\frac{2}{5}$ is a zero and the function can be factored as follows.

$$P(x) = \left(x - \frac{2}{5}\right)\left(5x^2 - 5x + 20\right)$$

$$P(x) = 5\left(x - \frac{2}{5}\right)\left(x^2 - x + 4\right)$$

$$P(x) = (5x - 2)\left(x^2 - x + 4\right)$$

Houghton Mifflin Company. All rights reserved.

51. The only possible rational zeros are factors of 7 divided by factors of 1:

$$\frac{\text{Factors of } 7}{\text{Factors of } 1} = \frac{\pm 1, \pm 7}{\pm 1} = \pm 1, \pm 7.$$

All feasible rational zeros are integers.

53. Graph the function

$$f(x) = x^4 + 2x^3 - 13x^2 - 14x + 24.$$

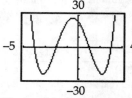

The graph indicates x-intercepts at $(-4, 0)$, $(-2, 0)$, $(1, 0)$, and $(3, 0)$. We check these possible zeros using synthetic division.

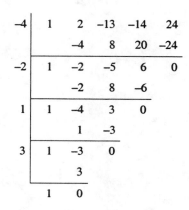

Thus, the solutions are -4, -2, 1, and 3.

55. Write the equation with 0 on one side.

$$10x^3 + 49x + 12 = 53x^2$$

$$10x^3 - 53x^2 + 49x + 12 = 0$$

Graph the function

$$f(x) = 10x^3 - 53x^2 + 49x + 12.$$

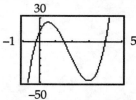

The graph indicates an x-intercept at $(4, 0)$. We check this possible zero using synthetic division.

$$\begin{array}{r|rrrr} 4 & 10 & -53 & 49 & 12 \\ & & 40 & -52 & -12 \\ \hline & 10 & -13 & -3 & 0 \end{array}$$

Thus, 4 is a zero and the equation can be written in factored form.

$$(x - 4)\left(10x^2 - 13x - 3\right) = 0$$

$$(x - 4)(5x + 1)(2x - 3) = 0$$

The remaining zeros are found by using the Zero Factor Property.

$$\begin{array}{ll} 5x + 1 = 0 & 2x - 3 = 0 \\ 5x = -1 & 2x = 3 \\ x = -\dfrac{1}{5} & x = \dfrac{3}{2} \end{array}$$

Thus the solutions are $-\dfrac{1}{5}$, $\dfrac{3}{2}$, and 4.

57. Write the equation with 0 on one side.

$$7x^2 + 6x = 20 - x^3$$

$$x^3 + 7x^2 + 6x - 20 = 0$$

Graph the function $f(x) = x^3 + 7x^2 + 6x - 20$.

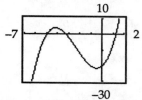

The graph indicates an x-intercept at $(-5, 0)$. We check this possible zero using synthetic division.

$$\begin{array}{r|rrrr} & 1 & 7 & 6 & -20 \\ & & -5 & -10 & 20 \\ \hline & 1 & 2 & -4 & 0 \end{array}$$

Thus, -5 is a zero and the equation can be written in factored form.

$$(x + 5)\left(x^2 + 2x - 4\right) = 0$$

The remaining solutions are found using the Quadratic Formula.

$$x^2 + 2x - 4 = 0$$

$$x = \frac{-2 \pm \sqrt{2^2 - 4(1)(-4)}}{2(1)}$$

$$= \frac{-2 \pm \sqrt{20}}{2} = -1 \pm \sqrt{5}$$

Thus, the solutions are -5, $-1 - \sqrt{5} \approx -3.24$, and $-1 + \sqrt{5} \approx 1.24$.

© Houghton Mifflin Company. All rights reserved.

59. Write the equation with 0 on one side.

$$15 - 13x^2 = -4\left(x^3 + 2x\right)$$

$$15 - 13x^2 = -4x^3 - 8x$$

$$4x^3 - 13x^2 + 8x + 15 = 0$$

Graph the function $f(x) = 4x^3 - 13x^2 + 8x + 15$.

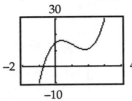

The graph indicates an x-intercept between -1, and 0. The list of possible rational zeros is

$$\pm 1, \pm 3, \pm 5, \pm 15, \pm \frac{1}{2}, \pm \frac{3}{2}, \pm \frac{5}{2}, \pm \frac{15}{2}, \pm \frac{1}{4}, \pm \frac{3}{4}$$

$$\pm \frac{5}{4}, \pm \frac{15}{4}.$$

Of these, $-\frac{1}{2}, -\frac{1}{4},$ and $-\frac{3}{4}$ lie between -1 and 0. We check the possible zeros using synthetic division.

$$
\begin{array}{r|rrrr}
-\dfrac{1}{2} & 4 & -13 & 8 & 15 \\
 & & -2 & \dfrac{15}{2} & -\dfrac{31}{4} \\
\hline
 & 4 & -15 & \dfrac{31}{2} & \dfrac{29}{4}
\end{array}
$$

$$
\begin{array}{r|rrrr}
-\dfrac{1}{4} & 4 & -13 & 8 & 15 \\
 & & -1 & \dfrac{7}{2} & -\dfrac{23}{8} \\
\hline
 & 4 & -14 & \dfrac{23}{2} & \dfrac{97}{8}
\end{array}
$$

$$
\begin{array}{r|rrrr}
-\dfrac{3}{4} & 4 & -13 & 8 & 15 \\
 & & -3 & 12 & -15 \\
\hline
 & 4 & -16 & 20 & 0
\end{array}
$$

Thus, $-\dfrac{3}{4}$ is a zero and the equation can be written in factored form.

$$\left(x + \frac{3}{4}\right)\left(4x^2 - 16x + 20\right) = 0$$

$$4\left(x + \frac{3}{4}\right)\left(x^2 - 4x + 5\right) = 0$$

The remaining solutions are found using the Quadratic Formula.

$$x^2 - 4x + 5 = 0$$

$$x = \frac{-(-4) \pm \sqrt{(-4)^2 - 4(1)(5)}}{2(1)} = \frac{4 \pm \sqrt{-4}}{2}$$

This produces no real zeros. The only solution is $-\dfrac{3}{4}$.

Houghton Mifflin Company. All rights reserved.

61. Write the equation with 0 on one side.

$$3\left(2x^4 - 23x - 20\right) = -\left(23x^3 + 2x^2\right)$$

$$6x^4 - 69x - 60 = -23x^3 - 2x^2$$

$$6x^4 + 23x^3 + 2x^2 - 69x - 60 = 0$$

Graph the function

$$f(x) = 6x^4 + 23x^3 + 2x^2 - 69x - 60.$$

According to the Rational Zero Theorem, the only rational number zeros are $\pm1, \pm2, \pm3, \pm4, \pm5,$ $\pm6, \pm10, \pm12, \pm15, \pm20, \pm30, \pm60,$

$$\pm\frac{1}{2}, \pm\frac{3}{2}, \pm\frac{5}{2}, \pm\frac{15}{2}, \pm\frac{1}{3}, \pm\frac{2}{3}, \pm\frac{4}{3}, \pm\frac{5}{3},$$

$$\pm\frac{10}{3}, \pm\frac{20}{3}, \pm\frac{1}{6}, \pm\frac{5}{6}$$

The graph suggests that $-\frac{3}{2}, -\frac{5}{2}, -\frac{4}{3}$ and $-\frac{5}{3}$ are feasible possibilities.

We check each of these using synthetic division.

```
-3/2 |  6   23    2    -69    -60
     |           -9   -21   57/2  243/4
     |  6   14  -19  -81/2   3/4
```

```
-5/2 |  6   23    2   -69    -60
     |          -15  -20    45    60
     |  6    8  -18  -24     0
```

Thus, $-\frac{5}{2}$ is a zero and we can now reduce the table by one column.

```
-4/3 |  6    8   -18   -24
     |           -8     0    24
     |  6    0   -18     0
```

Thus $-\frac{4}{3}$ is a zero and the equation can now be written in factored form.

$$\left(x + \frac{5}{2}\right)\left(x + \frac{4}{3}\right)\left(6x^2 - 18\right) = 0$$

$$6\left(x + \frac{5}{2}\right)\left(x + \frac{4}{3}\right)\left(x^2 - 3\right) = 0$$

The remaining solutions are found using the Zero Factor Property.

$$x^2 - 3 = 0$$
$$x^2 = 3$$
$$x = \pm\sqrt{3}$$

Thus, the solutions are $-\frac{5}{2}, -\frac{4}{3}, -\sqrt{3} \approx -1.73,$ and $\sqrt{3} \approx 1.73.$

© Houghton Mifflin Company. All rights reserved.

63. The inequality $x^3 - 21x - 20 > 0$ is true for all values of x for which the graph of

$y = x^3 - 21x - 20$ is above the x-axis.

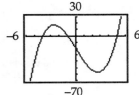

Referring to the graph, the inequality is satisfied by all values of x in the interval $(-4, -1)$ or in the interval $(5, \infty)$.

65. The inequality $2x^3 - x^2 - 21x \le 0$ is true for all values of x for which the graph of

$y = 2x^3 - x^2 - 21x$ is on or below the x-axis.

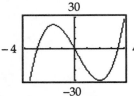

Referring to the graph, the inequality is satisfied by all values of x in the interval $(-\infty, -3]$ or in the

interval $\left[0, \dfrac{7}{2}\right]$.

67. The inequality $x^4 - 16x^2 \le 0$ is true for all values of x for which the graph of $y = x^4 - 16x^2$ is on or below the x-axis.

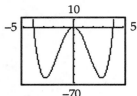

Referring to the graph, the inequality is satisfied for all values of x in the interval $[-4, 4]$.

69. The inequality $4x^4 + 15x^3 - 7x^2 - 12x > 0$ is true for all values of x for which the graph of

$y = 4x^4 + 15x^3 - 7x^2 - 12x$ is above the x-axis.

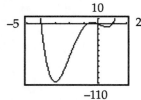

Referring to the graph, the inequality is satisfied for all values of x in the interval $(-\infty, -4)$ or in the

interval $\left(-\dfrac{3}{4}, 0\right)$ or in the interval $(1, \infty)$.

71. The inequality $x^5 + 3x^4 + 3x^3 + x^2 \le 0$ is true for all values of x for which the graph of

$y = x^5 + 3x^4 + 3x^3 + x^2$ is on or below the x-axis.

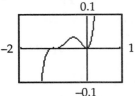

Adjust the window of your graphing calculator until you can see the detail near the origin. Referring to the graph, -1 and 0 are intercepts. It also indicates -1 is a zero of multiplicity 3 (or more) while 0 is a zero of multiplicity 2 (or more). The inequality is satisfied for all values of x in the interval $(-\infty, -1]$ or by the single value $\{0\}$.

73. Using synthetic division, we have the following table.

-2	1	0	k	-4	
		-2	4	-8	$-2k$
	1	-2	$4 + k$	-8	

Therefore, $-12 - 2k = -8$

$$-2k = 4$$
$$k = -2$$

75. If -2, 1, and 4 are zeros, $x + 2$, $x - 1$, and $x - 4$ are factors. So $P(x) = a_n(x+2)(x-1)(x-4)$.
If the graph contains $(3, 30)$, then

$$P(3) = 30$$
$$a_n(3+2)(3-1)(3-4) = 30$$
$$5(2)(-1)a_n = 30$$
$$-10a_n = 30$$
$$a_n = -3$$

Therefore, $P(x) = -3(x+2)(x-1)(x-4)$.

77. If -4 (multiplicity 3) and 0 are zeros, $(x+4)^3$ and x are factors. So $P(x) = a_n(x+4)^3 x$.
If $P(-5) = 25$,

$$a_n(-5+4)^3(-5) = 25$$
$$(-1)^3(-5)a_n = 25$$
$$5a_n = 25$$
$$a_n = 5$$

Therefore, $P(x) = 5x(x+4)^3$.

Houghton Mifflin Company. All rights reserved.

79. Use synthetic division to divide $x^n - c^n$ by $x - c$. No matter what the value of n is, our table will appear as follows.

c	1	0	0	0	...	0	$-c^n$
		c	c^2	c^3		c^{n-1}	c^n
	1	c	c^2	c^3		c^{n-1}	0

Since the remainder is 0, $x - c$ is a factor of $x^n - c^n$.

81. a. $x + 3 = -2$
$x = -5$
$x + 3 = 4$
$x = 1$
$x + 3 = 7$
$x = 4$
The zeros of $P(x + 3)$ are −5, 1, and 4.

b. $2x = -2$
$x = -1$
$2x = 4$
$x = 2$
$2x = 7$
$x = \dfrac{7}{2}$

The zeros of $P(2x)$ are −1, 2 and $\dfrac{7}{2}$.

83. a. Letting x represent the number of years since 1993, the data values are (1, 0), (2, 5), (3, 3), and (4, 8). Using the cubic regression option, we obtain
$$y = 2.\overline{3}x^3 - 17.5x^2 + 41.1\overline{6}x - 26$$
Notice that $\dfrac{1}{6}\left(14x^3 - 105x^2 + 247x - 156\right)$
$$= \dfrac{7}{3}x^3 - \dfrac{35}{2}x^2 + \dfrac{247}{6}x - 26$$
$$= 2.\overline{3}x^3 - 17.5x^2 + 41.1\overline{6}x - 26.$$
Thus, the regression equation and $P(x)$ have the same graphs.

b.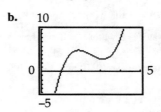

The x-intercept (1, 0) corresponds to the 1994 data.

c.

1	$\dfrac{7}{3}$	$-\dfrac{35}{2}$	$\dfrac{247}{6}$	−26
		$\dfrac{7}{3}$	$-\dfrac{91}{6}$	26
	$\dfrac{7}{3}$	$-\dfrac{91}{6}$	26	0

Because the remainder is 0, 1 is a zero of the

3.3 Exercises

1. A vertical asymptote occurs at a value of x for which the function is not defined.

3. Because $r(0) = -\dfrac{7}{9}$, the y-intercept is $\left(0, -\dfrac{7}{9}\right)$.
Because there is no value which will make the numerator equal to 0, there is no x-intercept.
Because $\dfrac{9}{2}$ is a restricted value, the line $x = \dfrac{9}{2}$ is a vertical asymptote.
The degree of the numerator is less than the degree of the denominator, so the horizontal asymptote is $y = 0$.

© Houghton Mifflin Company. All rights reserved.

5. Because $g(0) = -\frac{1}{5}$, the y-intercept is $\left(0, -\frac{1}{5}\right)$.

The numerator is 0 when $x = -\frac{1}{2}$, so the

x-intercept is $\left(-\frac{1}{2}, 0\right)$. Because 5 is a restricted

value, the line $x = 5$ is a vertical asymptote. The degree of the numerator is equal to the degree of the denominator, so the horizontal asymptote is the horizontal line equal to the ratio of their

leading coefficients which is $y\frac{2}{1} = 2$.

7. Because $f(0) = 0$, the y-intercept is (0,0).

The numerator is 0 when $x = 0$ or when $x = -\frac{5}{3}$,

so the x-intercepts are (0,0) and $\left(-\frac{5}{3}, 0\right)$.

Because $-\frac{1}{2}$ and 1 are restricted values, the lines

$x = -\frac{1}{2}$ and $x = 1$ are vertical asymptotes. The

degree of the numerator is equal to the degree of the denominator, so the horizontal asymptote is the horizontal line equal to the ratio of their

leading coefficients which is $y = \frac{3}{2}$.

9. C

11. D

13. A zero of the function Q pertains to a vertical asymptote of the function R.

15. Because $f(0) = -\frac{1}{2}$, the y-intercept is $\left(0, -\frac{1}{2}\right)$.

Because there is no value which will make the numerator equal to 0, there is no x-intercept. Because 6 is a restricted value, the line $x = 6$ is a vertical asymptote. The degree of the numerator is less than the degree of the denominator, so $y = 0$ is the horizontal asymptote.

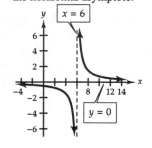

17. Because $g(0) = -\frac{1}{4}$, the y-intercept is $\left(0, -\frac{1}{4}\right)$.

Because there is no value which will make the numerator equal to 0, there is no x-intercept. Because -4 is a restricted value, the line $x = -4$ is a vertical asymptote. The degree of the numerator is less than the degree of the denominator, so $y = 0$ is the horizontal asymptote.

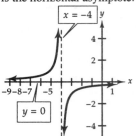

19. $f(0)$ is undefined, so there is no y-intercept. The numerator is 0 when $x = -3$, so the x-intercept is $(-3, 0)$. Because 0 is a restricted value, the line $x = 0$ is a vertical asymptote. The degree of the numerator is equal to the degree of the

denominator, so $y = \frac{1}{1} = 1$, is the horizontal

asymptote.

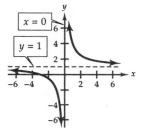

21. Because $h(0) = \frac{1}{2}$, the y-intercept is $\left(0, \frac{1}{2}\right)$.

The numerator is 0 when $x = 1$, so the x-intercept is $(1, 0)$. Because -2 is a restricted value, the line $x = -2$ is a vertical asymptote. The degree of the numerator is equal to the degree of the

denominator, so $y = \frac{-1}{1} = -1$ is the horizontal

asymptote.

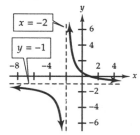

Houghton Mifflin Company. All rights reserved.

23. Because $r(0) = \dfrac{1}{4}$, the y-intercept is $\left(0, \dfrac{1}{4}\right)$.
Because there is no value which will make the numerator equal to 0, there is no x-intercept. There are no restricted values, so there are no vertical asymptotes. The degree of the numerator is less than the degree of the denominator, so $y = 0$ is the horizontal asymptote.

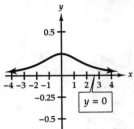

25. Because $h(0) = 0$, the y-intercept is $(0, 0)$. The numerator is 0 when $x = 0$, so the x-ntercept is $(0, 0)$. Because -1 is a restricted value, the line $x = -1$ is a vertical asymptote. The degree of the numerator is less than the degree of the denominator, so $y = 0$ is the horizontal asymptote.

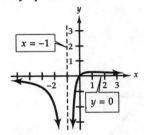

27. Because $g(0) = -5$, the y-intercept is $(0, -5)$.
The numerator is 0 when $x = \dfrac{5}{2}$, so the x-intercept is $\left(\dfrac{5}{2}, 0\right)$ Because -1 is a restricted value, the line $x = -1$ is a vertical asymptote. The degree of the numerator is equal to the degree of the denominator so $y = \dfrac{2}{1} = 2$ is the horizontal asymptote.

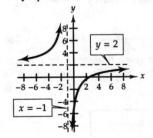

29. Because $w(0) = 0$, the y-intercept is $(0, 0)$. The numerator is 0 when $x = 0$, so the x-intercept is $(0, 0)$. There are no restricted values, so there are no vertical asymptotes. The degree of the numerator is equal to the degree of the denominator, so $y = \dfrac{1}{1} = 1$ is the horizontal asymptote.

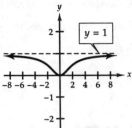

31. Because $h(0) = 0$, the y-intercept is $(0, 0)$. The numerator is 0 when $x = 0$, so the x-intercept is $(0, 0)$. Because -2 and 2 are restricted values, the lines $x = -2$ and $x = 2$ are vertical asymptotes. The degree of the numerator is less than the degree of the denominator, so $y = 0$ is the horizontal asymptote.

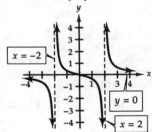

33. Because $r(0) = \dfrac{2}{3}$, the y-intercept is $\left(0, \dfrac{2}{3}\right)$.
The numerator is 0 when $x = 2$, so the x-intercept is $(2, 0)$. Because -1 and -3 are restricted values, the lines $x = -1$ and $x = -3$ are vertical asymptotes.
The degree of the numerator is less than the degree of the denominator, so $y = 0$ is the horizontal asymptote.

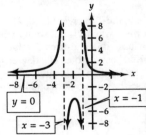

© Houghton Mifflin Company. All rights reserved.

35. Because $f(0) = -\dfrac{1}{3}$, the y-intercept is $\left(0, -\dfrac{1}{3}\right)$.

Using the Quadratic Formula, the numerator is 0

when $x = \dfrac{1 \pm \sqrt{5}}{2}$, so the x-intercepts are

approximately $(-0.62, 0)$ and $(1.62, 0)$. There are no restricted values so there are no vertical asymptotes. The degree of the numerator is equal

to the degree of the denominator, so $y = \dfrac{1}{1} = 1$ is

the horizontal asymptote.

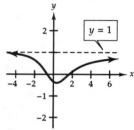

37. i $f(x)$ has vertical asymptotes $x = -2$ and $x = 3$.

ii $g(x)$ has horizontal asymptote $y = 0$.

iii $h(x)$ satisfies the conditions.

39. Because $h(0)$ is undefined, there is no y-intercept. The numerator is 0 when $x = -2$ and $x = 1$, so the x-intercepts are $(-2, 0)$ and $(1, 0)$. Because 0 is a restricted value, the line $x = 0$ is a vertical asymptote. Re-writing the function,

$h(x) = x + 1 + \dfrac{2}{x}$. thus, $y = x + 1$ is a slant asymptote.

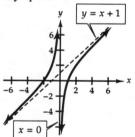

41. Because $f(0) = 2$, the y-intercept is $(0, 2)$. The numerator is 0 when $x = -2$, so the x-intercept is $(-2, 0)$. Because -4 is a restricted value, the line $x = -4$ is a vertical asymptote. The degree of the numerator is greater than the degree of the denominator, so there is no horizontal asymptote.

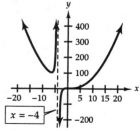

43. Because $g(0) = 0$, the y-intercept is $(0, 0)$. The numerator is 0 when $x = -8$ and $x = 0$, so the x-intercepts are $(-8, 0)$ and $(0, 0)$. Because -4 and 2 are restricted values, the lines $x = 4$ and $x = 2$ are vertical asymptotes. Re-writing the function,

$g(x) = x + 6 + \dfrac{-4x + 48}{x^2 + 2x - 8}$. Thus, $y = x + 6$ is a slant asymptote.

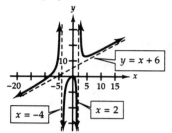

45. Because $f(0)$ is undefined, there is no y-intercept. The numerator is 0 when $x = \sqrt[3]{2}$, so the x-intercept is approximately $(1.26, 0)$. Because 0 is a restricted value, the line $x = 0$ is a vertical asymptote. Re-writing the function,

$f(x) = -x + \dfrac{2}{x^2}$. Thus $y = -x$ is a slant asymptote.

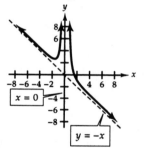

47. For the function B the degree of the numerator is equal to the degree of the denominator, so the horizontal asymptote is determined by the ratio of the leading coefficients: $y = \dfrac{1}{1} = 1$. The values of B which lie above the horizontal asymptote are

solutions of the inequality $\dfrac{x^2}{x^2 - 1} > 1$.

49. The function $y = \dfrac{2}{x - 5}$ has restricted value 5 and no zeros. The graph lies above the x-axis to the right of the line $x = 5$. Thus, the solution set is $(5, \infty)$.

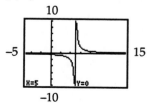

Houghton Mifflin Company. All rights reserved.

51. First, write the inequality with 0 on one side.

$$\frac{x+1}{4-x}-3\le 0$$

The function $y=\dfrac{x+1}{4-x}-3$ has a zero at $\dfrac{11}{4}$ and restricted value 4. The graph of the function is below the x-axis to the left of $\dfrac{11}{4}$ and to the right of 4. Thus, the solution set is $\left(-\infty,\dfrac{11}{4}\right]\cup(4,\infty)$.

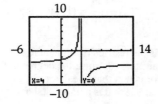

53. First, write the inequality with 0 on one side.

$$\frac{x+1}{x+5}-\frac{x}{x-1}\ge 0$$

The function $y=\dfrac{x+1}{x+5}-\dfrac{x}{x-1}$ has a zero at $-\dfrac{1}{5}$ and restricted values -5 and 1. The graph of the function is above the x-axis to the left of the line $x=-5$ and between the zero $-\dfrac{1}{5}$ and the line $x=1$. Thus, the solution set is $(-\infty,-5]\cup\left[-\dfrac{1}{5},1\right)$.

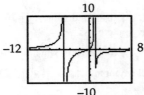

55. First, write the inequality with 0 on one side.

$$3+\frac{2-x}{x}-\frac{x+1}{x-2}>0$$

The function $y=3+\dfrac{2-x}{x}-\dfrac{x+1}{x-2}$ has zeros at -1 and 4 and restricted values 0 and 2. The graph of the function lies above the x-axis to the left of the zero -1, between the lines $x=0$ and $x=2$, and to the right of the zero 4. Thus, the solution set is $(-\infty,-1)\cup(0,2)\cup(4,\infty)$.

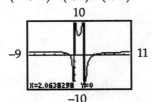

57. First, write the inequality with 0 on one side.

$$x+1-\frac{1}{x}<0$$

The function $y=x+1-\dfrac{1}{x}$ has zeros at approximately -1.62 and 0.62 and restricted value 0. The graph of the function is below the x-axis to the left of the zero -1.62 and between the line $x=0$ and the zero 0.62. Thus, the solution set is $(-\infty,-1.62)\cup(0,0.62)$.

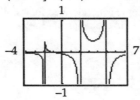

59. First, write the inequality with 0 on one side.

$$\frac{1}{x^2-4}-\frac{2}{x^2-3x-10}\ge 0$$

The function $y=\dfrac{1}{x^2-4}-\dfrac{2}{x^2-3x-10}$ has a zero at -1 and restricted values -2, 2, and 5. The graph of the function is above the x-axis between the line $x=-2$ and the zero -1 and between the line $x=2$ and the line $x=5$. Thus, the solution set is $(-2,-1]\cup(2,5)$.

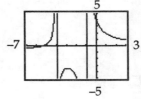

61. Write the inequality with 0 on one side.

$$\frac{4}{x+1}-\frac{x}{x^2+5x+4}<0$$

The function $y=\dfrac{4}{x+1}-\dfrac{x}{x^2+5x+4}$ has a zero at $-\dfrac{16}{3}$ and restricted values -1 and -4. The graph of the function is below the x-axis to the left of the zero $-\dfrac{16}{3}$ and between the line $x=-4$ and the line $x=-1$. thus, the solution set is $\left(-\infty,-\dfrac{16}{3}\right)\cup(-4,-1)$.

© Houghton Mifflin Company. All rights reserved.

63. The function $y = \dfrac{x^2 + x - 6}{x - 4}$ has zeros at -3 and 2 and restricted value 4. The graph of the function is below the x-axis to the left of the zero -3 and between the zero 2 and the line $x = 4$. Thus, the solution set is $(-\infty, -3] \cup [2, 4)$.

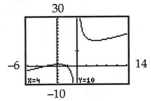

65. **a.** Simplify the rational expression $R(x)$.

$$R(x) = \frac{x^2 - 10x}{x - 10} = \frac{x(x - 10)}{x - 10} = x$$

The graph of $y = x$ is a line.

b. The function R is not defined for $x = 10$.

c. The domain of R is $\{x \mid x \neq 10\}$ and the domain of f is $\mathbb{R}$.

67. Simplify the rational expression.

$$H(x) = \frac{x^3 - 3x^2}{3 - x} = \frac{x^2(x - 3)}{3 - x} = -x^2$$

the graph of H is the same as the graph of

$y = -x^2$ except H is not defined for $x = 3$. Thus, there is a "hole" in the graph at $(3, -9)$.

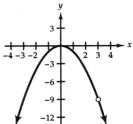

69. Simplify the rational expression.

$$\begin{aligned} f(x) &= \frac{x^3 + 4x^2 + x - 6}{x + 2} \\ &= \frac{(x + 3)(x + 2)(x - 1)}{x + 2} = (x + 3)(x - 1) \end{aligned}$$

The graph of f is the same as the graph of

$y = (x + 3)(x - 1) = x^2 + 2x - 3$ except f is not defined for $x = -2$. thus, there is a "hole" in the graph at $(-2, -3)$.

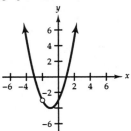

71. **a.** $h(x) = \dfrac{11x - 17}{x}$

$$h(5) = \frac{11(5) - 17}{5} = \frac{38}{5} = 7.6$$

The expected height after 5 years is 7.6 feet.

b. For values of x near the vertical asymptote $x = 0$, the model indicates that the height is negative.

c. As $x \to \infty$, $h(x) \to \dfrac{11}{1} = 11$. thus, the limiting height is 11 feet.

73 **a.** Since 0 is the restricted value there is a vertical asymptote at $x = 0$. Since the degree of the numerator is less than the degree of the denominator, $y = 0$ is a horizontal asymptote.

b. The function is not defined for $x = 0$.

c. A high school graduate has completed 12 years of formal schooling.

$$p(12) = \frac{200}{12} = 16.\overline{6}$$

thus, a recent high school graduate has probability of about 16.67% of being functionally illiterate.

d. As the number of years x increases, the probability approaches 0. The model appears to be reasonable.

75 **a.** Since 100 is a restricted value, $r_2 = 100$ is a vertical asymptote. As

$$r_2 \to \infty,\ r_1 \to \frac{100}{1} = 100. \text{ Thus, } r_1 = 100 \text{ is a}$$

horizontal asymptote.

b. If $r_2 < 100$, the denominator is negative so the function r is negative. Thus, the domain of r_1 is $(100, \infty)$.

c. As $r_2 \to 100$ from the right, $r_1 \to \infty$.

As $r_2 \to \infty$, $r_1 \to 100$.

Houghton Mifflin Company. All rights reserved.

77. a. Width: $w = x + 6$; height: $h = y + 8$

b. The area of the printed area is $x \cdot y$. Thus, $xy = 600$.

c. $xy = 600$

$$y = \frac{600}{x}$$

$$h = y + 8 = \frac{600}{x} + 8$$

d. Area $= w \cdot h$

$$A(x) = (x+6)\left(\frac{600}{x} + 8\right)$$

e.

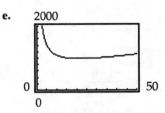

Using the calculator's *minimum* feature, the poster material will be a minimum when the width is 21 inches.

79.

20

0 ⌐ 20
0

Using the calculator's *minimum* feature, the cost will be a minimum when 10 refrigerators are manufactured.

3.4 Exercises

1. The definition requires that n be positive, so $-n$ is negative.

3. $\sqrt{-36} = i\sqrt{36} = 6i$

5. $-3 - \sqrt{-75} = -3 - i\sqrt{75}$
$\qquad = -3 - i\sqrt{25 \cdot 3}$
$\qquad = -3 - 5i\sqrt{3}$

7. $(2+3i) + (1-4i) = (2+1) + (3-4)i = 3 - i$

9. $(6i-4) - (7+2i) = (-4-7) + (6-2)i = -11 + 4i$

11. $5i - (-3+2i) = 3 + (5-2)i = 3 + 3i$

13. $3i(2i-1) - 3(2+i)$
$\quad = 6i^2 - 3i - 6 - 3i = 6(-1) - 6 - 6i$
$\quad = -12 - 6i$

15. $(1+3i)(2-i)$
$\quad = 2 - i + 6i - 3i^2 = 2 + 5i - 3(-1) = 5 + 5i$

17. $(5-4i)(4-5i)$
$\quad = 20 - 25i - 16i + 20i^2$
$\quad = 20 - 41i + 20(-1) = -41i$

19. $(3+5i)(3-5i) = 9 - 25i^2 = 9 + 25 = 34$

21. $\left(\sqrt{3} - 2i\right)\left(\sqrt{3} + 2i\right) = 3 - 4i^2 = 3 + 4 = 7$

23. $(1+3i)^2 = 1^2 + 2(1)(3i) + (3i)^2$
$\qquad = 1 + 6i + 9(-1) = -8 + 6i$

25. $i^9 = \left(i^4\right)^2 \cdot i = 1^2 \cdot i = i$

27. $i^{31} = \left(i^4\right)^7 \cdot i^3 = 1^7 \cdot -i = -i$

29. $\left(i^3\right)^5 = i^{15} = \left(i^4\right)^3 \cdot i^3 = 1^3 \cdot -i = -i$

31. $\sqrt{-9}\sqrt{-1} = i\sqrt{9} \cdot i\sqrt{1} = 3i \cdot i = 3i^2 = 3(-1) = -3$

33. $\sqrt{-2}\left(\sqrt{-8} + \sqrt{2}\right)$
$\quad = i\sqrt{2}\left(i\sqrt{8} + \sqrt{2}\right) = i\sqrt{2} \cdot i\sqrt{8} + i\sqrt{2} \cdot \sqrt{2}$
$\quad = i^2\sqrt{16} + i\sqrt{4} = 4i^2 + 2i = 4(-1) + 2i = -4 + 2i$

35. $\left(2+\sqrt{-3}\right)\left(2-\sqrt{-3}\right) = \left(2+i\sqrt{3}\right)\left(2-i\sqrt{3}\right)$
$\quad = 2^2 - \left(i\sqrt{3}\right)^2 = 4 - 3i^2 = 4 - 3(-1) = 7$

37. $\dfrac{-4+\sqrt{-12}}{2} = \dfrac{-4+i\sqrt{12}}{2} = \dfrac{-4+2i\sqrt{3}}{2} = -2 + i\sqrt{3}$

39. $(a+b)(a-b) = a^2 - b^2$
$\quad (a+bi)(a-bi) = a^2 - b^2 i^2$
$\quad = a^2 - b^2(-1) = a^2 + b^2$

41. The conjugate of $3 + 2i$ is $3 - 2i$.
$\quad (3+2i)(3-2i) = 9 - 4i^2 = 9 - 4(-1) = 13$

43. The conjugate of $i\sqrt{5} - 4$ is $-4 - i\sqrt{5}$.
$\quad \left(i\sqrt{5} - 4\right)\left(-4 - i\sqrt{5}\right)$
$\quad = \left(-4 + i\sqrt{5}\right)\left(-4 - i\sqrt{5}\right)$
$\quad = 16 - 5i^2 = 16 - 5(-1) = 21$

45. $\dfrac{1+i}{2-i} = \dfrac{1+i}{2-i} \cdot \dfrac{2+i}{2+i} = \dfrac{2+i+2i-1}{4+1}$
$\qquad = \dfrac{1+3i}{5} = \dfrac{1}{5} + \dfrac{3}{5}i$

© Houghton Mifflin Company. All rights reserved.

86   *Chapter 3: Polynomial and Rational Functions*

47. $\dfrac{2+i}{1+3i}=\dfrac{2+i}{1+3i}\cdot\dfrac{1-3i}{1-3i}=\dfrac{2-6i+i+3}{1+9}$

$=\dfrac{5-5i}{10}=\dfrac{1}{2}-\dfrac{1}{2}i$

49. $\dfrac{3+2i}{i}=\dfrac{3+2i}{i}\cdot\dfrac{-i}{-i}=\dfrac{-3i+2}{1}=2-3i$

51. $\dfrac{5i}{1+3i}=\dfrac{5i}{1+3i}\cdot\dfrac{1-3i}{1-3i}=\dfrac{5i+15}{1+9}=\dfrac{15+5i}{10}=\dfrac{3}{2}+\dfrac{1}{2}i$

53. $\dfrac{\sqrt3-i\sqrt2}{\sqrt2+i\sqrt3}=\dfrac{\sqrt3-i\sqrt2}{\sqrt2+i\sqrt3}\cdot\dfrac{\sqrt2-i\sqrt3}{\sqrt2-i\sqrt3}$

$=\dfrac{\sqrt6-3i-2i-\sqrt6}{2+3}=\dfrac{-5i}{5}=-i$

55. Since $i^4=1$, divide n by 4 to determine the remainder r, $r=0$, 1, 2, or 3. Since $i^n=i^r$, evaluate i^r.

57. $x^2+25=0$

$x^2=-25$

$x=\pm\sqrt{-25}$

$x=\pm5i$

59. $x^2-6x+13=0$

$x=\dfrac{-(-6)\pm\sqrt{(-6)^2-4(1)(13)}}{2(1)}$

$=\dfrac{6\pm\sqrt{-16}}{2}$

$=\dfrac{6\pm4i}{2}$

$=3\pm2i$

61. $16x=16x^2+5$

$-16x^2+16x-5=0$

$x=\dfrac{-16\pm\sqrt{16^2-4(-16)(-5)}}{2(-16)}$

$=\dfrac{-16\pm\sqrt{-64}}{-32}$

$=\dfrac{-16\pm8i}{-32}$

$=\dfrac{1}{2}\pm\dfrac{1}{4}i$

63. $x^2=8x-18$

$x^2-8x+18=0$

$x=\dfrac{-(-8)\pm\sqrt{(-8)^2-4(1)(18)}}{2(1)}$

$=\dfrac{8\pm\sqrt{-8}}{2}$

$=\dfrac{8\pm2i\sqrt2}{2}$

$=4\pm i\sqrt2\approx4\pm1.41i$

65. $5=4x(1-x)$

$5=4x-4x^2$

$4x^2-4x+5=0$

$x=\dfrac{-(-4)\pm\sqrt{(-4)^2-4(4)(5)}}{2(4)}$

$=\dfrac{4\pm\sqrt{-64}}{8}$

$=\dfrac{4\pm8i}{8}$

$=\dfrac{1}{2}\pm i$

67. $i^4+i^8+i^{12}+i^{16}$

$=i^4+\left(i^4\right)^2+\left(i^4\right)^3+\left(i^4\right)^4$

$=1+1^2+1^3+1^4$

$=4$

69. $(a+3i)-(5-bi)=4i$

$(a-5)+(3+b)i=0+4i$

We must have $a-5=0$, so that $a=5$; and $3+b=4$, so $b=1$.

71. $\dfrac{a+bi}{2i}=3-\dfrac{3}{2}i$

First, perform the division.

$\dfrac{a+bi}{2i}=\dfrac{a+bi}{2i}\cdot\dfrac{-2i}{-2i}=\dfrac{-2ai-2bi^2}{4}=\dfrac{b}{2}-\dfrac{a}{2}i$

Solve the equation.

$\dfrac{b}{2}-\dfrac{a}{2}i=3-\dfrac{3}{2}i$

We must have $-\dfrac{a}{2}=-\dfrac{3}{2}$, So $a=3$; and $\dfrac{b}{2}=3$, so $b=6$.

73. $x^2-2x+5=(1-2i)^2-2(1-2i)+5$

$=\left[1^2-2(1)(2i)-4\right]-2+4i+5$

$=[-3-4i]+3+4i$

$=0$

Houghton Mifflin Company. All rights reserved.

75.

$$x^2 + 4x + 9 = \left(-2 + i\sqrt{5}\right)^2 + 4\left(-2 + i\sqrt{5}\right) + 9$$

$$= \left[4 - 2(2)\left(i\sqrt{5}\right) - 5\right] - 8 + 4i\sqrt{5} + 9$$

$$= \left[-1 - 4i\sqrt{5}\right] + 1 + 4i\sqrt{5}$$

$$= 0$$

77. $(x + 2i)(x - 2i) = x^2 + 4$

79. $\left[x - (3 + i)\right]\left[x - (3 - i)\right]$

$$= x^2 - (3 - i)x - (3 + i)x + (9 + 1)$$

$$= x^2 - 6x + 10$$

81. $\dfrac{1}{a + bi} = \dfrac{1}{a + bi} \cdot \dfrac{a - bi}{a - bi}$

$$= \dfrac{a - bi}{a^2 + b^2} = \dfrac{a}{a^2 + b^2} - \dfrac{b}{a^2 + b^2}i$$

83. $ix + 3 - 4i = i$

$$ix = i - (3 - 4i)$$

$$ix = -3 + 5i$$

$$x = \dfrac{-3 + 5i}{i}$$

$$= \dfrac{-3 + 5i}{i} \cdot \dfrac{-i}{-i}$$

$$= \dfrac{-5i^2 + 3i}{-i^2}$$

$$= 5 + 3i$$

85.

$$(5 + 2i)x + i = 2i + (1 + i)x$$

$$(5 + 2i)x - (1 + i)x = 2i - i$$

$$(4 + i)x = i$$

$$x = \dfrac{i}{4 + i}$$

$$x = \dfrac{i}{4 + i} \cdot \dfrac{4 - i}{4 - i}$$

$$x = \dfrac{4i - i^2}{16 + 1}$$

$$x = \dfrac{1}{17} + \dfrac{4}{17}i$$

87. a. $i^{-3} = \dfrac{1}{i^3} = \dfrac{1}{i^3} \cdot \dfrac{i}{i} = \dfrac{i}{i^4} = \dfrac{i}{1} = i$

b. $i^{-2} = \dfrac{1}{i^2} = \dfrac{1}{i^2} \cdot \dfrac{i^2}{i^2} = \dfrac{i^2}{1} = -1$

c. $(1 + 2i)^{-1} = \dfrac{1}{1 + 2i} = \dfrac{1}{1 + 2i} \cdot \dfrac{1 - 2i}{1 - 2i}$

$$= \dfrac{1 - 2i}{1 + 4} = \dfrac{1}{5} - \dfrac{2}{5}i$$

d. $(2 - 3i)^{-2} = \dfrac{1}{(2 - 3i)^2} = \dfrac{1}{4 - 2(2)(3i) - 9}$

$$= \dfrac{1}{-5 - 12i} = \dfrac{1}{-5 - 12i} \cdot \dfrac{-5 + 12i}{-5 + 12i}$$

$$= \dfrac{-5 + 12i}{25 + 144} = -\dfrac{5}{169} + \dfrac{12}{169}i$$

3.5 Exercises

1. No, complex roots of a polynomial function occur in conjugate pairs, so the function has an even number of complex roots. Thus, a polynomial whose degree is odd has at least one real number zero.

3. $(x - 6)\left(x - 1 - \sqrt{2}\right)\left(x - 1 + \sqrt{2}\right)$

$$= (x - 6)\left(x^2 - 2x - 1\right) = x^3 - 8x^2 + 11x + 6$$

5. $\left(x + \sqrt{3}\right)\left(x - \sqrt{3}\right)(x + 2i)(x - 2i)$

$$= \left(x^2 - 3\right)\left(x^2 + 4\right) = x^4 + x^2 - 12$$

7. Because $2 - i$ is a zero, its conjugate $2 + i$ is also a zero.

$$(x - 3)(x - 2 + i)(x - 2 - i)$$

$$= (x - 3)\left(x^2 - 4x + 5\right)$$

$$= x^3 - 7x^2 + 17x - 15$$

9. Because i is a zero, its conjugate $-i$ is also a zero.

$$(x - 1)(x + 2)(x - i)(x + i)$$

$$= \left(x^2 + x - 2\right)\left(x^2 + 1\right)$$

$$= x^4 + x^3 - x^2 + x - 2$$

11. Because $1 - \sqrt{2}$ is a zero, its conjugate $1 + \sqrt{2}$ is also a zero. Because $i\sqrt{3}$ is a zero, its conjugate $-i\sqrt{3}$ is also a zero.

$$\left(x - 1 + \sqrt{2}\right)\left(x - 1 - \sqrt{2}\right)\left(x - i\sqrt{3}\right)\left(x + i\sqrt{3}\right)$$

$$= \left(x^2 - 2x - 1\right)\left(x^2 + 3\right)$$

$$= x^4 - 2x^3 + 2x^2 - 6x - 3$$

13. The conjugate of the zero is also a zero of multiplicity 3. So the lowest possible degree is 6.

Houghton Mifflin Company. All rights reserved.

15. Because $1+\sqrt{2}$ is a zero, its conjugate $1-\sqrt{2}$ is also a zero.

$$Q(x) = \left(x - 1 - \sqrt{2}\right)\left(x - 1 + \sqrt{2}\right)P(x)$$

$$= \left(x^2 - 2x - 1\right)P(x)$$

Now find $P(x)$ using division.

$$P(x) = \frac{Q(x)}{x^2 - 2x - 1}$$

$$= \frac{x^4 - 2x^3 + 8x^2 - 18x - 9}{x^2 - 2x - 1}$$

$$= x^2 + 9$$

The remaining zeros of $Q(x)$ are the zeros of

$x^2 + 9$.

$$x^2 + 9 = 0$$

$$x^2 = -9$$

$$x = \pm 3i$$

The zeros are $1+\sqrt{2}, 1-\sqrt{2}, 3i$, and $-3i$.

17. Because $3 + 2i$ is a zero, its conjugate $3 - 2i$ is also a zero.

$$f(x) = (x - 3 - 2i)(x - 3 + 2i)Q(x)$$

$$= \left(x^2 - 6x + 13\right)Q(x)$$

Now find $Q(x)$ using division.

$$Q(x) = \frac{f(x)}{x^2 - 6x + 13}$$

$$= \frac{x^4 - 6x^3 + 8x^2 + 30x - 65}{x^2 - 6x + 13}$$

$$= x^2 - 5$$

The remaining zeros of $f(x)$ are the zeros of

$x^2 - 5$.

$$x^2 - 5 = 0$$

$$x^2 = 5$$

$$x = \pm\sqrt{5}$$

The zeros are $3 + 2i, 3 - 2i, \sqrt{5}$, and $-\sqrt{5}$.

19. Because $-1 - i$ is a zero, its conjugate $-1 + i$ is also a zero.

$$P(x) = (x + 1 + i)(x + 1 - i)Q(x)$$

$$= \left(x^2 + 2x + 2\right)Q(x)$$

Now find $Q(x)$ using division.

$$Q(x) = \frac{P(x)}{x^2 + 2x + 2}$$

$$= \frac{x^4 + 6x^3 + 11x^2 + 10x + 2}{x^2 + 2x + 2}$$

$$= x^2 + 4x + 1$$

The remaining zeros of $P(x)$ are the zeros of

$x^2 + 4x + 1$.

$$x^2 + 4x + 1 = 0$$

$$x = \frac{-4 \pm \sqrt{4^2 - 4(1)(1)}}{2(1)}$$

$$= \frac{-4 \pm \sqrt{12}}{2} = -2 \pm \sqrt{3}$$

The zeros are $-1-i, -1+i, -2+\sqrt{3}$, and $-2-\sqrt{3}$.

21. Because $-1+\sqrt{3}$ is a zero, its conjugate $-1-\sqrt{3}$ is also a zero.

$$q(x) = \left(x + 1 - \sqrt{3}\right)\left(x + 1 + \sqrt{3}\right)Q(x)$$

$$= \left(x^2 + 2x - 2\right)Q(x)$$

Now find $Q(x)$ using division.

$$Q(x) = \frac{q(x)}{x^2 + 2x - 2}$$

$$= \frac{x^4 - 3x^2 + 10x - 6}{x^2 + 2x - 2}$$

$$= x^2 - 2x + 3$$

The remaining zeros of $q(x)$ are the zeros of

$x^2 - 2x + 3$.

$$x^2 - 2x + 3 = 0$$

$$x = \frac{-(-2) \pm \sqrt{(-2)^2 - 4(1)(3)}}{2(1)}$$

$$= \frac{2 \pm \sqrt{-8}}{2}$$

$$= 1 \pm i\sqrt{2}$$

The zeros are $-1+\sqrt{3}, -1-\sqrt{3}, 1+i\sqrt{2}$, and $1-i\sqrt{2}$.

Houghton Mifflin Company. All rights reserved.

23. Because $3i$ is a zero, its conjugate $-3i$ is also a zero.

$$f(x) = (x-3i)(x+3i)Q(x)$$

$$= \left(x^2+9\right)Q(x)$$

Now find $Q(x)$ using division.

$$Q(x) = \frac{f(x)}{x^2+9}$$

$$= \frac{x^4+4x^3+22x^2+36x+117}{x^2+9}$$

$$= x^2+4x+13$$

The remaining zeros of $f(x)$ are the zeros of $x^2+4x+13$.

$$x^2+4x+13=0$$

$$x = \frac{-4\pm\sqrt{4^2-4(1)(13)}}{2(1)}$$

$$= \frac{-4\pm\sqrt{-36}}{2}$$

$$= -2\pm3i$$

The zeros are $3i$, $-3i$, $-2+3i$, and $-2-3i$.

25. Because $-2+i$ is a zero, its conjugate $-2-i$ is also a zero.

$$h(x) = (x+2-i)(x+2+i)Q(x)$$

$$= \left(x^2+4x+5\right)Q(x)$$

Now find $Q(x)$ using division.

$$Q(x) = \frac{h(x)}{x^2+4x+5}$$

$$= \frac{x^4+2x^3-x^2-2x+10}{x^2+4x+5} = x^2-2x+2$$

The remaining zeros of $h(x)$ are the zeros of x^2-2x+2.

$$x^2-2x+2=0$$

$$x = \frac{-(-2)\pm\sqrt{(-2)^2-4(1)(2)}}{2(1)}$$

$$= \frac{2\pm\sqrt{-4}}{2} = 1\pm i$$

The zeros are $-2+i$, $-2-i$, $1+i$, and $1-i$.

27. No, $-i$ is not a zero. Because the polynomial does not have real coefficients, the result does not violate the Conjugate Zero Theorem.

29. To factor $P(x) = 2x^3+9x^2+2x-30$, we produce the graph and locate the x-intercepts.

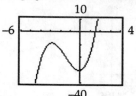

One x-intercept appears to be $\left(\frac{3}{2}, 0\right)$. We verify this using synthetic division.

$$\begin{array}{r|rrrr} \frac{3}{2} & 2 & 9 & 2 & -30 \\ & & 3 & 18 & 30 \\ \hline & 2 & 12 & 20 & 0 \end{array}$$

Therefore, $\frac{3}{2}$ is a zero and we write $P(x)$ in factored form.

$$P(x) = \left(x-\frac{3}{2}\right)\left(2x^2+12x+20\right)$$

$$= 2\left(x-\frac{3}{2}\right)\left(x^2+6x+10\right)$$

Now use the Quadratic Formula to find the remaining zeros.

$$x = \frac{-6\pm\sqrt{6^2-4(1)(10)}}{2(1)} = \frac{-6\pm\sqrt{-4}}{2} = -3\pm i$$

Thus, the zeros are $\frac{3}{2}$, $-3+i$, and $-3-i$.

Houghton Mifflin Company. All rights reserved.

31. To factor $f(x) = x^4 - 6x^3 + 10x^2 + 42x + 25$, we produce the graph and locate the *x*-intercepts.

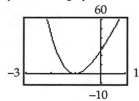

One *x*-intercept appears to be $(-1, 0)$; in fact, -1 appears to be a zero of multiplicity 2. We verify this using synthetic division.

$$
\begin{array}{r|rrrrr}
-1 & 1 & -6 & 10 & 42 & 25 \\
   &   & -1 & 7 & -17 & -25 \\
\hline
-1 & 1 & -7 & 17 & 25 & 0 \\
   &   & -1 & 8 & -25 \\
\hline
   & 1 & -8 & 25 & 0
\end{array}
$$

Therefore, -1 is a zero of multiplicity 2 and we write $f(x)$ in factored form.

$$f(x) = (x+1)^2 \left(x^2 - 8x + 25 \right)$$

Now we use the Quadratic Formula to find the remaining zeros.

$$x = \frac{-(-8) \pm \sqrt{(-8)^2 - 4(1)(25)}}{2(1)}$$

$$= \frac{8 \pm \sqrt{-36}}{2} = 4 \pm 3i$$

Thus, the zeros are -1 (multiplicity 2), $4 + 3i$, and $4 - 3i$.

33. To factor

$$q(x) = 4x^6 - 4x^5 - 9x^4 + 8x^3 + 6x^2 - 4x - 1,$$ we produce the graph and locate the *x*-intercepts.

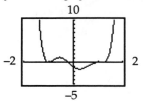

One *x*-intercept appears to be $(-1, 0)$; in fact, -1 appears to be a zero of multiplicity 2. Another *x*-intercept appears to be $(1, 0)$; at first glance it appears to be a zero of multiplicity 3. We verify these estimates using synthetic division.

$$
\begin{array}{r|rrrrrrr}
-1 & 4 & -4 & -9 & 8 & 6 & -4 & -1 \\
   &   & -4 & 8 & 1 & -9 & 3 & 1 \\
\hline
-1 & 4 & -8 & -1 & 9 & -3 & -1 & 0 \\
   &   & -4 & 12 & -11 & 2 & 1 \\
\hline
1 & 4 & -12 & 11 & -2 & -1 & 0 \\
  &   & 4 & -8 & 3 & 1 \\
\hline
1 & 4 & -8 & 3 & 1 & 0 \\
  &   & 4 & -4 & -1 \\
\hline
1 & 4 & -4 & -1 & 0 \\
  &   & 4 & 0 \\
\hline
  & 4 & 0 & -1
\end{array}
$$

Although 1 is not a zero of multiplicity 3, it is a zero of multiplicity 2. We write $q(x)$ in factored form.

$$q(x) = (x+1)^2 (x-1)^2 \left(4x^2 - 4x - 1 \right)$$

Now we use the Quadratic Formula to find the remaining zeros.

$$x = \frac{-(-4) \pm \sqrt{(-4)^2 - 4(4)(-1)}}{2(4)}$$

$$= \frac{4 \pm \sqrt{32}}{8} = \frac{1}{2} \pm \frac{\sqrt{2}}{2}$$

Thus, the zeros are -1 (multiplicity 2), 1 (multiplicity 2), $\frac{1}{2} + \frac{\sqrt{2}}{2}$, and $\frac{1}{2} - \frac{\sqrt{2}}{2}$.

Houghton Mifflin Company. All rights reserved.

35. Neither ± 1 nor ± 2 are zeros of $P(x)$. No, we can't conclude that $P(x)$ has four complex zeros. In fact, $P(x)$ has a conjugate pair of irrational zeros $\left(\sqrt{2} \text{ and } -\sqrt{2}\right)$ and a conjugate pair of complex zeros (i and $-i$).

$$x^4 - x^2 - 2 = 0$$
$$(x^2 - 2)(x^2 + 1) = 0$$
$$x^2 - 2 = 0 \qquad x^2 + 1 = 0$$
$$x^2 = 2 \qquad x^2 = -1$$
$$x = \pm\sqrt{2} \qquad x = \pm i$$

37. To solve the equation, we produce the graph of $y = x^3 - 10x^2 + 35x - 44$ and locate the x-intercepts.

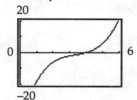

One x-intercept appears to be $(4, 0)$. We verify this using synthetic division.

$$\begin{array}{r|rrrr} 4 & 1 & -10 & 35 & -44 \\ & & 4 & -24 & 44 \\ \hline & 1 & -6 & 11 & 0 \end{array}$$

Therefore, 4 is a zero and we write the equation in factored form.

$$x^3 - 10x^2 + 35x - 44 = 0$$
$$(x - 4)(x^2 - 6x + 11) = 0$$

We use the Quadratic Formula to find the remaining zeros.

$$x = \frac{-(-6) \pm \sqrt{(-6)^2 - 4(1)(11)}}{2(1)}$$
$$= \frac{6 \pm \sqrt{-8}}{2} = 3 \pm i\sqrt{2}$$

Thus, the solutions are 4, $3 + i\sqrt{2} \approx 3 + 1.41i$, $3 - i\sqrt{2} \approx 3 - 1.41i$.

39. To solve the equation, we produce the graph of $y = 2x^4 - 7x^3 - 9x^2 + 2x + 2$ and locate the x-intercepts.

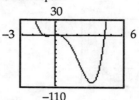

$(-1, 0)$ and $\left(\frac{1}{2}, 0\right)$ appear to be x-intercepts. We verify this using synthetic division.

$$\begin{array}{r|rrrrr} -1 & 2 & -7 & -9 & 2 & 2 \\ & & -2 & 9 & 0 & -2 \\ \hline \frac{1}{2} & 2 & -9 & 0 & 2 & 0 \\ & & 1 & -4 & -2 & \\ \hline & 2 & -8 & -4 & 0 & \end{array}$$

Therefore, -1 and $\frac{1}{2}$ are zeros and we write the equation in factored form.

$$2x^4 - 7x^3 - 9x^2 + 2x + 2 = 0$$
$$(x + 1)\left(x - \frac{1}{2}\right)(2x^2 - 8x - 4) = 0$$
$$2(x + 1)\left(x - \frac{1}{2}\right)(x^2 - 4x - 2) = 0$$

We use the Quadratic Formula to find the remaining zeros.

$$x = \frac{-(-4) \pm \sqrt{(-4)^2 - 4(1)(-2)}}{2(1)}$$
$$= \frac{4 \pm \sqrt{24}}{2} = 2 \pm \sqrt{6}$$

Thus, the solutions are -1, $\frac{1}{2}$, $2 + \sqrt{6} \approx 4.45$, and $2 - \sqrt{6} \approx -0.45$.

Houghton Mifflin Company. All rights reserved.

41. To solve the equation, we produce the graph of
$y = x^5 - 4x^4 + 7x^3 - 4x^2 - 8x + 8$ and locate the
x-intercepts.

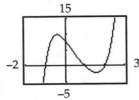

$(-1, 0)$, $(1, 0)$, and $(2, 0)$ appear to be x-intercepts.
We verify this using synthetic division.

$$
\begin{array}{r|rrrrrr}
-1 & 1 & -4 & 7 & -4 & -8 & 8 \\
   &   & -1 & 5 & -12 & 16 & -6 \\
\hline
1 & 1 & -5 & 12 & -12 & 8 & 0 \\
  &   & 1 & -4 & 8 & -8 & \\
\hline
2 & 1 & -4 & 8 & -8 & 0 & \\
  & 1 & -2 & 4 & 8 & & \\
\end{array}
$$

Therefore, $-1, 1$, and 2 are zeros and we write the
equation in factored form.

$$x^5 - 4x^4 + 7x^3 - 4x^2 - 8x + 8 = 0$$

$$(x+1)(x-1)(x-2)\left(x^2 - 2x + 4\right) = 0$$

We use the Quadratic Formula to find the
remaining zeros.

$$x = \frac{-(-2) \pm \sqrt{(-2)^2 - 4(1)(4)}}{2(1)}$$

$$= \frac{2 \pm \sqrt{-12}}{2} = 1 \pm i\sqrt{3}$$

Thus, the solutions are $-1, 1, 2,$
$1 + i\sqrt{3} \approx 1 + 1.73i$, and $1 - i\sqrt{3} \approx 1 - 1.73i$.

43. The zeros of $P(x)$ are the same as the solutions of
the equation.

45. Because $x - 3 + \sqrt{5}$ is a factor, $x - 3 - \sqrt{5}$ must
also be a factor. Therefore

$$F(x) = \left(x - 3 + \sqrt{5}\right)\left(x - 3 - \sqrt{5}\right)Q(x)$$

$$= \left(x^2 - 6x + 4\right)Q(x)$$

for some quotient polynomial $Q(x)$. We now solve
for $Q(x)$.

$$Q(x) = \frac{F(x)}{x^2 - 6x + 4}$$

$$= \frac{x^4 - 8x^3 + 15x^2 - 2x - 4}{x^2 - 6x + 4} = x^2 - 2x - 1$$

The remaining zeros of $F(x)$ are the zeros of $Q(x)$
which we find using the Quadratic Formula.

$$x = \frac{-(-2) \pm \sqrt{(-2)^2 - 4(1)(-1)}}{2(1)}$$

$$= \frac{2 \pm \sqrt{8}}{2} = 1 \pm \sqrt{2}$$

Thus,

$$F(x) = \left(x - 3 + \sqrt{5}\right)\left(x - 3 - \sqrt{5}\right)\left(x - 1 + \sqrt{2}\right)\left(x - 1 - \sqrt{2}\right).$$

47. Because $x + 2 - 4i$ is a factor, $x + 2 + 4i$ must
also be a factor. Therefore,

$$R(x) = (x + 2 - 4i)(x + 2 + 4i)Q(x)$$

$$= \left(x^2 + 4x + 20\right)Q(x)$$

for some quotient polynomial $Q(x)$. We now solve
for $Q(x)$.

$$Q(x) = \frac{R(x)}{x^2 + 4x + 20}$$

$$= \frac{x^4 + 6x^3 + 26x^2 + 32x - 40}{x^2 + 4x + 20}$$

$$= x^2 + 2x - 2$$

The remaining zeros of $R(x)$ are the zeros of $Q(x)$
which we find using the Quadratic Formula.

$$x = \frac{-2 \pm \sqrt{2^2 - 4(1)(-2)}}{2(1)}$$

$$= \frac{-2 \pm \sqrt{12}}{2} = -1 \pm \sqrt{3}$$

Thus,

$$R(x) = (x + 2 - 4i)(x + 2 + 4i)\left(x + 1 + \sqrt{3}\right)\left(x + 1 - \sqrt{3}\right).$$

Houghton Mifflin Company. All rights reserved.

49. Because $x - 2 + i\sqrt{5}$ is a factor, $x - 2 - i\sqrt{5}$ must also be a factor. Therefore,

$$G(x) = \left(x - 2 + i\sqrt{5}\right)\left(x - 2 - i\sqrt{5}\right)Q(x)$$

$$= \left(x^2 - 4x + 9\right)Q(x)$$

for some quotient polynomial $Q(x)$. We now solve for $Q(x)$.

$$Q(x) = \frac{G(x)}{x^2 - 4x + 9}$$

$$= \frac{x^4 - 6x^3 + 27x^2 - 58x + 90}{x^2 - 4x + 9}$$

$$= x^2 - 2x + 10$$

The remaining zeros of $G(x)$ are the zeros of $Q(x)$ which we find using the Quadratic Formula.

$$x = \frac{-(-2) \pm \sqrt{(-2)^2 - 4(1)(10)}}{2(1)}$$

$$= \frac{2 \pm \sqrt{-36}}{2} = 1 \pm 3i$$

Therefore,

$$G(x) = \left(x - 2 + i\sqrt{5}\right)\left(x - 2 - i\sqrt{5}\right)(x - 1 + 3i)(x - 1 - 3i).$$

51. To locate the zeros, we produce the graph of $f(x)$ and find the x-intercepts.

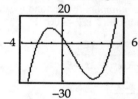

One x-intercept appears to be $(5, 0)$. We verify this result using synthetic division.

$$
\begin{array}{r|rrrr}
5 & 1 & -3 & -11 & 5 \\
  &   & 5 & 10 & -5 \\
\hline
  & 1 & 2 & -1 & 0
\end{array}
$$

Therefore, 5 is a zero. We write the function in factored form.

$$f(x) = (x - 5)\left(x^2 + 2x - 1\right)$$

We use the Quadratic Formula to find the remaining zeros.

$$x = \frac{-2 \pm \sqrt{2^2 - 4(1)(-1)}}{2(1)}$$

$$= \frac{-2 \pm \sqrt{8}}{2} = -1 \pm \sqrt{2}$$

Thus, $f(x) = (x - 5)\left(x + 1 + \sqrt{2}\right)\left(x + 1 - \sqrt{2}\right).$

53. To locate the zeros, we produce the graph of $p(x)$ and find the x-intercepts.

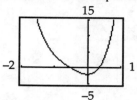

$\left(-\frac{1}{2}, 0\right)$ and $\left(\frac{1}{3}, 0\right)$ appear to be x-intercepts. We verify these results using synthetic division.

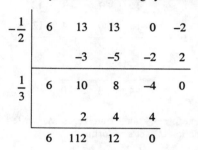

Therefore, $-\dfrac{1}{2}$ and $\dfrac{1}{3}$ are zeros. We write the function in factored form.

$$p(x) = \left(x + \frac{1}{2}\right)\left(x - \frac{1}{3}\right)\left(6x^2 + 12x + 12\right)$$

$$= 6\left(x + \frac{1}{2}\right)\left(x - \frac{1}{3}\right)\left(x^2 + 2x + 2\right)$$

$$= (2x + 1)(3x - 1)\left(x^2 + 2x + 2\right)$$

We use the Quadratic Formula to find the remaining zeros.

$$x = \frac{-2 \pm \sqrt{2^2 - 4(1)(2)}}{2(1)} = \frac{-2 \pm \sqrt{-4}}{2} = -1 \pm i$$

Thus, $p(x) = (2x + 1)(3x - 1)(x + 1 + i)(x + 1 - i).$

Houghton Mifflin Company. All rights reserved.

55. To locate the zeros, we produce the graph of $q(x)$ and find the x-intercepts.

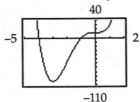

$(-4, 0)$ and $(-1, 0)$ appear to be x-intercepts. We verify these results using synthetic division.

$$
\begin{array}{r|rrrrr}
-4 & 4 & 16 & -1 & -1 & 12 \\
   &   & -16 & 0 & 4 & -12 \\
\hline
-1 & 4 & 0 & -1 & 3 & 0 \\
   &   & -4 & 4 & -3 & \\
\hline
   & 4 & -4 & 3 & 0 &
\end{array}
$$

Therefore, -4 and -1 are zeros. We write the function in factored form.

$$q(x) = (x+4)(x+1)\left(4x^2 - 4x + 3\right)$$

We use the Quadratic Formula to find the remaining zeros.

$$x = \frac{-(-4) \pm \sqrt{(-4)^2 - 4(4)(3)}}{2(4)}$$

$$= \frac{4 \pm \sqrt{-32}}{8} = \frac{1}{2} \pm \frac{\sqrt{2}}{2} i$$

Thus,

$$q(x) = (x+4)(x+1)\left(x - \frac{1}{2} + \frac{\sqrt{2}}{2}i\right)\left(x - \frac{1}{2} - \frac{\sqrt{2}}{2}i\right)$$

57. $x^2 - 5ix - 6 = 0$

$$x = \frac{-(-5i) \pm \sqrt{(-5i)^2 - 4(1)(-6)}}{2(1)}$$

$$x = \frac{5i \pm \sqrt{-25 + 24}}{2}$$

$$x = \frac{5i \pm i}{2}$$

$$x = 2i, 3i$$

Review Exercises

1.

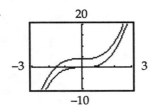

The graph of f is the graph of y shifted upward 4 units.

3. $-x^4$; since the degree is even and the coefficient is negative, the graph falls on the left and falls on the right.

5.
$$
\begin{array}{lll}
2x - 1 = 0 & x + 1 = 0 & x - 2 = 0 \\
2x = 1 & x = -1 & x = 2 \\
x = \dfrac{1}{2} & &
\end{array}
$$

$\dfrac{1}{2}$ (multiplicity 1), -1 (multiplicity 3), and 2 (multiplicity 2).

7. If n is odd, the graph crosses the x-axis at $(c, 0)$. If n is even, the graph is tangent to the x-axis at $(c, 0)$.

9. Enter the data points $(1, -6)$, $(3, -12)$, $(4, 0)$, and $(6, 84)$. Use cubic regression to obtain
$$y = x^3 - 3x^2 - 4x.$$

11. Produce the graph of the function.

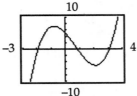

We use the calculator's *maximum* feature and locate a relative maximum at $(-0.79, 7.21)$ We use the calculator's *minimum* feature and locate a relative minimum at $(2.12, -5.06)$.

13.

$$
\require{enclose}
\begin{array}{r}
2x^2 + 4x + 3 \\
2x - 1 \enclose{longdiv}{4x^3 + 6x^2 + 2x + 5} \\
\underline{4x^3 - 2x^2} \\
8x^2 + 2x \\
\underline{8x^2 - 4x} \\
6x + 5 \\
\underline{6x - 3} \\
8
\end{array}
$$

$$P(x) = (2x - 1)\left(2x^2 + 4x + 3\right) + 8$$

Houghton Mifflin Company. All rights reserved.

15.

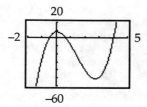

$$\left(-\frac{1}{2},0\right),\left(\frac{2}{3},0\right),(4,0) \text{ are apparent } x\text{-intercepts.}$$

We use synthetic division to verify the results.

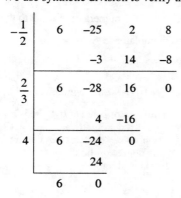

Thus, $-\frac{1}{2}, \frac{2}{3}$, and 4 are the rational zeros of P.

17.

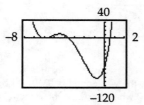

$(-6, 0)$ is an apparent x-intercept; in fact, -6 appears to be a zero of multiplicity 2. We use synthetic division to verify the result.

```
-6 | 1   15    70    84   -72
   |      -6   -54   -96    72
-6 | 1    9    16    -2     0
   |      -6   -18    12
     1    3    -2     0
```

We find the remaining zeros using the Quadratic Formula.

$$x = \frac{-3 \pm \sqrt{3^2 - 4(1)(-2)}}{2(1)} = \frac{-3 \pm \sqrt{17}}{2}$$

Thus, the zeros are -6 (multiplicity 2),

$$\frac{-3+\sqrt{17}}{2} \approx 0.56 \text{ and } \frac{-3-\sqrt{17}}{2} \approx -3.56.$$

19.

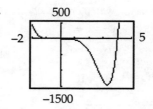

$(0, 0)$ is an apparent x-intercept; in fact, 0 appears to be a 0 of multiplicity 3. Also, $(4, 0)$ is an apparent x-intercept. Since x is a factor of $H(x)$, simplify the function first.

$$H(x) = 4x^6 - 12x^5 - 15x^4 - 4x^3$$
$$= x^3\left(4x^3 - 12x^2 - 15x - 4\right)$$

Therefore 0 is a zero of multiplicity 3. Now verify that 4 is a zero using synthetic division of $4x^3 - 12x^2 - 15x - 4$.

```
4 | 4   -12   -15   -4
  |      16    16    4
    4    4     1     0
```

We write $H(x)$ in factored form.

$$H(x) = x^3(x-4)\left(4x^2 + 4x + 1\right)$$
$$= x^3(x-4)(2x+1)^2$$

Houghton Mifflin Company. All rights reserved.

21. Write the equation with 0 on one side.

$$5x^3 + 10x + 6 = 2x^4 + x^2$$

$$2x^4 - 5x^3 + x^2 - 10x - 6 = 0$$

Graph the function

$$f(x) = 2x^4 - 5x^3 + x^2 - 10x - 6.$$

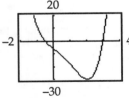

According to the Rational Zero Theorem, the only rational number zeros are

$\pm 1, \pm 2, \pm 3, \pm 6, \pm \dfrac{1}{2}, \pm \dfrac{3}{2}$. The graph suggests $-\dfrac{1}{2}$ and 3 are feasible possibilities. We check using synthetic division.

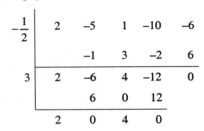

We find the remaining solutions using the Zero Factor Property.

$$2x^2 + 4 = 0$$

$$x^2 = -2$$

$$x = \pm i\sqrt{2}$$

Thus, the real number solutions are $-\dfrac{1}{2}$ and 3.

23. The inequality $x^4 + 8x^3 + 15x^2 \geq 0$ is true for all values of x for which the graph of

$y = x^4 + 8x^3 + 15x^2$ is on or above the x-axis.

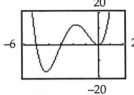

Referring to the graph, the inequality is satisfied by all values of x to the left of the zero -5 and to the right of the zero -3. Thus, the solution set is $(-\infty, -5] \cup [-3, \infty)$.

25. Because $h(0) = 0$, the y-intercept is $(0, 0)$. The numerator is 0 when $x = 0$, so the x-intercept is $(0, 0)$. There are no restricted values, so there are no vertical asymptotes. The degree of the numerator is less than the degree of the denominator so $y = 0$ is the horizontal asymptote.

27. Because $f(0)$ *is* undefined, there is no y-intercept. The numerator is 0 when $x = 3$ and $x = -1$, so the x-intercepts are $(3, 0)$ and $(-1, 0)$. Because 0 and -2 are restricted values, the lines $x = 0$ and $x = -2$ are vertical asymptotes.

The degree of the numerator is equal to the degree of the denominator so $y = \dfrac{1}{1} = 1$ is the horizontal asymptote.

29. Because $f(0) = -\dfrac{2}{3}$, the y-intercept is $\left(0, -\dfrac{2}{3}\right)$. Since the numerator is never 0, there is no x-intercept. -3 is a restricted value so $x = -3$ is a vertical asymptote. The degree of the numerator is less than the degree of the denominator, so $y = 0$ is the horizontal asymptote.

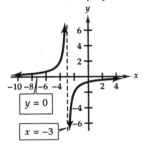

31. Because $g(0)$, the y-intercept is $(0, 0)$. The numerator is 0 when $x = 0$, so the x-intercept is $(0, 0)$. There are no restricted values, so there are no vertical asymptotes. The degree of the numerator is less than the degree of the denominator, so $y = 0$ is the horizontal asymptote.

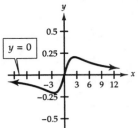

Houghton Mifflin Company. All rights reserved.

33. Write the inequality with 0 on one side.

$$\frac{x-2}{x+3} \geq -4$$

$$\frac{x-2}{x+3} + 4 \geq 0$$

The inequality is true for all values of x for which

the graph of $y = \dfrac{x-2}{x+3} + 4$ is on or above the

x-axis.

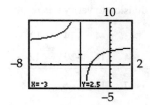

Referring to the graph, the inequality is satisfied by all values of x to the left of the vertical asymptote $x = -3$ and to the right of the zero -2. Thus, the solution set is $(-\infty, -3) \cup [-2, \infty)$.

35. Factor the function.

$$y = \frac{x^2 - 2x}{x-2} = \frac{x(x-2)}{x-2} = x$$

We see it simplifies to $y = x$. So, as x approaches 2, y approaches 2, not ∞.

37. a. 4

b. -7

c. $4 + 7i$

39. $(3i - 2) + (2 + i) = (-2 + 2) + (3 + 1)i = 4i$

41. $3(i - 5) - i(2 + i) = 3i - 15 - 2i - i^2$

$$= (-15 + 1) + (3 - 2)i$$

$$= -14 + i$$

43. $(3 - 4i)^2 = (3)^2 - 2(3)(4i) + (4i)^2$

$$= (9 - 16) - 24i$$

$$= -7 - 24i$$

45. $\dfrac{2-i}{1-3i} = \dfrac{2-i}{1-3i} \cdot \dfrac{1+3i}{1+3i}$

$$= \frac{2 + 6i - i - 3i^2}{1^2 + 3^2}$$

$$= \frac{5 + 5i}{10} = \frac{1}{2} + \frac{1}{2}i$$

47. $x^2 + 2x + 5 = 0$

$$x = \frac{-2 \pm \sqrt{2^2 - 4(1)(5)}}{2(1)}$$

$$= \frac{-2 \pm \sqrt{-16}}{2} = -1 \pm 2i$$

The solutions are $-1 + 2i$ and $-1 - 2i$.

49. By the Conjugate Zero Theorem, $-4 + 5i$ is also a zero.

51. $f(x) = (x - 1 + \sqrt{2})(x - 1 - \sqrt{2})(x - 5i)(x + 5i)$

$$= (x^2 - 2x - 1)(x^2 + 25)$$

$$= x^4 - 2x^3 + 24x^2 - 50x - 25$$

53. Because $i\sqrt{7}$ is a zero, its conjugate $-i\sqrt{7}$ is a zero. We write $F(x)$ in factored form.

$$F(x) = (x - i\sqrt{7})(x + i\sqrt{7})Q(x)$$

$$= (x^2 + 7)Q(x)$$

for some quotient polynomial $Q(x)$. We find $Q(x)$ using division.

$$Q(x) = \frac{F(x)}{x^2 + 7}$$

$$= \frac{x^4 - 4x^3 + 6x^2 - 28x - 7}{x^2 + 7}$$

$$= x^2 - 4x - 1$$

The remaining zeros of $F(x)$ are the zeros of $Q(x)$. We solve for the zeros using the Quadratic Formula.

$$x = \frac{-(-4) \pm \sqrt{(-4)^2 - 4(1)(-1)}}{2(1)}$$

$$= \frac{4 \pm \sqrt{20}}{2} = 2 \pm \sqrt{5}$$

Thus, the zeros are $i\sqrt{7}$, $-i\sqrt{7}$, $2 + \sqrt{5} \approx 4.24$, and $2 - \sqrt{5} \approx -0.24$.

Houghton Mifflin Company. All rights reserved.

55.

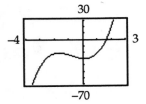

$\left(\dfrac{5}{3}, 0\right)$ is an apparent x-intercept. We use synthetic division to verify this result.

$$\dfrac{5}{3} \,\bigg|\; \begin{array}{cccc} 3 & 7 & -2 & -30 \\ & 5 & 20 & 30 \\ \hline 3 & 12 & 18 & 0 \end{array}$$

Therefore, $\dfrac{5}{3}$ is a zero. Write $f(x)$ in factored form.

$$F(x) = \left(x - \dfrac{5}{3}\right)\!\left(3x^2 + 12x + 18\right)$$

$$= 3\left(x - \dfrac{5}{3}\right)\!\left(x^2 + 4x + 6\right)$$

remaining zeros.

$$x = \dfrac{-4 \pm \sqrt{4^2 - 4(1)(6)}}{2(1)}$$

$$= \dfrac{-4 \pm \sqrt{-8}}{2} = -2 \pm i\sqrt{2}$$

Thus, the zeros are $\dfrac{5}{3}$, $-2 + i\sqrt{2} \approx -2 + 1.41i$, and $-2 - i\sqrt{2} \approx -2 - 1.41i$.

57.

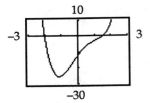

$(-2, 0)$, and $\left(\dfrac{3}{2}, 0\right)$ are apparent x-intercepts. We use synthetic division to verify this result.

$$\begin{array}{r} -2 \\ \\ \dfrac{3}{2} \\ \\ \\ \end{array} \bigg|\; \begin{array}{ccccc} 2 & -3 & -4 & 14 & -12 \\ & -4 & 14 & -20 & 12 \\ \hline 2 & -7 & 10 & -6 & 0 \\ & & 3 & -6 & 6 \\ \hline 2 & -4 & 4 & 0 & 0 \end{array}$$

Therefore, -2 and $\dfrac{3}{2}$ are zeros. Write $f(x)$ in factored form.

$$f(x) = (x + 2)\left(x - \dfrac{3}{2}\right)\!\left(2x^2 - 4x + 4\right)$$

$$= 2(x + 2)\left(x - \dfrac{3}{2}\right)\!\left(x^2 - 2x + 2\right)$$

$$= (x + 2)(2x - 3)\left(x^2 - 2x + 2\right)$$

We use the Quadratic Formula to find the remaining zeros.

$$x = \dfrac{-(-2) \pm \sqrt{(-2)^2 - 4(1)(2)}}{2(1)} = \dfrac{2 \pm \sqrt{-4}}{2} = 1 \pm i$$

Thus, $f(x) = (x + 2)(2x - 3)(x - 1 + i)(x - 1 - i)$.

Houghton Mifflin Company. All rights reserved.

59. Write the equation with 0 on one side.

$$2x^3 + 50x = 5x^2 + 125$$

$$2x^3 - 5x^2 + 50x - 125 = 0$$

The solutions of the equation are the zeros of the

function $y = 2x^3 - 5x^2 + 50x - 125$. Graph the

function and locate the x-intercepts.

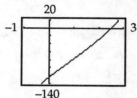

$\left(\dfrac{5}{2}, 0\right)$ is an apparent x-intercept. We use

synthetic division to verify this result.

$\dfrac{5}{2}$	2	−5	50	−125
		5	0	125
	2	0	50	0

Therefore, $\dfrac{5}{2}$ is a zero. Write the equation in

factored form.

$$\left(x - \frac{5}{2}\right)\left(2x^2 + 50\right) = 0$$

The remaining zeros are the zeros of $2x^2 + 50$.

$$2x^2 + 50 = 0$$

$$x^2 = -25$$

$$x = \pm 5i$$

Thus, the solutions are $\dfrac{5}{2}$, $5i$ and $-5i$.

Houghton Mifflin Company. All rights reserved.

Chapter 4: Exponential and Logarithmic Functions

4.1 Exercises

1. Either (1) determine the product of $f(x)$ and $g(x)$ and evaluate the resulting function for $x = -2$, or (2) evaluate $f(-2)$ and $g(-2)$ and multiply the results. Method (2) is easier.

3. a.
$$(f+g)(1) = f(1) + g(1)$$
$$= 3(1) + (1 - 1^2)$$
$$= 3 + 0$$
$$= 3$$

b.
$$(f-g)(2) = f(2) - g(2)$$
$$= 3(2) - (1 - 2^2)$$
$$= 6 - (-3)$$
$$= 9$$

c.
$$(fg)(-1) = f(-1) \cdot g(-1)$$
$$= 3(-1) \cdot [1 - (-1)^2]$$
$$= -3(0)$$
$$= 0$$

d.
$$\left(\frac{f}{g}\right)(0) = \frac{f(0)}{g(0)} = \frac{3(0)}{1 - 0^2} = \frac{0}{1} = 0$$

5. a.
$$(f+g)(1) = f(1) + g(1)$$
$$= 2^1 + \frac{1}{1-3}$$
$$= 2 - \frac{1}{2}$$
$$= \frac{3}{2}$$

b.
$$(f-g)(2) = f(2) - g(2)$$
$$= 2^2 - \frac{1}{2-3}$$
$$= 4 - \frac{1}{-1}$$
$$= 5$$

c.
$$(fg)(-1) = f(-1) \cdot g(-1)$$
$$= 2^{-1} \cdot \frac{1}{-1-3}$$
$$= \frac{1}{2} \cdot \left(-\frac{1}{4}\right)$$
$$= -\frac{1}{8}$$

d.
$$\left(\frac{f}{g}\right)(0) = \frac{f(0)}{g(0)} = \frac{2^0}{\frac{1}{0-3}} = \frac{1}{-\frac{1}{3}} = -3$$

7. a.
$$(f+g)(0) = f(0) + g(0)$$
$$= \sqrt{1 - 2(0)} + (|2(0)| - 1)$$
$$= 1 + (-1)$$
$$= 0$$

b.
$$(f-g)(0.5) = f(0.5) - g(0.5)$$
$$= \sqrt{1 - 2(0.5)} - (2|0.5| - 1)$$
$$= 0 - 0$$
$$= 0$$

c.
$$\left(\frac{f}{g}\right)(-4) = \frac{f(-4)}{g(-4)} = \frac{\sqrt{1 - 2(-4)}}{2|-4| - 1} = \frac{3}{7}$$

d.
$$(ff)(-2) = f(-2) \cdot f(-2)$$
$$= \sqrt{1 - 2(-2)} \cdot \sqrt{1 - 2(-2)}$$
$$= \sqrt{5} \cdot \sqrt{5}$$
$$= 5$$

9.
$$(f+g)(x) = f(x) + g(x)$$
$$= (x+1) + (x^2 + 2x + 1)$$
$$= x^2 + 3x + 2$$
$$(f-g)(x) = f(x) - g(x)$$
$$= (x+1) - (x^2 + 2x + 1)$$
$$= -x^2 - x$$
$$(fg)(x) = f(x) \cdot g(x)$$
$$= (x+1) \cdot (x^2 + 2x + 1)$$
$$= x^3 + 3x^2 + 3x + 1$$
$$\left(\frac{f}{g}\right)(x) = \frac{f(x)}{g(x)} = \frac{x+1}{x^2 + 2x + 1} = \frac{x+1}{(x+1)^2} = \frac{1}{x+1}$$

Domain of $\dfrac{f}{g} = \{x | x \neq 1\}$.

© Houghton Mifflin Company. All rights reserved.

11. $(f+g)(x) = f(x) + g(x)$

$$= \frac{x}{x-1} + \frac{1-x}{x+2}$$

$$= \frac{x}{x-1} \cdot \frac{x+2}{x+2} + \frac{1-x}{x+2} \cdot \frac{x-1}{x-1}$$

$$= \frac{4x-1}{(x-1)(x+2)}$$

$(f-g)(x) = f(x) - g(x)$

$$= \frac{x}{x-1} - \frac{1-x}{x+2}$$

$$= \frac{2x^2+1}{(x-1)(x+2)}$$

$(fg)(x) = f(x) \cdot g(x)$

$$= \frac{x}{x-1} \cdot \frac{1-x}{x+2}$$

$$= \frac{x}{x-1} \cdot \frac{-(x-1)}{x+2}$$

$$= -\frac{x}{x+2}$$

$$\left(\frac{f}{g}\right)(x) = \frac{f(x)}{g(x)} = \frac{\frac{x}{x-1}}{\frac{1-x}{x+2}} = \frac{x}{x-1} \cdot \frac{x+2}{1-x} = -\frac{x^2+2x}{(x-1)^2}$$

Domain of $\frac{f}{g} = \{x|x \neq 1, -2\}$.

13. Parts (ii) and (iii) are possible definitions of f and g. In (i), both functions are defined for $x = -1$, so $(fg)(-1)$ is also defined.

15. a. $(f+g)(x) = 2x^2$; Domain = **R**

b. $(f-g)(x) = -8$; Domain = **R**

c. $(fg)(x) = x^4 - 16$; Domain = **R**

d. $\left(\frac{f}{g}\right)(x) = \frac{x^2-4}{x^2+4}$; Domain = **R**

17. a. $(f+g)(x) = \frac{x^2-6}{x(x-3)}$; Domain = $\{x|x \neq 0, 3\}$

b. $(f-g)(x) = \frac{-x^2+2x+6}{x(x-3)}$;

Domain = $\{x|x \neq 0, 3\}$

c. $(fg)(x) = \frac{x+2}{x(x-3)}$; Domain = $\{x|x \neq 0, 3\}$

d. $\left(\frac{f}{g}\right)(x) = \frac{x}{(x-3)(x+2)}$;

Domain = $\{x|x \neq -2, 0, 3\}$

19. a. $(f+g)(x) = \sqrt{x} + \sqrt{x-3}$; Domain = $\{x|x \geq 3\}$

b. $(f-g)(x) = \sqrt{x} - \sqrt{x-3}$; Domain = $\{x|x \geq 3\}$

c. $(fg)(x) = \sqrt{x(x-3)}$; Domain = $\{x|x \geq 3\}$

d. $\left(\frac{f}{g}\right)(x) = \sqrt{\frac{x}{x-3}}$; Domain = $\{x|x \geq 3\}$

21. Because $g(0) = -1$ and -1 is not in the domain of f, 0 is not in the domain of $f \circ g$.

23. a. $(f \circ g)(1) = f(g(1))$

$$= f([1^2 - 1])$$
$$= f(0)$$
$$= 2(0) - 3$$
$$= -3$$

b. $(g \circ f)(2) = g(f(2))$

$$= g([2(2) - 3])$$
$$= g(1)$$
$$= 1^2 - 1$$
$$= 0$$

c. $(f \circ g)(-3) = f(g(-3))$

$$= f([(-3)^2 - 1])$$
$$= f(8)$$
$$= 2(8) - 3$$
$$= 13$$

d. $(g \circ f)(0) = g(f(0))$

$$= g([2(0) - 3])$$
$$= g(-3)$$
$$= (-3)^2 - 1$$
$$= 8$$

© Houghton Mifflin Company. All rights reserved.

25. a. $(f \circ g)(0) = f(g(0))$

$$= f(0+2)$$
$$= f(2)$$
$$= \sqrt[3]{3-2}$$
$$= 1$$

b. $(g \circ f)(-5) = g(f(-5))$

$$= g\left(\sqrt[3]{3-(-5)}\right)$$
$$= g(2)$$
$$= 2+2$$
$$= 4$$

c. $(f \circ f)(4) = f(f(4))$

$$= f\left(\sqrt[3]{3-4}\right)$$
$$= f(-1)$$
$$= \sqrt[3]{3-(-1)}$$
$$= \sqrt[3]{4}$$
$$\approx 1.59$$

d. $(g \circ g)(-7) = g(g(-7))$

$$= g(-7+2)$$
$$= g(-5)$$
$$= -5+2$$
$$= -3$$

27. $(f \circ g)(x) = f(g(x)) = f(3-x) = (3-x)^2$
$(g \circ f)(x) = g(f(x)) = g(x^2) = 3-x^2$
$(f \circ f)(x) = f(f(x)) = f(x^2) = (x^2)^2 = x^4$

29. $(f \circ g)(x) = f(g(x))$

$$= f\left(\sqrt[3]{x+1}\right)$$
$$= \left(\sqrt[3]{x+1}\right)^3 - 2$$
$$= (x+1) - 2$$
$$= x - 1$$

$(g \circ f)(x) = g(f(x))$

$$= g(x^3 - 2)$$
$$= \sqrt[3]{(x^3-2)+1}$$
$$= \sqrt[3]{x^3 - 1}$$

31. $(f \circ g)(x) = f(g(x))$

$$= f(3-2x)$$
$$= \frac{1}{(3-2x)+1}$$
$$= \frac{1}{4-2x}$$

$(g \circ f)(x) = g(f(x))$

$$= g\left(\frac{1}{x+1}\right)$$
$$= 3 - 2\left(\frac{1}{x+1}\right)$$
$$= 3 - \frac{2}{x+1}$$

$(f \circ f)(x) = f(f(x)) = f\left(\frac{1}{x+1}\right) = \frac{1}{\frac{1}{x+1}+1} = \frac{x+1}{x+2}$

33. $(f \circ g)(x) = f(g(x)) = f(x^{3/2}) = (x^{3/2})^4 = x^6$
$(g \circ f)(x) = g(f(x)) = g(x^4) = (x^4)^{3/2} = x^6$
$(f \circ f)(x) = f(f(x)) = f(x^4) = (x^4)^4 = x^{16}$

35. $(f \circ g)(x) = f(g(x))$

$$= f(x^3 + 2x^2 - x - 1)$$
$$= (x^3 + 2x^2 - x - 1) + 3$$
$$= x^3 + 2x^2 - x + 2$$

$(g \circ f)(x) = g(f(x))$

$$= g(x+3)$$
$$= (x+3)^3 + 2(x+3)^2 - (x+3) - 1$$

$(f \circ f)(x) = f(f(x))$

$$= f(x+3)$$
$$= (x+3) + 3$$
$$= x + 6$$

37. $(f \circ g)(x) = f(g(x))$

$$= f\left(\frac{1}{2}x + 3\right)$$
$$= 2\left(\frac{1}{2}x + 3\right) - 6$$
$$= x + 6 - 6$$
$$= x$$

$(g \circ f)(x) = g(f(x))$

$$= g(2x - 6)$$
$$= \frac{1}{2}(2x - 6) + 3$$
$$= x - 3 + 3$$
$$= x$$

© Houghton Mifflin Company. All rights reserved.

39. $(f \circ g)(x) = f(g(x))$

$$= f\left(\frac{x}{x+1}\right)$$

$$= \frac{\frac{x}{x+1}}{1 - \frac{x}{x+1}}$$

$$= \frac{\frac{x}{x+1}}{\frac{1}{x+1}}$$

$$= x$$

$(g \circ f)(x) = g(f(x))$

$$= g\left(\frac{x}{1-x}\right)$$

$$= \frac{\frac{x}{1-x}}{\frac{x}{1-x}+1}$$

$$= \frac{\frac{x}{1-x}}{\frac{1}{1-x}}$$

$$= x$$

41. $(f \circ g)(x) = f(g(x))$

$$= f((x-1)^5)$$

$$= \sqrt[5]{(x-1)^5} + 1$$

$$= (x-1)+1$$

$$= x$$

$(g \circ f)(x) = g(f(x))$

$$= g(\sqrt[5]{x}+1)$$

$$= \left[(\sqrt[5]{x}+1)-1\right]^5$$

$$= \sqrt[5]{x}^5$$

$$= x$$

43. a. $(f \circ g)(x) = f(g(x))$

$$= f\left(\frac{x}{3}\right)$$

$$= 3\left(\frac{x}{3}\right)+1$$

$$= x+1$$

Domain = **R**

b. $(g \circ f)(x) = g(f(x))$

$$= g(3x+1)$$

$$= \frac{3x+1}{3}$$

Domain = **R**

45. a. $(f \circ g)(x) = f(g(x)) = f(\sqrt{x}) = 3 - \sqrt{x}$

Domain = $\{x | x \geq 0\}$

b. $(g \circ f)(x) = g(f(x)) = g(3-x) = \sqrt{3-x}$

Domain = $\{x | x \leq 3\}$

47. Let $f(x) = \sqrt{x}$ and $g(x) = 2x+1$. Then

$h(x) = (f \circ g)(x) = f(g(x)) = f(2x+1) = \sqrt{2x+1}$.

49. Let $f(x) = \frac{2}{x}$ and $g(x) = x^2 - x$. Then

$h(x) = (f \circ g)(x) = f(g(x)) = f(x^2 - x) = \frac{2}{x^2 - x}$.

51. Let $f(x) = \frac{x-4}{x+4}$ and $g(x) = x^2$. Then

$h(x) = (f \circ g)(x) = f(g(x)) = f(x^2) = \frac{x^2 - 4}{x^2 + 4}$.

53. Let $f(x) = x^3 + 3x^2 + 3x + 5$ and $g(x) = x - 1$. Then

$h(x) = (f \circ g)(x)$

$= f(g(x))$

$= f(x-1)$

$= (x-1)^3 + 3(x-1)^3 + 3(x-1) + 5$

55. Let $f(x) = |2-x|$ and $g(x) = |2-x|$. Then

$h(x) = (f \circ g)(x)$

$= f(g(x))$

$= f(|2-x|)$

$= |2 - |2-x||$

57. a. $(f+g)(2t) = f(2t) + g(2t)$

$= [(2t)^2 - 2(2t)] + [(2t) - 4]$

$= (4t^2 - 4t) + (2t - 4)$

$= 4t^2 - 2t - 4$

b. $(f-g)(t+1) = f(t+1) - g(t+1)$

$= [(t+1)^2 - 2(t+1)] - [(t+1) - 4]$

$= (t^2 - 1) - (t-3)$

$= t^2 - t + 2$

c. $\left(\frac{g}{f}\right)(a^2) = \frac{g(a^2)}{f(a^2)}$

$$= \frac{a^2 - 4}{(a^2)^2 - 2(a^2)}$$

$$= \frac{a^2 - 4}{a^4 - 2a^2}$$

© Houghton Mifflin Company. All rights reserved.

59. a.

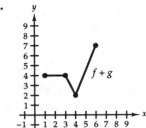

b. Domain $= \{x|1 \le x \le 6\}$

61. a. $\left(\dfrac{f}{g}\right)(1) = \dfrac{f(1)}{g(1)} = \dfrac{3}{1} = 3$

b. Domain $= \{x|1 \le x < 2 \cup 2 < x < 4\}$

63. a. $(f \circ g)(4) = f(g(4)) = f(0) = 3$

b. $(g \circ f)(3) = g(f(3)) = g(5) = 0$

65. Because f is even, $f(-x) = f(x)$. Because g is odd, $g(-x) = -g(x)$. So, we have
$$(fg)(-x) = f(-x) \cdot g(-x)$$
$$= f(x) \cdot [-g(x)]$$
$$= -f(x)g(x)$$
$$= -(fg)(x)$$
Thus, fg is odd.

67. a. $P(x) = R(x) - C(x)$
$$P(x) = 100,000\sqrt{x} - 30,000x$$

b. $R(4) - C(4) = 100,000\sqrt{4} - 30,000(4)$
$$R(4) - C(4) = 80,000$$

c. $P(4) = 100,000\sqrt{4} - 30,000(4)$
$$P(4) = 80,000$$
Thus, $P(4) = R(4) - C(4)$.

69. a. Since distance = rate × time, time $= \dfrac{\text{distance}}{\text{rate}}$.

We have $h(d) = \dfrac{d}{50}$

b. There were 18 gallons at the beginning of the trip and the car consumes 20 miles per gallons. After h hours at 50 miles per hour, the car will use $\dfrac{50}{20}h = 2.5h$ gallons.
Thus, $f(h) = 18 - 2.5h$.

c. $(f \circ h)(d) = f(h(d))$
$$= f\left(\dfrac{d}{50}\right)$$
$$= 18 - 2.5\left(\dfrac{d}{50}\right)$$
$$= 18 - \dfrac{d}{20}$$
The composition $f \circ h$ gives the amount of gasoline remaining after the car has traveled d miles.

d. $(f \circ h)(200) = 18 - \dfrac{200}{20} = 18 - 10 = 8$
After the car travels 200 miles, 8 gallons of gasoline remain.

71. a. $r(t) = 1.5t$

b. $A(r) = \pi r^2$

c. $f(t) = (A \circ r)(t)$
$$= A(r(t))$$
$$= A(1.5t)$$
$$= \pi(1.5t)^2$$
$$= 2.25\pi t^2$$

d. $f(2.5) = 2.25\pi(2.5)^2 = 14.0625\pi \approx 44.2$ square feet

73. a. $(C + N)(x) = C(x) + N(x)$
$$= (574x + 5676) + (512x + 4824)$$
$$= 1086x + 10,500$$
The function represents the total number of agencies.

b.

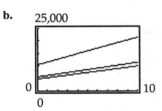

Referring to the graphs,
$C(4) = 7972$
$N(4) = 6872$
$(C + N)(4) = 14,844$
Thus, $(C + N)(4) = C(4) + N(4)$.

4.2 Exercises

1. The Vertical Line Test guarantees that the relation is a function and the Horizontal Line Test guarantees that the function is a one-to-one function.

3. $\{(4, -3), (-1, 0), (9, 5)\}$

5. $y = \sqrt{9 - x^2}$

7. $xy + x = 3$

© Houghton Mifflin Company. All rights reserved.

9. The inverse is obtained by reversing the coordinates of the ordered pairs. So its graph is in Quadrant IV.

11. No, the inverse is not a function.

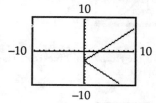

13. Yes, the inverse is a function.

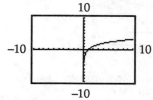

15. Yes, the inverse is a function.

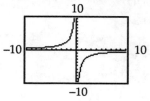

17. a. $D: \{x | x \neq 4\}, \ R: \{y | y \neq 2\}$

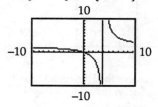

  b. $D: \{x | x \neq 2\}, \ R: \{y | y \neq 4\}$

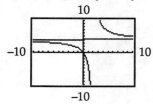

19. a. $D: \{x | x \leq 3\}, \ R: \{y | y \geq 0\}$

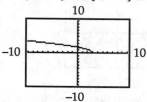

  b. $D: \{x | x \geq 0\}, \ R: \{y | y \leq 3\}$

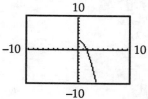

21. a. $D: \{x | x \geq 2\}, \ R: \{y | y \geq 0\}$

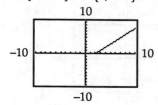

  b. $D: \{x | x \geq 0\}, \ R: \{y | y \geq 2\}$

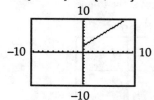

23. a. $D: \mathbf{R}, R: \mathbf{R}$

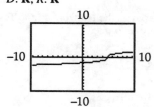

  b. $D: \mathbf{R}, R: \mathbf{R}$

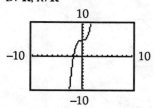

25. Yes, the function is one-to-one.

27. No, the function is not one-to-one.

© Houghton Mifflin Company. All rights reserved.

29.

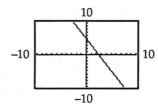

The graph passes the Horizontal Line Test.
The function is one-to-one.

$$f(x) = 5 - 2x$$
$$y = 5 - 2x$$
$$x = 5 - 2y$$
$$x - 5 = -2y$$
$$y = \frac{x - 5}{-2}$$
$$= -\frac{1}{2}x + \frac{5}{2}$$
$$f^{-1}(x) = -\frac{1}{2}x + \frac{5}{2}$$

31.

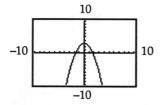

The graph fails the Horizontal Line Test. The function is not one-to-one.

33.

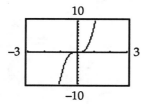

The graph passes the Horizontal Line Test. The function is one-to-one.

$$g(x) = 8x^3$$
$$y = 8x^3$$
$$x = 8y^3$$
$$\frac{x}{8} = y^3$$
$$y = \sqrt[3]{\frac{x}{8}}$$
$$= \frac{\sqrt[3]{x}}{2}$$
$$g^{-1}(x) = \frac{\sqrt[3]{x}}{2}$$

35.

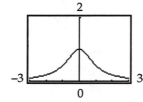

The graph fails the Horizontal Line Test. The function is not one-to-one.

37.

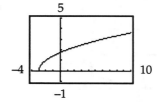

The graph passes the Horizontal Line Test. The function is one-to-one.

$$p(x) = \sqrt{x + 3}$$
$$y = \sqrt{x + 3}$$
$$x = \sqrt{y + 3}$$
$$x^2 = y + 3$$
$$y = x^2 - 3$$
$$p^{-1}(x) = x^2 - 3, \ x \geq 0$$

39.

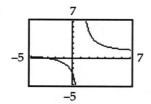

The graph passes the Horizontal Line Test. The function is one-to-one.

$$f(x) = \frac{x + 3}{x - 1}$$
$$y = \frac{x + 3}{x - 1}$$
$$x = \frac{y + 3}{y - 1}$$
$$xy - x = y + 3$$
$$xy - y = x + 3$$
$$y(x - 1) = x + 3$$
$$y = \frac{x + 3}{x - 1}$$
$$f^{-1}(x) = \frac{x + 3}{x - 1}$$

© Houghton Mifflin Company. All rights reserved.

41.

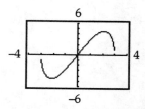

The graph fails the Horizontal Line Test. The function is not one-to-one.

43.

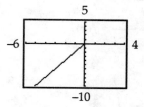

The graph fails the Horizontal Line Test. The function is not one-to-one.

45. $(f \circ f^{-1})(x) = f(f^{-1}(x)) = f\left(\dfrac{3x+1}{2}\right)$

$= \dfrac{2\left(\frac{3x+1}{2}\right)-1}{3}$

$= \dfrac{(3x+1)-1}{3}$

$= \dfrac{3x}{3}$

$= x$

$(f^{-1} \circ f)(x) = f^{-1}(f(x))$

$= f^{-1}\left(\dfrac{2x-1}{3}\right)$

$= \dfrac{3\left(\frac{2x-1}{3}\right)+1}{2}$

$= \dfrac{(2x-1)+1}{2}$

$= \dfrac{2x}{2}$

$= x$

47. $(f \circ f^{-1})(x) = f(f^{-1}(x)) = f\left(\dfrac{1}{x-1}\right)$

$= \dfrac{\frac{1}{x-1}+1}{\frac{1}{x-1}}$

$= \dfrac{\frac{1+(x-1)}{x-1}}{\frac{1}{x-1}}$

$= \dfrac{x}{1}$

$= x$

$(f^{-1} \circ f)(x) = f^{-1}(f(x)) = f^{-1}\left(\dfrac{x+1}{x}\right)$

$= \dfrac{1}{\frac{x+1}{x}-1}$

$= \dfrac{1}{\frac{(x+1)-x}{x}}$

$= \dfrac{1}{\frac{1}{x}}$

$= x$

49. $f(x) = \dfrac{x+2}{5}$

$y = \dfrac{x+2}{5}$

$x = \dfrac{y+2}{5}$

$5x = y+2$

$y = 5x-2$

$f^{-1}(x) = 5x-2$

51.

$f(x) = \sqrt[3]{x}-7$

$y = \sqrt[3]{x}-7$

$x = \sqrt[3]{y}-7$

$x+7 = \sqrt[3]{y}$

$y = (x+7)^3$

$f^{-1}(x) = (x+7)^3$

53.

$f(x) = -\dfrac{3}{x+5}$

$y = -\dfrac{3}{x+5}$

$x = -\dfrac{3}{y+5}$

$x(y+5) = -3$

$xy+5x = -3$

$xy = -3-5x$

$y = \dfrac{-3-5x}{x}$

$y = -\dfrac{3}{x}-5$

$f^{-1}(x) = -\dfrac{3}{x}-5$

© Houghton Mifflin Company. All rights reserved.

55. $f(x) = 2 - \sqrt{x}$

$y = 2 - \sqrt{x}$

$x = 2 - \sqrt{y}$

$x - 2 = -\sqrt{y}$

$\sqrt{y} = 2 - x$

$y = (2 - x)^2, \ x \le 2$

$f^{-1}(x) = (2 - x)^2, \ x \le 2$

57. $f(x) = (x - 1)^2, \ x \ge 1$

With the given domain for f, the range is $\{y | y \ge 0\}$, which will be the domain of the inverse function.

$y = (x - 1)^2$

$x = (y - 1)^2$

$\sqrt{x} = y - 1$

$y = \sqrt{x} + 1$

$f^{-1}(x) = \sqrt{x} + 1, \ x \ge 0$

59. $g(x) = 4 - x^2, \ x \ge 0$

With the given domain for g, the range is $\{y | y \le 4\}$, which will be the domain of the inverse function.

$y = 4 - x^2$

$x = 4 - y^2$

$y^2 = 4 - x$

$y = \sqrt{4 - x}$

$g^{-1}(x) = \sqrt{4 - x}, \ x \le 4$

61. $h(x) = (x + 2)^2, \ x \le -2$

With the given domain for h, the range is $\{y | y \ge 0\}$, which will be the domain of the inverse function.

$y = (x + 2)^2$

$x = (y + 2)^2$

$-\sqrt{x} = y + 2$

$-\sqrt{x} - 2 = y$

$h^{-1}(x) = -\sqrt{x} - 2, \ x \ge 0$

63. A constant function is not one-to-one and so does not have an inverse.

65. $f(x) = \dfrac{x}{x - 1}$

$y = \dfrac{x}{x - 1}$

$x = \dfrac{y}{y - 1}$

$xy - x = y$

$xy - y = x$

$y(x - 1) = x$

$y = \dfrac{x}{x - 1}$

Thus, $f^{-1}(x) = \dfrac{x}{x - 1} = f(x)$.

67.

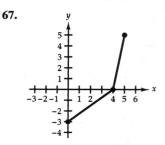

Yes, the inverse relation is a funciton.

69.

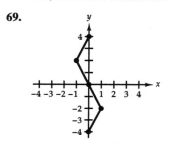

No, the inverse relation is not a function.

71. $(f \circ g)(x) = f(g(x)) = f\left(x^{1/5}\right) = \left(x^{1/5}\right)^5 = x$

$(g \circ f)(x) = g(f(x)) = g(x^5) = (x^5)^{1/5} = x$

Tracing the graphs of each composition indicates g is the inverse of f.

73. $(f \circ g)(x) = f(g(x))$

$= f(x^2 + 1)$

$= (x^2 + 1)^2 - 1$

$= x^4 + 2x^2$

Tracing the graph of this composition indicates g is not the inverse of f.

© Houghton Mifflin Company. All rights reserved.

75. Yes, f is one-to-one.

For $x \le 0$: $\quad y = -x^2$
$$x = -y^2$$
$$y^2 = -x$$
$$y = -\sqrt{-x}$$

For $x > 0$: $\quad y = x^2$
$$x = y^2$$
$$y = \sqrt{x}$$

$$f^{-1}(x) = \begin{cases} -\sqrt{-x}, & x \le 0 \\ \sqrt{x}, & x > 0 \end{cases}$$

77. Let $y = f(x)$. If f is an even function, then
$f(-x) = f(x) = y$.

Thus, $f^{-1}(y) = x$ and $f^{-1}(y) = -x$. Therefore f is not a function.

79. a. No, R is not a function. Two pairs have the same first coordinate: $(2, A)$ and $(2, B)$.

b. Yes, R^{-1} is a function. Each letter is associated with exactly one number.

c. Yes, as illustrated in parts (a) and (b).

81. a. $r(m) = 800 - 3m$

b. $0 \le m \le 100$

c. $r = 800 - 3m$
$$3m = 800 - r$$
$$m = \frac{800 - r}{3}$$

Thus, $m = \dfrac{800 - r}{3}$ gives the number m of members in attendance for total revenue r.

83. a. Let t be the time for both painters working together to complete the job.

$$\frac{t}{3} + \frac{t}{x} = 1$$
$$t\left(\frac{1}{3} + \frac{1}{x}\right) = 1$$
$$t\left(\frac{1}{3} \cdot \frac{x}{x} + \frac{1}{x} \cdot \frac{3}{3}\right) = 1$$
$$t\frac{x+3}{3x} = 1$$
$$t = \frac{3x}{x+3}$$

Thus, $f(x) = \dfrac{3x}{x+3}$

b. $f(4.5) = \dfrac{3(4.5)}{4.5+3} = 1.8$

The answer indicates that if the second painter can paint the house in 4.5 days, then working together, the two painters can paint the house in 1.8 days.

c.

Yes, f is a one-to-one function.
$$y = \frac{3x}{x+3}$$
$$x = \frac{3y}{y+3}$$
$$xy + 3x = 3y$$
$$xy - 3y = -3x$$
$$y(x-3) = -3x$$
$$y = -\frac{3x}{x-3}$$
$$y = \frac{3x}{3-x}$$

Thus, $f^{-1}(x) = \dfrac{3x}{3-x}$.

d. $f^{-1}(2) = \dfrac{3(2)}{3-2} = 6$

The value indicates that if working together the painters can complete the job in 2 days, the second painter could paint the house in 6 days.

© Houghton Mifflin Company. All rights reserved.

85. a. Using the data values (2, 5), (4, 15), (6, 40), (8, 100), (10, 300), and (12,570), we use the calculator's cubic regression and obtain

$$f(x) = 0.68x^3 - 5.41x^2 + 15.02x - 6.67$$

b. $f(7) = 0.68(7)^3 - 5.41(7)^2 + 15.02(7) - 6.67$
$= 66.62$
The value indicates that in (1988 + 7 =) 1995 the number of cards was 66.62 million.

c. $f^{-1}(66.62) = 7$

d. Domain = $\{x | 5 \le x \le 570\}$

4.3 Exercises

1. Since $g(x) = \left(\dfrac{1}{2}\right)^{-x} = \left(\dfrac{1}{\frac{1}{2}}\right)^x = 2^x = f(x)$, the functions are equivalent.

3. $f(3.2) = 5^{3.2} = 172.47$

5. $g(-\pi) = \pi^{-\pi} = 0.03$

7. $h(-\sqrt{2}) = 6.8^{-\sqrt{2}} = 0.07$

9. $f(3) = e^3 = 20.09$

11. $f\left(-\dfrac{3}{4}\right) = e^{-\frac{3}{4}} = 0.47$

13. $f(-0.3) = 0.74$

15. $f(x) = 2x$ is a linear function, $g(x) = 2^x$ is an exponential function, and $h(x) = x^2$ is a quadratic function.

17. a. $4^a = 1$, so $a = 0$

b. $4^b = 16$, so $b = 2$

c. $4^3 = c$, so $c = 64$

d. $4^{-1} = d$, so $d = \dfrac{1}{4} = 0.25$

19. a. $e^a = e$, so $a = 1$

b. $e^b = 1$, so $b = 0$

c. $e^c = e^3$, so $c = 3$

d. $e^d = \dfrac{1}{e}$, so $d = -1$

21. Since e represents a specific number, (i) and (ii) are not variable expressions. Only (iii) is a variable expression.

23.

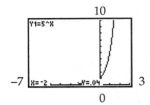

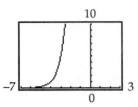

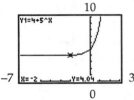

a. The graph of f is the graph of y shifted to the left 4 units.

b. The graph of g is the graph of y shifted upward 4 units.

25.

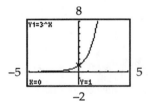

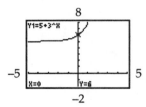

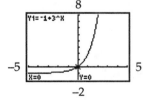

a. The graph of f is the graph of y shifted upward 5 units.

b. The graph of g is the graph of y shifted downward 1 unit.

© Houghton Mifflin Company. All rights reserved.

27. If the graph $f(x) = b^x$ contains the point $\left(-2, \dfrac{1}{9}\right)$, then $f(-2) = b^{-2} = \dfrac{1}{9} = \dfrac{1}{3^2} = 3^{-2}$. Thus $b = 3$.

29.

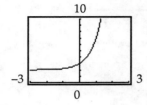

  a. The y-intercept is (0, 3). There is no x-intercept.

  b. The horizontal asymptote is $y = 2$.

  c. The range is $\{y | y > 2\}$.

31.

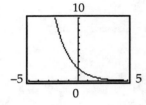

  a. The y-intercept is (0, 2). There is no x-intercept.

  b. The horizontal asymptote is $y = 0$.

  c. The range is $\{y | y > 0\}$.

33.

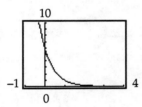

  a. The y-intercept is (0, 6). There is no x-intercept.

  b. The horizontal asymptote is $y = 0$.

  c. The range is $\{y | y > 0\}$.

35.

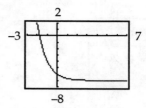

  a. The y-intercept is (0, –6). The x-intercept is (–1.77, 0).

  b. The horizontal asymptote is $y = -7$.

  c. The range is $\{y | y > -7\}$.

37.

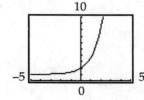

  a. The y-intercept is (0, 2). There is no x-intercept.

  b. The horizontal asymptote is $y = 1$.

  c. The range is $\{y | y > 1\}$.

39.

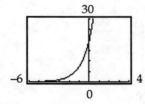

  a. The y-intercept is (0, 20.09). There is no x-intercept.

  b. The horizontal asymptote is $y = 0$.

  c. The range is $\{y | y > 0\}$.

41.

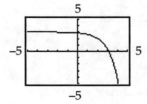

  a. The y-intercept is (0, 2.86). The x-intercept is (3.10, 0).

  b. The horizontal asymptote is $y = 3$.

  c. The range is $\{y | y < 3\}$.

© Houghton Mifflin Company. All rights reserved.

43.

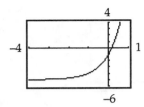

   a. The y-intercept is $(0, -1)$.
   The x-intercept is $(0.16, 0)$.

   b. The horizontal asymptote is $y = -5$.

   c. The range is $\{y|y > -5\}$.

45. B

47. F

49. C

51.

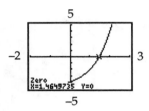

   The zero is 1.46.

53.

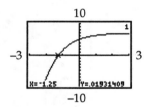

   The zero is -1.25.

55.

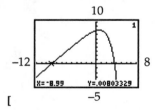

   [

   The zeros are -9.00 and 3.66.

57. i. The graph of g is the graph of f shifted
   downward k units.

   ii. The graph of h is the graph of f shifted to the
   right k units.

59.

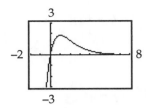

   a. The relative maximum is $(1, 1.84)$.

   b. The horizontal asymptote is $y = 0$.

   c. The range is $\{y|y \leq 1.84\}$.

61.

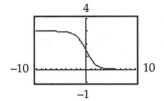

   a. There are no relative extrema.

   b. The horizontal asymptotes are $y = 3$ and $y = 0$.

   c. The range is $\{y|0 < y < 3\}$.

63. i. $f(m+n) = b^{m+n} = b^m \cdot b^n = f(m) \cdot f(n)$

   ii. $f(kx) = b^{kx} = \left(b^x\right)^k = \left[f(x)\right]^k$

65. Use $A = P\left(1+\dfrac{r}{n}\right)^{nt}$ with $P = 10{,}000$, $r, = 0.07$,
   $n = 12$, and $t = 18$.
   $$A = 10{,}000\left(1+\dfrac{0.07}{12}\right)^{12 \cdot 18} = 35{,}125.39$$
   The value of the fund when the child turned 18 is
   \$35,125.39.

67. a. $D = P_0(0.8)^x = 22{,}000(0.8)^3 = 11{,}264$
   The value of the automobile after 3 years is
   \$11,264.

   b. If the value of a is decreased, the rate of
   depreciation increases.

69. a. The year 1998 corresponds to
   $x = 1998 - 1990 = 8$.
   $$F(8) = 54(1.36)^8 = 631.98$$
   The estimated funding for 1998 was \$632
   million.

   b. Because the graph rises very quickly, the
   model is unreliable for future years.

© Houghton Mifflin Company. All rights reserved.

71. Using $A = Pe^{rt}$ with $A = 10{,}517.36$, r, $= 0.08$, and $t = 5$, we have $10{,}517.36 = Pe^{0.08 \cdot 5}$

$$P = \frac{10{,}517.36}{e^{0.4}} = 7050.00$$

The employee earned a profit of $10{,}517.36 - 7050.00 = 3467.36$, or \$3467.36.

73. Using $A = P\left(1 + \dfrac{r}{n}\right)^{nt}$ for each investment, we have

$$A = 120{,}000\left(1 + \frac{0.065}{12}\right)^{20 \cdot 12} + 80{,}000\left(1 + \frac{0.0675}{12}\right)^{20 \cdot 12}$$

$$= 438{,}773.60 + 307{,}428.93$$

$$= 746{,}202.53$$

The total combined worth was \$746,202.53.

75. We produce the graphs of the two functions $f(x) = 18{,}000e^{0.06x}$ and $g(x) = 30{,}900$ and determine their point of intersection.

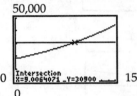

Because $f(9) \approx 30{,}900$, it will take about 9 years for the person to afford the property.

77.

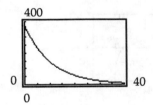

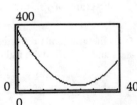

If the number of strikes is expected to decrease, the exponential model is better. If the number of strikes is expected to increase, the quadratic model is better.

4.4 Exercises

1. Either $\log a = 0$ or $\log b = 0$. Thus either $a = 1$ or $b = 1$.

3. $9^2 = 81$

5. $t^3 = x$

7. $\log_7 \dfrac{1}{49} = -2$

9. $\ln P = rt$

11. Write the logrithmic form in the equivalent exponential form.

$$\log_3 81 = y$$
$$3^y = 81$$
$$3^y = 3^4$$
$$y = 4$$

Therefore, $\log_3 81 = 4$.

13. $\log_4 \dfrac{1}{16} = y$

$$4^y = \frac{1}{16}$$
$$4^y = 4^{-2}$$
$$y = -2$$

Therefore, $\log_4 \dfrac{1}{16} = -2$.

15. $\log_{1/2} \dfrac{1}{16} = y$

$$\left(\frac{1}{2}\right)^y = \frac{1}{16}$$
$$\left(\frac{1}{2}\right)^y = \left(\frac{1}{2}\right)^4$$
$$y = 4$$

Therefore, $\log_{1/2} \dfrac{1}{16} = 4$.

17. $\ln e^4 = y$

$$e^y = e^4$$
$$y = 4$$

Therefore, $\ln e^4 = 4$.

19. $\log_7 \sqrt[3]{49} = y$

$$7^y = \sqrt[3]{49}$$
$$7^y = \left(7^2\right)^{1/3}$$
$$7^y = 7^{2/3}$$
$$y = \frac{2}{3}$$

Therefore, $\log_7 \sqrt[3]{49} = \dfrac{2}{3}$.

© Houghton Mifflin Company. All rights reserved.

21. $\log_8 0.25 = y$

$$8^y = 0.25$$

$$\left(2^3\right)^y = \frac{1}{4}$$

$$2^{3y} = 2^{-2}$$

$$3y = -2$$

$$y = -\frac{2}{3}$$

Therefore, $\log_8 0.25 = -\frac{2}{3}$.

23. $\ln 2 + \ln\sqrt{2} = 0.693 + 0.347 = 1.040$, or 1.04.

25. $\dfrac{\ln\sqrt{2}}{\ln 5} = \dfrac{0.347}{1.609} = 0.216$, or 0.22.

27. $\log_7 4 = \dfrac{\ln 4}{\ln 7} = \dfrac{1.386}{1.946} = 0.712$, or 0.71.

29. $\log_5 \sqrt[3]{324} = \dfrac{\log \sqrt[3]{324}}{\log 5} = \dfrac{0.837}{0.699} = 1.197$, or 1.20.

31. $\log_4 3^5 = \dfrac{\ln 3^5}{\ln 4} = \dfrac{5.493}{1.386} = 3.963$, or 3.96

33.

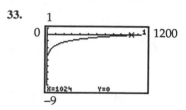

a. The x-intercept is $(1024, 0)$.
There is no y-intercept.

b. The vertical asymptote is $x = 0$.

c. The domain is $\{x|x > 0\}$.

35.

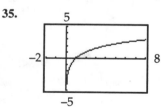

a. The x-intercept is $(1, 0)$.
There is no y-intercept.

b. The vertical asymptote is $x = 0$.

c. The domain is $\{x|x > 0\}$.

37.

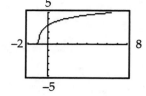

a. The x-intercept is $(-0.95, 0)$.
The y-intercept is $(0, 3)$.

b. The vertical asymptote is $x = -1$.

c. The domain is $\{x|x > -1\}$

39. The vertical asymptote of each graph is $x = 0$.

41.

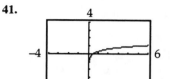

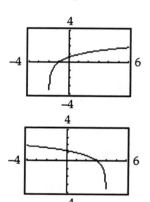

a. the graph of f is the graph of y shifted to the left 2 units.

b. The graph of g is the graph of y reflected across the y-axis and shifted to the right 4 units.

© Houghton Mifflin Company. All rights reserved.

43.

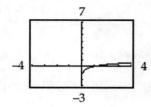

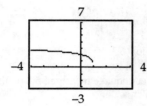

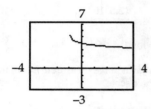

    a. The graph of f is the graph of y reflected across the y-axis and then shifted to the right 1 unit and upward 2 units.

    b. The graph of g is the graph of y reflected across the x-axis and then shifted to the left 1 unit and upward 4 units.

45. E

47. A

49. F

51.
$$y = 5^x$$
$$x = 5^y$$
$$\log_5 x = y$$
$$f^{-1}(x) = \log_5 x$$

53.
$$y = \log_{1/2}(x+1)$$
$$x = \log_{1/2}(y+1)$$
$$\left(\frac{1}{2}\right)^x = y+1$$
$$y = \left(\frac{1}{2}\right)^x - 1$$
$$g^{-1}(x) = \left(\frac{1}{2}\right)^x - 1$$

55.
$$y = 4^{x-2}$$
$$x = 4^{y-2}$$
$$\log_4 x = y - 2$$
$$y = 2 + \log_4 x$$
$$f^{-1}(x) = 2 + \log_4 x$$

57.
$$y = 1 - \ln x$$
$$x = 1 - \ln y$$
$$\ln y = 1 - x$$
$$y = e^{1-x}$$
$$r^{-1}(x) = e^{1-x}$$

59. We can use the Change of Base Formula to write
$$\log_2 x \text{ as } \frac{\log_4 x}{\log_4 2}.$$

61. The argument must be positive.
$$2x - 7 > 0$$
$$2x > 7$$
$$x > \frac{7}{2}$$
The domain is $\left\{x \middle| x > \frac{7}{2}\right\}$.

63. $|x - 2| > 0$
This inequality is true for all values of except where $|x - 2| = 0$, or $x = 2$. The domain is $\{x | x \neq 2\}$.

65.

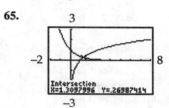

The graphs intersect at (1.31, 0.27). The graphs of $y = e^x$ and $y = \ln x$ do not intersect.

67. No, the graphs are not the same. The domain of $y = \ln e^x$ is **R** whereas the domain of $y = e^{\ln x}$ is $\{x | x > 0\}$.

69.

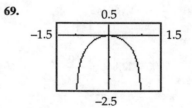

    a. The relative maximum is (0, 0).

    b. The y-intercept is (0, 0).
        The x-intercept is (0, 0).

© Houghton Mifflin Company. All rights reserved.

71.

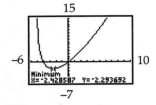

a. The relative minimum is (–2.43, –2.29).

b. The y-intercept is (0, 0).
The x-intercepts are (–4, 0) and (0, 0).

73. To find f^{-1}, use our normal procedure.

$$y = \log(\log x)$$

$$x = \log(\log y)$$

$$10^x = \log y$$

$$10^{10^x} = y$$

Thus, $f^{-1}(x) = 10^{10^x}$

75. **a. (i)**
$$A(680) = 22,860 \ln \frac{760}{680}$$
$$= 22,860(0.111226)$$
$$= 2542.63$$
The altitude is 2543 meters.

(ii)
$$A(650) = 22,860 \ln \frac{760}{650}$$
$$= 22,860(0.156346)$$
$$= 3574.07$$
The altitude is 3574 meters.

b. (i)
$$3500 = 22,860 \ln \frac{760}{x}$$
$$\ln \frac{760}{x} = 0.153106$$
$$\frac{760}{x} = e^{0.153106}$$
$$\frac{760}{x} = 1.165448$$
$$x = 652.11 \text{ mm Hg}$$

(ii)
$$5000 = 22,860 \ln \frac{760}{x}$$
$$\ln \frac{760}{x} = 0.218723$$
$$\frac{760}{x} = e^{0.218723}$$
$$\frac{760}{x} = 1.244486$$
$$x = 610.69 \text{ mm Hg}$$

77. a. (i)
$$R = \log \frac{(4.0 \cdot 10^8)I_0}{I_0}$$
$$= \log(4.0 \cdot 10^8)$$
$$= 8.6$$
The earthquake measured 8.6 on the Richter scale.

(ii)
$$R = \log \frac{(6.3 \cdot 10^5)I_0}{I_0}$$
$$= \log(6.3 \cdot 10^5)$$
$$= 5.8$$
The earthquake measured 5.8 on the Richter scale.

(iii)
$$R = \log \frac{(2.5 \cdot 10^7)I_0}{I_0}$$
$$= \log(2.5 \cdot 10^7)$$
$$= 7.4$$
The earthquake measured 7.4 on the Richter scale.

b. (i)
$$6.2 = \log \frac{I}{I_0}$$
$$\frac{I}{I_0} = 10^{6.2}$$
$$\frac{I}{I_0} = 1.6 \cdot 10^6$$
$$I = (1.6 \cdot 10^6)I_0$$
The intensity of the earthquake was $(1.6 \cdot 10^6)I_0$.

(ii)
$$8.4 = \log \frac{I}{I_0}$$
$$\frac{I}{I_0} = 10^{8.4}$$
$$\frac{I}{I_0} = 2.5 \cdot 10^8$$
$$I = (2.5 \cdot 10^8)I_0$$
The intensity of the earthquake was $(2.5 \cdot 10^8)I_0$.

(iii)
$$7.6 = \log \frac{I}{I_0}$$
$$\frac{I}{I_0} = 10^{7.6}$$
$$\frac{I}{I_0} = 4.0 \cdot 10^7$$
$$I = (4.0 \cdot 10^7)I_0$$
The intensity of the earthquake was $(4.0 \cdot 10^7)I_0$.

© Houghton Mifflin Company. All rights reserved.

79. a. $S(0) = 58 + 16\ln(0+1)$

$\qquad = 58 + 16(0)$

$\qquad = 58$

The initial average score was 58.

b. $S(4) = 58 + 16\ln(4+1)$

$\qquad = 58 + 16(1.61)$

$\qquad = 58 + 25.75$

$\qquad = 83.75$

After 4 weeks, the score was 83.75.

c. To determine when the score exceeded 90, we solve the equation

$90 = 58 + 16\ln(t+1)$

$16\ln(t+1) = 32$

$\ln(t+1) = 2$

$t+1 = e^2$

$t = -1 + e^2$

$t = 6.39$

The score exceeded 90 after 6.39 weeks.

81. a. $R(1) = 142 + 111\ln(1)$

$\qquad = 142 + 111(0)$

$\qquad = 142$

The first-year revenue was \$142,000.

b. To determine when first-year revenue doubled, we solve the equation

$284 = 142 + 111\ln t$

$142 = 111\ln t$

$\ln t = \dfrac{142}{111}$

$t = e^{142/111}$

$t = 3.59$

The first-year revenue doubled after 3.59 years.

c. $R(10) = 142 + 111\ln(10)$

$\qquad = 142 + 255.59$

$\qquad = 397.587$

The anticipated revenue after 10 years is \$397,587.

4.5 Exercises

1. Because $\dfrac{\log 100}{\log 10} = \dfrac{2}{1} = 2$ and

$\log 100 - \log 10 = 2 - 1 = 1$, it is clear that

$\dfrac{\log M}{\log N} \neq \log M - \log N$.

3. $10^{-\log 3} = \dfrac{1}{10^{\log 3}} = \dfrac{1}{3}$

5. $8^{-\log_2 5} = \dfrac{1}{8^{\log_2 5}}$

$\qquad = \dfrac{1}{(2^3)^{\log_2 5}}$

$\qquad = \dfrac{1}{(2^{\log_2 5})^3}$

$\qquad = \dfrac{1}{5^3}$

$\qquad = \dfrac{1}{125}$

7. $e^{\ln 6 - \ln 2} = \dfrac{e^{\ln 6}}{e^{\ln 2}} = \dfrac{6}{2} = 3$

9. $\log_7(\log_3 3) = \log_7(1) = 0$

11. $\log_5 5^{x-2} = x - 2$

13. $e^{\ln(3x)} = 3x$

15. $\ln 3(x-1) = \ln 3 + \ln(x-1)$

17. $\log x + \log\sqrt{x+3} = \log x\sqrt{x+3}$

19. $\log_5 \dfrac{t}{t-2} = \log_5 t - \log_5(t-2)$

21. $\ln(t^2-1) - \ln\sqrt{t} = \ln\dfrac{t^2-1}{\sqrt{t}}$

23. $\ln a^{-3} = -3\ln a$

25. $-\log_3 w = \log_3 w^{-1} = \log_3 \dfrac{1}{w}$

27. $\log_9 \dfrac{xy^2}{z} = \log_9 xy^2 - \log_9 z$

$\qquad = \log_9 x + \log_9 y^2 - \log_9 z$

$\qquad = \log_9 x + 2\log_9 y - \log_9 z$

29. $\ln x(y+3)^2 = \ln x + \ln(y+3)^2$

$\qquad = \ln x + 2\ln(y+3)$

31. $\log\sqrt[3]{\dfrac{a}{2b}} = \log\left(\dfrac{a}{2b}\right)^{1/3}$

$\qquad = \dfrac{1}{3}\log\dfrac{a}{2b}$

$\qquad = \dfrac{1}{3}[\log a - \log(2b)]$

$\qquad = \dfrac{1}{3}[\log a - (\log 2 + \log b)]$

$\qquad = \dfrac{1}{3}\log a - \dfrac{1}{3}\log 2 - \dfrac{1}{3}\log b$

© Houghton Mifflin Company. All rights reserved.

33. $\log_4 \dfrac{\sqrt{z+2}}{x^3 y} = \log_4 \sqrt{z+2} - \log_4(x^3 y)$

$\qquad\qquad = \log_4(z+2)^{1/2} - (\log_4 x^3 + \log_4 y)$

$\qquad\qquad = \dfrac{1}{2}\log_4(z+2) - 3\log_4 x - \log_4 y$

35. $\log_7 \dfrac{x^2-4}{x^3} = \log_7(x^2-4) - \log_7 x^3$

$\qquad\qquad = \log_7(x+2)(x-2) - 3\log_7 x$

$\qquad\qquad = \log_7(x+2) + \log_7(x-2) - 3\log_7 x$

37. $\ln(x^2+6x+9) = \ln(x+3)^2$

$\qquad\qquad = 2\ln(x+3)$

39. $\log(100x - 10x^2) = \log 10x(10-x)$

$\qquad\qquad = \log 10 + \log x + \log(10-x)$

$\qquad\qquad = 1 + \log x + \log(10-x)$

41. $2\log_6 y + 3\log_6 x = \log_6 y^2 + \log_6 x^3$

$\qquad\qquad = \log_6 x^3 y^2$

43. $\dfrac{1}{2}\log_3 a - \log_3 b = \log_3 a^{1/2} - \log_3 b$

$\qquad\qquad = \log_3 \sqrt{a} - \log_3 b$

$\qquad\qquad = \log_3 \dfrac{\sqrt{a}}{b}$

45. $3\log x - \dfrac{1}{2}\log y + \log z = \log x^3 - \log y^{1/2} + \log z$

$\qquad\qquad = \log x^3 - \log\sqrt{y} + \log z$

$\qquad\qquad = \log x^3 + \log z - \log\sqrt{y}$

$\qquad\qquad = \log \dfrac{x^3 z}{\sqrt{y}}$

47. $\dfrac{2}{3}\log(x+1) + \dfrac{1}{3}\log(x-2) = \dfrac{1}{3}[2\log(x+1) + \log(x-2)]$

$\qquad\qquad = \dfrac{1}{3}[\log(x+1)^2 + \log(x-2)]$

$\qquad\qquad = \dfrac{1}{3}\log(x+1)^2(x-2)$

$\qquad\qquad = \log\sqrt[3]{(x+1)^2(x-2)}$

49. $\ln(a+1) + 2\ln b - 3\ln(c-3) = \ln(a+1) + \ln b^2 - \ln(c-3)^3$

$\qquad\qquad = \ln(a+1)b^2 - \ln(c-3)^3$

$\qquad\qquad = \ln \dfrac{(a+1)b^2}{(c-3)^3}$

© Houghton Mifflin Company. All rights reserved.

51.
$$3 + 2\log_5 y = 3\log_5 5 + 2\log_5 y$$
$$= \log_5 5^3 + \log_5 y^2$$
$$= \log_5 5^3 y^2$$
$$= \log_5 125 y^2$$

53.
$$6x + \ln y = 6x \ln e + \ln y$$
$$= \ln e^{6x} + \ln y$$
$$= \ln y e^{6x}$$

55.

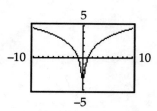

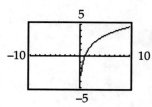

The domain of $y = \ln x^2$ is $\{x \mid x \neq 0\}$, whereas the domain of $y = 2\ln x$ is $\{x \mid x > 0\}$.

57. Since $\log_{1/b} b = \log_{1/b}\left(\dfrac{1}{b}\right)^{-1} = -1$, the statement is true.

59. Since $\ln \dfrac{7}{3} = \ln 7 - \ln 3 \neq \dfrac{\ln 7}{\ln 3}$, the statement is false.

61. $\log(xy) = \log x + \log y \neq (\log x)(\log y)$, the statement is false.

63. Since $\log_{15} 6 = \dfrac{\log 6}{\log 15}$ or $\log_{15} 6 = \dfrac{\ln 6}{\ln 15}$, the statement is true.

65.
$$\log_4 24 - \log_4 3 = \log_4 \dfrac{24}{3}$$
$$= \log_4 8$$
$$= \log_4 4 \cdot 2$$
$$= \log_4 4 + \log_4 2$$
$$= 1 + \log_4 \sqrt{4}$$
$$= 1 + \log_4 4^{1/2}$$
$$= 1 + \dfrac{1}{2}$$
$$= \dfrac{3}{2}$$

67.
$$\log \sqrt{10} + 10^{\log 3} = \log 10^{1/2} + 3$$
$$= \dfrac{1}{2} + 3$$
$$= 3.5$$

69.
$$\log_b \dfrac{2}{15} = \log_b 2 - \log_b 15$$
$$= \log_b 2 - \log_b 3 \cdot 5$$
$$= \log_b 2 - (\log_b 3 + \log_b 5)$$
$$= \log_b 2 - \log_b 3 - \log_b 5$$
$$= A - B - C$$

71.
$$\log_b \dfrac{\sqrt{b}}{60} = \log_b \sqrt{b} - \log_b 60$$
$$= \log_b b^{1/2} - \log_b 2^2 \cdot 3 \cdot 5$$
$$= \dfrac{1}{2}\log_b b - (\log_b 2^2 + \log_b 3 + \log_b 5)$$
$$= \dfrac{1}{2} - 2\log_b 2 - \log_b 3 - \log_b 5$$
$$= \dfrac{1}{2} - 2A - B - C$$

73.
$$\ln x\sqrt{yz} = \ln x(yz)^{1/2}$$
$$= \ln x + \ln(yz)^{1/2}$$
$$\ln x + \dfrac{1}{2}\ln(yz)$$
$$= \ln x + \dfrac{1}{2}(\ln y + \ln z)$$
$$= \ln x + \dfrac{1}{2}\ln y + \dfrac{1}{2}\ln z$$
$$= a + \dfrac{1}{2}b + \dfrac{1}{2}c$$

75.
$$\ln(exz) - \ln\left(\dfrac{x}{e}\right) = (\ln e + \ln x + \ln z) - (\ln x - \ln e)$$
$$= 2\ln e + \ln z$$
$$= 2 + c$$

77.
$$\log_B \dfrac{1}{A} = \log_B A^{-1}$$
$$= -\log_B A$$
$$\log_{1/B} A = \dfrac{\log_B A}{\log_B \dfrac{1}{B}}$$
$$= \dfrac{\log_B A}{\log_B 1 - \log_B B}$$
$$= \dfrac{\log_B A}{-\log_B B}$$
$$= -\log_B A$$

Thus, $\log_B \dfrac{1}{A} = \log_{1/B} A$.

© Houghton Mifflin Company. All rights reserved.

79.

x	2.3	23	230	2300
$\log x$	0.3617	1.3617	2.3617	3.3617

Because each number is 10 times the preceeding number, each logarithmic value is 1 more than the preceding value.

81. We want to prove $\log_b \dfrac{M}{N} = \log_b M - \log_b N$. Let $\log_b M = r$ and $\log_b N = s$.

$b^r = M$ and $b^s = N$ Convert each logarithmic form to exponential form.

$\dfrac{M}{N} = \dfrac{b^r}{b^s}$ Write the quotient of M and N.

$\dfrac{M}{N} = b^{r-s}$ Quotient Rule for Exponents

$\log_b \dfrac{M}{N} = r - s$ Convert to logarithmic form.

$\log_b \dfrac{M}{N} = \log_b M - \log_b N$ $r = \log_b M$ and $s = \log_b N$

83. We first solve for the intensity of the 140 decibel sound.

$$140 = 10\log \frac{I}{I_0}$$

$$\log \frac{I}{I_0} = 14$$

$$\frac{I}{I_0} = 10^{14}$$

$$I = 10^{14} I_0$$

Now solve for the intensity of the 130 decibel sound.

$$130 = 10\log \frac{I}{I_0}$$

$$\log \frac{I}{I_0} = 13$$

$$\frac{I}{I_0} = 10^{13}$$

$$I = 10^{13} I_0$$

Therefore, the dangerous sound level is more intense than a rock concert by a factor of $\dfrac{10^{14} I_0}{10^{13} I_0} = 10$.

© Houghton Mifflin Company. All rights reserved.

4.6 Exercises

1. We can also solve the equation by equating the logarithms of the expressions.

3. $3^{2x} = 81$
$3^{2x} = 3^4$
$2x = 4$
$x = 2$

5. $4^{-3x} = 32^{1-x}$
$(2^2)^{-3x} = (2^5)^{1-x}$
$2^{-6x} = 2^{5-5x}$
$-6x = 5 - 5x$
$-x = 5$
$x = -5$

7. $\dfrac{1}{25^{x+1}} = 125^{2-x}$
$25^{-(x+1)} = 125^{2-x}$
$(5^2)^{-x-1} = (5^3)^{2-x}$
$5^{-2x-2} = 5^{6-3x}$
$-2x - 2 = 6 - 3x$
$x = 8$

9. $8 \cdot 2^{x^2} = 4^{-2x}$
$2^3 \cdot 2^{x^2} = (2^2)^{-2x}$
$2^{3+x^2} = 2^{-4x}$
$3 + x^2 = -4x$
$x^2 + 4x + 3 = 0$
$(x+3)(x+1) = 0$
$x = -3$ or $x = -1$

11. $9^{x^2} = 3^x$
$(3^2)^{x^2} = 3^x$
$3^{2x^2} = 3^x$
$2x^2 = x$
$2x^2 - x = 0$
$x(2x-1) = 0$
$x = 0$ or $x = \dfrac{1}{2}$

13. $\left(\dfrac{4}{3}\right)^x = \dfrac{9}{16}$
$\left(\dfrac{4}{3}\right)^x = \left(\dfrac{3}{4}\right)^2$
$\left(\dfrac{4}{3}\right)^x = \left(\dfrac{4}{3}\right)^{-2}$
$x = -2$

15. $3^{2x} = 16$
$\ln 3^{2x} = \ln 16$
$2x \cdot \ln 3 = \ln 16$
$x = \dfrac{\ln 16}{2 \cdot \ln 3}$
$x \approx 1.26$

17. $3^x = 4^{x+1}$
$\ln 3^x = \ln 4^{x+1}$
$x \ln 3 = (x+1)\ln 4$
$x \ln 3 = x \ln 4 + \ln 4$
$x \ln 3 - x \ln 4 = \ln 4$
$x(\ln 3 - \ln 4) = \ln 4$
$x = \dfrac{\ln 4}{\ln 3 - \ln 4}$
$x \approx -4.82$

19. $5 \cdot 3^{1-2x} = 14$
$3^{1-2x} = \dfrac{14}{5}$
$\ln 3^{1-2x} = \ln \dfrac{14}{5}$
$(1-2x)\ln 3 = \ln 14 - \ln 5$
$1 - 2x = \dfrac{\ln 14 - \ln 5}{\ln 3}$
$2x = 1 - \dfrac{\ln 14 - \ln 5}{\ln 3}$
$x = \dfrac{1}{2}\left(1 - \dfrac{\ln 14 - \ln 5}{\ln 3}\right)$
$x \approx 0.03$

21. $3e^x + 7 = 12$
$3e^x = 5$
$e^x = \dfrac{5}{3}$
$\ln e^x = \ln \dfrac{5}{3}$
$x = \ln 5 - \ln 3$
$x \approx 0.51$

23. $3 + \log_5 x = 5$
$\log_5 x = 2$
$x = 5^2$
$x = 25$

© Houghton Mifflin Company. All rights reserved.

25. $-2 + 5\ln(3x) = 0$

$$5\ln(3x) = 2$$

$$\ln(3x) = \frac{2}{5}$$

$$3x = e^{2/5}$$

$$x = \frac{1}{3}e^{2/5}$$

$$x \approx 0.50$$

27. $\log_7 49^x = 3$

$$\log_7 (7^2)^x = 3$$

$$\log_7 (7^{2x}) = 3$$

$$2x = 3$$

$$x = \frac{3}{2}$$

29. $\log_2 \sqrt[3]{x} = 4$

$$\sqrt[3]{x} = 2^4$$

$$\sqrt[3]{x} = 16$$

$$x = 16^3$$

$$x = 4096$$

31. $\ln x(x+3) = 0$

$$x(x+3) = e^0$$

$$x^2 + 3x = 1$$

$$x^2 + 3x - 1 = 0$$

We solve the equation using the Quadratic Formula.

$$x = \frac{-3 \pm \sqrt{3^2 - 4(1)(-1)}}{2(1)}$$

$$x = \frac{-3 \pm \sqrt{13}}{2}$$

$$x \approx -3.30 \text{ or } x \approx 0.30$$

33. $\ln(2x-5) = \ln(x+1)$

$$2x - 5 = x + 1$$

$$x = 6$$

35. $\frac{1}{2}\log_2(2x-1) = \log_2 3$

$$\log_2(2x-1)^{1/2} = \log_2 3$$

$$(2x-1)^{1/2} = 3$$

$$2x - 1 = 3^2$$

$$2x - 1 = 9$$

$$2x = 10$$

$$x = 5$$

37. The expression on the left is easy to simplify, however the Change of Base Formula is needed to evaluate the expression on the right.

39. $\log_3 x + \log_3(x+2) = 1$

$$\log_3 x(x+2) = 1$$

$$x(x+1) = 3^1$$

$$x^2 + 2x = 3$$

$$x^2 + 2x - 3 = 0$$

$$(x+3)(x-1) = 0$$

$$x = -3 \text{ or } x = 1$$

Because x must be positive, -3 is an extraneous solution. Thus, the only solution is 1.

41. $\log_2(x+3) - \log_2(x-3) = 2$

$$\log_2 \frac{x+3}{x-3} = 2$$

$$\frac{x+3}{x-3} = 2^2$$

$$x + 3 = 4(x-3)$$

$$x + 3 = 4x - 12$$

$$15 = 3x$$

$$x = 5$$

43. $1 - \log_6 \frac{x}{3} = \log_6(x+3)$

$$1 = \log_6(x+3) + \log_6 \frac{x}{3}$$

$$6^1 = \frac{x(x+3)}{3}$$

$$18 = x^2 + 3x$$

$$x^2 + 3x - 18 = 0$$

$$(x+6)(x-3) = 0$$

$$x = -6 \text{ or } x = 3$$

Because $x + 3$ must be positive, -6 is an extraneous solution. Thus, the only solution is 3.

45. $\ln(4-x) - \ln(x+1) = \ln 2$

$$\ln \frac{4-x}{x+1} = \ln 2$$

$$\frac{4-x}{x+1} = 2$$

$$4 - x = 2x + 2$$

$$2 = 3x$$

$$x = \frac{2}{3}$$

47. $\log x = \log(x+6) - \log(x+2)$

$$\log x = \log \frac{x+6}{x+2}$$

$$x = \frac{x+6}{x+2}$$

$$x^2 + 2x = x + 6$$

$$x^2 + x - 6 = 0$$

$$(x+3)(x-2) = 0$$

$$x = -3 \text{ or } x = 2$$

Because $x + 2$ must be positive, -3 is an extraneous solution. Thus, the only solution is 2.

© Houghton Mifflin Company. All rights reserved.

49. $\log x + \log(x+3) - \log(x+1) = \log(x-3)$

$$\log\frac{x(x+3)}{x+1} = \log(x-3)$$

$$\frac{x(x+3)}{x+1} = x-3$$

$$x(x+3) = (x-3)(x+1)$$

$$x^2 + 3x = x^2 - 2x - 3$$

$$5x = -3$$

$$x = -\frac{3}{5}$$

Because x must be positive, $-\frac{3}{5}$ is an extraneous solution. Thus, there is no solution.

51. Both -5 and 3 are solutions of (i). In (ii), x must be positive, so 3 is the only solution.

53. $\log_4|x| = 2$

$$|x| = 4^2$$

$$|x| = 16$$

$$x = -16 \text{ or } x = 16$$

55. $(\log x)^2 - \log x^2 - 8 = 0$

$$(\log x)^2 - 2\log x - 8 = 0$$

$$(\log x - 4)(\log x + 2) = 0$$

$$\log x - 4 = 0 \qquad \text{or} \quad \log x + 2 = 0$$

$$\log x = 4 \qquad\qquad \log x = -2$$

$$x = 10^4 \qquad\qquad x = 10^{-2}$$

$$x = 10,000 \qquad\qquad x = \frac{1}{100}$$

57. $x^{\log x} = 1000 x^2$

$$\log x^{\log x} = \log 1000 x^2$$

$$(\log x)(\log x) = \log 1000 + \log x^2$$

$$(\log x)^2 = \log 10^3 + 2\log x$$

$$(\log x)^2 = 3 + 2\log x$$

$$(\log x)^2 - 2\log x - 3 = 0$$

$$(\log x + 1)(\log x - 3) = 0$$

$$\log x + 1 = 0 \qquad \text{or} \quad \log x - 3 = 0$$

$$\log x = -1 \qquad\qquad \log x = 3$$

$$x = 10^{-1} \qquad\qquad x = 10^3$$

$$x = \frac{1}{10} \qquad\qquad x = 1000$$

59. $\log(\ln x) = 0$

$$\ln x = 10^0$$

$$\ln x = 1$$

$$x = e^1$$

$$x = e$$

61.

$$-2 + \log x = \sqrt{-2 + \log x}$$

$$(-2 + \log x)^2 = -2 + \log x$$

$$4 - 4\log x + (\log x)^2 = -2 + \log x$$

$$(\log x)^2 - 5\log x + 6 = 0$$

$$(\log x - 2)(\log x - 3) = 0$$

$$\log x - 2 = 0 \qquad \text{or} \quad \log x - 3 = 0$$

$$\log x = 2 \qquad\qquad \log x = 3$$

$$x = 10^2 \qquad\qquad x = 10^3$$

$$x = 100 \qquad\qquad x = 1000$$

63. $\ln[\ln(\ln x)] = 1$

$$\ln(\ln x) = e$$

$$\ln x = e^e$$

$$x = e^{e^e}$$

65. $7^{|x-1|} = 6$

$$|x-1|\ln 7 = \ln 6$$

$$|x-1| = \frac{\ln 6}{\ln 7}$$

$$x - 1 = \frac{\ln 6}{\ln 7} \qquad \text{or} \quad x - 1 = -\frac{\ln 6}{\ln 7}$$

$$x = 1 + \frac{\ln 6}{\ln 7} \qquad\qquad x = 1 - \frac{\ln 6}{\ln 7}$$

$$x \approx 1.92 \qquad\qquad x \approx 0.08$$

67. $e^x + 3e^{-x} = 4$

To solve the equation, multiply both sides by e^x.

$$e^x(e^x + 3e^{-x}) = e^x(4)$$

$$(e^x)^2 + 3 = 4e^x$$

$$(e^x)^2 - 4e^x + 3 = 0$$

$$(e^x - 1)(e^x - 3) = 0$$

$$e^x = 1 \qquad \text{or} \quad e^x = 3$$

$$x = \ln 1 \qquad\qquad x = \ln 3$$

$$x = 0 \qquad\qquad x \approx 1.10$$

69. $e^{2\ln x^3 - 3\ln x} = \ln e^8$

$$\frac{e^{2\ln x^3}}{e^{3\ln x}} = \ln e^8$$

$$\frac{e^{\ln(x^3)^2}}{e^{\ln x^3}} = \ln e^8$$

$$\frac{x^6}{x^3} = 8$$

$$x^3 = 8$$

$$x = 2$$

© Houghton Mifflin Company. All rights reserved.

71.
$$\frac{e^x + e^{-x}}{2} = 1$$
$$e^x + e^{-x} = 2$$
$$e^x(e^x + e^{-x}) = e^x(2)$$
$$e^x \cdot e^x + e^x \cdot e^{-x} = 2e^x$$
$$(e^x)^2 + 1 = 2e^x$$
$$(e^x)^2 - 2e^x + 1 = 0$$
$$(e^x - 1)^2 = 0$$
$$e^x - 1 = 0$$
$$e^x = 1$$
$$x = 0$$

73.
$$6^{\log_6(3x)} = 2$$
$$3x = 2$$
$$x = \frac{2}{3}$$

75. (i)
$$\log x^2 = \log x$$
$$2\log x = \log x$$
$$\log x = 0$$
$$x = 10^0$$
$$x = 1$$

(ii)
$$(\log x)^2 = \log x$$
$$(\log x)^2 - \log x = 0$$
$$\log x(\log x - 1) = 0$$
$$\log x = 0 \quad \text{or} \quad \log x - 1 = 0$$
$$x = 10^0 \qquad\qquad \log x = 1$$
$$x = 1 \qquad\qquad x = 10^1$$
$$\qquad\qquad\qquad x = 10$$

(iii)
$$2\log x = \log x$$
$$\log x = 0$$
$$x = 10^0$$
$$x = 1$$

77.

x	0.47	0.047	0.0047	0.00047
$\log x$	-0.3279	-1.3279	-2.3279	-3.279

Because each value of x is $\frac{1}{10}$ the preceding value of x, the value of each logarithm is 1 less than the preceding value.

79.

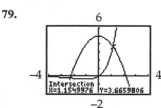

The points of intersection are $(-2.29, -0.23)$ and $(1.15, 3.67)$. Thus, the solutions are -2.29 and 1.15.

81.

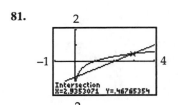

The points of intersection are $(0.11, -0.94)$ and $(2.94, 0.47)$.
Therefore, the solutions are 0.11 and 2.94.

83.

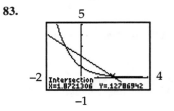

The points of intersection are $(-1, 3)$ and $(1.87, 0.13)$. The inequality is satisfied for x equal to or between the zeros -1 and 1.87. Thus, the solution is $[-1, 1.87]$.

85.

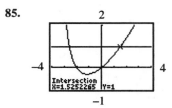

The points of intersection are $(-2.59, 1)$ and $(1.53, 1)$. The inequality is satisfied for x equal to or between the zeros -2.59 and 1.53. Thus, the solution is $[-2.59, 1.53]$.

87. a. Yes. Because the base is positive, f is defined on $(0, \infty)$.

b. The function is defined for some values such as -3 because $(-3)^{-3}$ is a real number. The function is not defined for values such as $-\frac{1}{2}$ because $\left(-\frac{1}{2}\right)^{-\frac{1}{2}}$ is not a real number.

c.

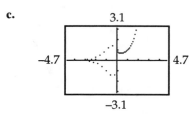

On $(-\infty, 0)$, the graph has "holes" because the function is not defined for all values in the interval.

© Houghton Mifflin Company. All rights reserved.

4.7 Exercises

1.
$$M = M_0 e^{rt}$$
$$3.6 \cdot 10^{-7} = 1.2 \cdot 10^{-8} e^{r \cdot 3}$$
$$e^{3r} = \frac{3.6 \cdot 10^{-7}}{1.2 \cdot 10^{-8}} = 30$$
$$3r = \ln 30$$
$$r = \frac{\ln 30}{3}$$
$$r \approx 1.13$$

3. If the initial mass is M_0, when it triples its mass will be $3M_0$.
$$3M_0 = M_0 e^{1.04t}$$
$$e^{1.04t} = 3$$
$$1.04t = \ln 3$$
$$t = \frac{\ln 3}{1.04}$$
$$t \approx 1.06$$
It takes 1.06 hours or 63.38 minutes for the mass to triple.

5.
$$100 = 300(2)^{-t/8.065}$$
$$2^{-t/8.065} = \frac{1}{3}$$
$$-\frac{t}{8.065} \ln 2 = \ln \frac{1}{3}$$
$$t = -8.065 \cdot \frac{\ln \frac{1}{3}}{\ln 2}$$
$$t \approx 12.78$$
It will take about 13 days for the level to drop to 100, on or about June 14.

7.
$$1 = 3(2)^{-t/15}$$
$$2^{-t/15} = \frac{1}{3}$$
$$-\frac{t}{15} \ln 2 = \ln \frac{1}{3}$$
$$t = -15 \frac{\ln \frac{1}{3}}{\ln 2}$$
$$t \approx 23.77$$
The experiment could begin after 24 minutes, or at 9:24 A.M.

9. We use the formula for continuous compounding, $A = Pe^{rt}$, and solve for r.
$$2 \cdot 10,000 = 10,000 e^{r \cdot 9}$$
$$e^{9r} = 2$$
$$9r = \ln 2$$
$$r = \frac{\ln 2}{9}$$
$$r \approx 0.0770$$
The investment must earn 7.7%.

11. We first solve for k. In 1987, $t = 1987 - 1967 = 20$ and $N = 1418$.
$$1418 = 1e^{k \cdot 20}$$
$$e^{20k} = 1418$$
$$20k = \ln 1418$$
$$k = \frac{\ln 1418}{20}$$
$$k \approx 0.3629$$
We now use the equation $N = e^{0.3629t}$ to solve for t when $N = 40,000$.
$$40,000 = e^{0.3629t}$$
$$0.3629t = \ln 40,000$$
$$t = \frac{\ln 40,000}{0.3629}$$
$$t \approx 29.20$$
40,000 transplants were predicted in 1967 + 29, or the year 1996.

13. We first need to solve for k in the model. Let t be the number of years after 1980. Then
$$N = N_0 e^{kt}$$
$$1.11 = 0.875 e^{k \cdot 10}$$
$$e^{10k} = \frac{1.11}{0.875}$$
$$10k = \ln \frac{1.11}{0.875}$$
$$k = \frac{1}{10} \ln \frac{1.11}{0.875}$$
$$k \approx 0.0238$$
Therefore, the growth model is $N = 0.875 e^{0.0238t}$. Now, solve for t when $N = 1.3$.
$$1.3 = 0.875 e^{0.0238t}$$
$$e^{0.0238t} = \frac{1.3}{0.875}$$
$$0.0238t = \ln \frac{1.3}{0.875}$$
$$t = \frac{1}{0.0238} \ln \frac{1.3}{0.875}$$
$$t \approx 16.63$$
The population is projected to reach 1.3 million in 1980 + 16, or in the year 1996.

15. $v(t) = 40(1 - 2e^{-t})$
$$v(1) = 40(1 - 2e^{-1})$$
$$v(1) \approx 10.57$$
After 1 second, the velocity is 10.57 feet per second.
$$v(2) = 40(1 - 2e^{-2})$$
$$v(2) \approx 29.17$$
After 2 seconds, the velocity is 29.17 feet per second.

As t increases, e^{-t} decreases, so the velocity approaches 40 feet per second.

© Houghton Mifflin Company. All rights reserved.

17. $T = 203$, $R = 70$, and $t = 0$.

$$\ln(T - R) = kt + C$$
$$\ln(203 - 70) = k \cdot 0 + C$$
$$C = \ln 133 \approx 4.89$$

After 1 minute, $T = 194$, $R = 70$, and $t = 1$.

$$\ln(194 - 70) = k \cdot 1 + 4.89$$
$$k = \ln 124 - 4.89$$
$$k \approx -0.07$$

We now solve for t when $T = 149$.

$$\ln(149 - 70) = -0.07t + 4.89$$
$$\ln 79 = -0.07t + 4.89$$
$$0.07t = 0.52$$
$$t \approx 7.43$$

It will take about 7.43 minutes.

19. At first, $T = 600$, $R = 40$, and $t = 0$.

$$\ln(T - R) = kt + C$$
$$\ln(600 - 40) = k \cdot 0 + C$$
$$C = \ln 560$$
$$C \approx 6.33$$

After 20 minutes, $T = 400$, $R = 40$, and $t = 20$.

$$\ln(400 - 40) = k \cdot 20 + 6.33$$
$$\ln 360 = 20k + 6.33$$
$$20k \approx -0.44$$
$$k \approx -0.022$$

We now solve for t if $T = 400 - 100 = 300$.

$$\ln(300 - 40) = -0.022t + 6.33$$
$$\ln 260 = -0.022t + 6.33$$
$$0.022t \approx 0.77$$
$$t \approx 35$$

It took another 35−20, or 15 minutes.

21. At the beginning, $t = 5$ and $n = 1$.
$C = \log 5 \approx 0.6990$.
After 10 practice trials, $t = 2$ and $n = 10$.

$$\log 2 = k \cdot \log 10 + 0.6990$$
$$k = \log 2 - 0.6990$$
$$k \approx -0.3979$$

Now solve for n for $t = 1$.

$$\log 1 = -0.3979 \cdot \log n + 0.6990$$
$$0.3979 \log n = 0.6990$$
$$\log n = \frac{0.6990}{0.3979}$$
$$\log n \approx 1.76$$
$$n \approx 10^{1.76}$$
$$n \approx 57.54$$

The reaction time is reduced to 1 second after 57 trials.

23. $\log t = k \cdot \log n + \log c$

$$\log t = \log n^k \cdot c$$

$$t = n^k \cdot c$$

Because $n \geq 1$ and $c > 0$, t is never zero.

25. $\ln N = k \ln A + 2.834$

$$\ln N = k \ln A + \ln e^{2.834}$$
$$\ln N = \ln A^k + \ln e^{2.834}$$
$$\ln N = \ln A^k e^{2.834}$$
$$N = A^k e^{2.834}$$

27. $\ln N = 0.28 \cdot \ln 0.5 + 2.834$

$$\ln N \approx 2.6399$$
$$N = e^{2.6399}$$
$$N = 14$$

The predicted number of species is 14.

29. For one crying baby, $D = 30$. Solve for I.

$$30 = 10 \log \frac{I}{10^{-12}}$$

$$\log \frac{I}{10^{-12}} = 3$$

$$\frac{I}{10^{-12}} = 10^3$$

$$I = 10^{-12} \cdot 10^3$$

$$I = 10^{-9}$$

Therefore, for seven crying babies, the intensity is $7 \cdot 10^{-9}$.

$$D = 10 \log \frac{7 \cdot 10^{-9}}{10^{-12}}$$
$$= 10 \log 7000$$
$$\approx 38.45$$

The loudness of crying septuplets is 38.45 decibels.

31. First, solve for B_1 the brightness of the sun with $M = -26.8$.

$$-26.8 = -2.5 \log \frac{B_1}{B_0}$$

$$\log \frac{B_1}{B_0} = 10.72$$

$$\frac{B_1}{B_0} = 10^{10.72}$$

$$B_1 \approx 5.25 \cdot 10^{10} B_0$$

Now, solve for B_2, the brightness of Sirius with $M = -1.5$.

$$-1.5 = -2.5 \log \frac{B_2}{B_0}$$

$$\log \frac{B_2}{B_0} = 0.6$$

$$\frac{B_2}{B_0} = 10^{0.6}$$

$$B_2 = 3.98 \cdot B_0$$

The sun is $\dfrac{B_1}{B_2} = \dfrac{5.25 \cdot 10^{10} B_0}{3.98 B_0} = 1.32 \cdot 10^{10}$ times as bright as Sirius.

© Houghton Mifflin Company. All rights reserved.

33. The initial population is the value of B for $t = 0$.

$$B(0) = \frac{580}{1+8.3e^{-0.12(0)}}$$
$$\approx 62.36$$

The initial population was 62 black bears.
The initial population will double when
$B(t) = 2 \cdot 62 = 124$.

$$124 = \frac{580}{1+8.3e^{-0.12t}}$$
$$1+8.3e^{-0.12t} = \frac{580}{124}$$
$$8.3e^{-0.12t} = \frac{580}{124} - 1$$
$$8.3e^{-0.12t} = 3.6774$$
$$e^{-0.12t} = \frac{3.6774}{8.3}$$
$$e^{-0.12t} = 0.4431$$
$$-0.12t = \ln 0.4431$$
$$t = \frac{\ln 0.4431}{-0.12}$$
$$t \approx 6.78$$

The initial population will double in about 6.8 years.

35. After 2 years, $P(t) = 30$, $M = 200$, $P_0 = 20$, and $t = 2$.

$$30 = \frac{200 \cdot 20}{20+(200-20)e^{-k \cdot 2}}$$
$$30 = \frac{4000}{20+180e^{-2k}}$$
$$20+180e^{-2k} = \frac{4000}{30}$$
$$20+180e^{-2k} = 133.33$$
$$180e^{-2k} = 113.33$$
$$e^{-2k} = \frac{113.33}{180}$$
$$e^{-2k} = 0.630$$
$$-2k = \ln 0.63$$
$$k = \frac{\ln 0.63}{-2}$$
$$k \approx 0.23$$

Now, solve for t for when $P(t) = 100$.

$$100 = \frac{4000}{20+180e^{-0.23t}}$$
$$20+180e^{-0.23t} = \frac{4000}{100} = 40$$
$$180e^{-0.23t} = 20$$
$$e^{-0.23t} = \frac{1}{9}$$
$$-0.23t = \ln \frac{1}{9}$$
$$t = \frac{-2.20}{-0.23}$$
$$t = 9.6$$

The population will reach 100 after 9.6 years.

© Houghton Mifflin Company. All rights reserved.

37. Let t be the number of years after 1994, when $t = 0$, $P(0) = P_0 = 5.643$. After one year, $P(t) = 1.015 \cdot 5.643 = 5.728$.

$$5.728 = \frac{10 \cdot 5.643}{5.643 + (10 - 5.643)e^{-k \cdot 1}}$$

$$5.643 + 4.357e^{-k} = \frac{10 \cdot 5.643}{5.728}$$

$$5.643 + 4.357e^{-k} = 9.852$$

$$4.357e^{-k} = 4.209$$

$$e^{-k} = \frac{4.209}{4.357}$$

$$e^{-k} = 0.966$$

$$-k = \ln(0.966) = -0.035$$

$$k = 0.035$$

Now solve for t when $P(t) = 9$

$$9 = \frac{10 \cdot 5.643}{5.632 + (10 - 5.643)e^{-0.035t}}$$

$$5.632 + 4.357e^{-0.035t} = \frac{10 \cdot 5.643}{9}$$

$$5.643 + 4.357e^{-09.035t} = 6.27$$

$$4.357e^{-0.035t} = 0.627$$

$$e^{-0.035t} = 0.1439$$

$$-0.035t = \ln 0.1439$$

$$t = \frac{\ln 0.1439}{-0.035}$$

$$t \approx 55.4$$

The population will reach 9 billion after 55 years, or in the year 2049.

39. a. The exponential regression equation is
$y = 105.48(0.63)^x$.

b.

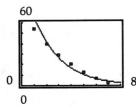

c. The model function never has a value of 0.

41. a.

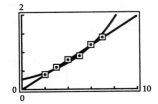

The linear regression model is $y = 0.197x - 0.0038$.
The exponential regression model is
$y = 0.274(1.273)^x$.

b. Because the rate of increase is approximately constant, the linear model appears to be better.

c. Because the expenditures are not likely to increase rapidly, a linear model is more reasonable.

43. a. The exponential regression equation is
$y = 0.034(1.037)^x$.

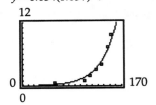

b. To determine when membership will be 10% of 260 million, or 26 million, we solve for x.

$$26 = 0.034(1.037)^x$$

$$(1.037)^x = 764.706$$

$$x \ln 1.037 = \ln 764.706$$

$$x = \frac{\ln 764.706}{\ln 1.037}$$

$$x \approx 182.7$$

10% of the population will be members after 182 years, or in the year 2026.

c. For the years 1950 – 80, the rate of increase is constant so a linear model is appropriate.

45. a. The natural logarithmic regression equation is $y = 11.2 + 15.2\ln x$.

b We solve for x when $y = 50$.

$$50 = 11.2 + 15.2 \ln x$$

$$15.2 \ln x = 38.8$$

$$\ln x = 2.55$$

$$x = e^{2.55}$$

$$x \approx 12.8$$

After 12.8 years, or in the year 2001, half the vehicles will be sport-utility vehicles.

© Houghton Mifflin Company. All rights reserved.

47. a.

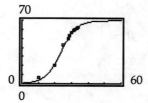

The trend is a rapid increase followed by a leveling off of the percentage.

b. The logistic regression equation is
$$y = \frac{69.586}{1 + 154.952e^{-0.215x}}.$$
For the year 2000, $x = 2000 - 1960 = 40$.
Using the trace feature, we obtain $y = 67.7$, or 67.7% with cable television.
For the year 2010, $x = 2010 - 1960 = 50$.
Using the trace feature, we obtain $y = 69.4\%$ with cable television.

c. As $x \to \infty$, $e^{-0.215x} \to 0$ so that $y \to 69.586$. The limiting value indicates that the percentage will level off at approximately 69.6%.

49. a. The logistic regression equation is
$$y = \frac{15.39}{1 + 1597.45e^{-0.82x}}$$

b. To determine when the number of accounts will reach 15 million, we solve for x.
$$15 = \frac{15.39}{1 + 1597.45e^{-0.82x}}$$
$$1 + 1597.45e^{-0.82x} = \frac{15.39}{15}$$
$$1597.45e^{-0.82x} = 1.026 - 1$$
$$e^{-0.82x} = \frac{.026}{1597.45}$$
$$e^{-0.82x} = .0000163$$
$$-0.82x = \ln.0000163$$
$$x = \frac{\ln.0000163}{-0.82}$$
$$x \approx 13.44$$
The number of accounts will reach 15 million after 13 years, or in the year 2003.

c. The natural logarithmic regression equation is $y = -42.97 + 23.1\ln x$.

d. To determine when the number of accounts will reach 15 million, according to this model, we solve for x.
$$15 = -42.97 + 23.1\ln x$$
$$57.97 = 23.1\ln x$$
$$\ln x = \frac{57.97}{23.1}$$
$$\ln x = 2.5095$$
$$x = e^{2.5095}$$
$$x \approx 12.30$$
The number of accounts will reach 15 million after 12 years, or in the year 2002.

e. The logistic model indicates that the number of accounts will level off and the logarithmic model indicates that the number will continue to increase.

© Houghton Mifflin Company. All rights reserved.

Review Exercises

1. a. $(f+g)(-2) = f(-2)+g(-2)$
$= [-3(-2)]+[(-2)^2-(-2)+1]$
$= 6+7$
$= 13$

b. $(f-g)(x) = f(x)-g(x)$
$= (-3x)-(x^2-x+1)$
$= -x^2-2x-1$

3. a. $(f \circ g)(1) = f(g(1))$
$= f(1^2-1+1)$
$= f(1)$
$= -3(1)$
$= -3$

b. $(g \circ f)(x) = g(f(x))$
$= g(-3x)$
$= (-3x)^2-(-3x)+1$
$= 9x^2+3x+1$

5. $\left(\dfrac{f}{g}\right)(x) = \dfrac{f(x)}{g(x)}$
$= \dfrac{\frac{4}{x-1}}{\frac{x+3}{1-x}}$
$= -\dfrac{4}{x+3}$
Domain: $\{x | x \neq -3, 1\}$

7. a. $(g \circ g)\left(\dfrac{3}{2}\right) = g\left(g\left(\dfrac{3}{2}\right)\right)$
$= g\left(\dfrac{1}{\frac{3}{2}-1}\right)$
$= \dfrac{1}{2-1}$
$= 1$

b. $(f \circ f)(-4) = f(f(-4))$
$= f(|-4+1|)$
$= f(3)$
$= |3+1|$
$= 4$

9. $(f \circ g)(x) = f(g(x))$
$= f\left(\dfrac{3}{2}x\right)$
$= \dfrac{2}{3}\left(\dfrac{3}{2}x\right)$
$= x$
$(g \circ f)(x) = g(f(x))$
$= g\left(\dfrac{2}{3}x\right)$
$= \dfrac{3}{2}\left(\dfrac{2}{3}x\right)$
$= x$

11. Let $f(x) = \dfrac{2}{x}$ and $g(x) = x^2+x$. Then
$H(x) = (f \circ g)(x) = f(g(x))$
$= f(x^2+x) = \dfrac{2}{x^2+x}$

13. $3xy-x^2 = 7$

15.

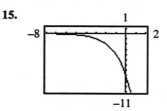

Yes. The graph passes the Horizontal Line Test. The inverse relation is a function.

17. a. $D: \{x | x \neq 2\}$, $R: \{y | y \neq 3\}$

b. $D: \{x | x \neq 3\}$, $R: \{y | y \neq 2\}$

19. Yes. The graph passes the Horizontal Line Test.
$y = \sqrt{2-x}$
$x = \sqrt{2-y}$
$x^2 = 2-y$
$y = 2-x^2$
$f^{-1}(x) = 2-x^2, \ x \geq 0$

21. Yes. The graph passes the Horizontal Line Test.
$y = \dfrac{x+1}{x}$
$x = \dfrac{y+1}{y}$
$xy = y+1$
$xy-y = 1$
$y(x-1) = 1$
$y = \dfrac{1}{x-1}$
$f^{-1}(x) = \dfrac{1}{x-1}$

© Houghton Mifflin Company. All rights reserved.

23.
$$y = x^2 - 1$$
$$x = y^2 - 1$$
$$y^2 = x + 1$$
$$y = \sqrt{x+1}$$
$$f^{-1}(x) = \sqrt{x+1}$$

25. a. $f(2.5) = e^{2.5}$
$$\approx 12.18$$

   b. $f(-1) = e^{-1}$
$$= \frac{1}{e}$$
$$\approx 0.37$$

27. The graph of f is the graph of $y = b^x$ shifted upward k units. The graph of g is the graph of $y = b^x$ shifted downward k units

29

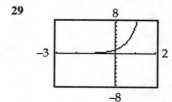

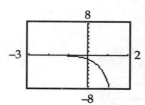

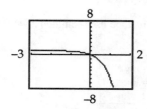

   a. The graph of f is the graph of y reflected across the x-axis.

   b. The graph of g is the graph of y reflected across the x-axis and then shifted upward 1 unit.

31.

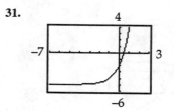

   a. The y-intercept is $(0, -2)$.
   The x-intercept is $(0.46, 0)$.

   b. The horizontal asymptote is $y = -5$.

   c. The range is $\{y | y > -5\}$.

33.

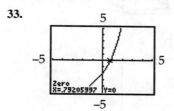

The zero is 0.79.

35. a. $\left(\dfrac{1}{4}\right)^{-2} = 16$

   b. $\ln A = -rt$

37. a. $\log_{\frac{2}{3}} \dfrac{27}{8} = \log_{\frac{2}{3}} \left(\dfrac{3}{2}\right)^3 = \log_{\frac{2}{3}} \left(\dfrac{2}{3}\right)^{-3} = -3$

   b. $\log_6 \sqrt[3]{36} = \log_6 (36)^{1/3}$
$$= \log_6 (6^2)^{1/3}$$
$$= \log_6 6^{2/3}$$
$$= \frac{2}{3}$$

39. Functions f and g are inverses, so their graphs are symmetric with respect to the line $y = x$.

© Houghton Mifflin Company. All rights reserved.

41.

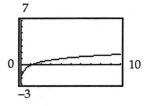

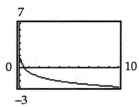

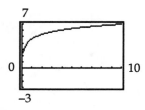

a. The graph of f is the graph of y reflected across the x-axis and shifted downward 1 unit.

b. The graph of g is the graph of y shifted upward 5 units.

43. a. $D = 10 \log \dfrac{10^{10} \cdot I_0}{I_0}$

$= 10 \log 10^{10}$

$= 10 \cdot 10$

$= 100$

A jackhammer has decibel level 100.

b. $65 = 10 \log \dfrac{I}{I_0}$

$\log \dfrac{I}{I_0} = 6.5$

$\dfrac{I}{I_0} = 10^{6.5}$

$I = 10^{6.5} \cdot I_0$

A dishwasher has a decibel level of $10^{6.5} \cdot I_0$.

45. $y = \log(x + 5)$

$x = \log(y + 5)$

$10^x = y + 5$

$y = 10^x - 5$

Thus, $g^{-1}(x) = 10^x - 5$.

47. a. $e^{-2 \ln 3} = (e^{\ln 3})^{-2} = (3)^{-2} = \dfrac{1}{9}$

b. $\log_3 3^7 = 7$

49. a. $\log_5(2ab) = \log_5 2 + \log_5 a + \log_5 b$

b. $\log t + \log(t + 2) = \log t(t + 2)$

51. a. $\log_2 z^{-5} = -5 \log_2 z$

b. $\dfrac{1}{3} \ln t = \ln t^{1/3} = \ln \sqrt[3]{t}$

53. $\log_7 \dfrac{5a^2}{b} = \log_7(5a^2) - \log_7 b$

$= \log_7 5 + \log_7 a^2 - \log_7 b$

$= \log_7 5 + 2 \log_7 a - \log_7 b$

55. $\ln 2x(x + 4)^3 = \ln 2 + \ln x + \ln(x + 4)^3$

$= \ln 2 + \ln x + 3 \ln(x + 4)$

57. $2 \log_9 t + 3 \log_9 (t^2 + 1) = \log_9 t^2 + \log_9 (t^2 + 1)^3$

$= \log_9 t^2 (t^2 + 1)^3$

59. $8^{x-2} = 16^{x+1}$

$(2^3)^{x-2} = (2^4)^{x+1}$

$2^{3x-6} = 2^{4x+4}$

$3x - 6 = 4x + 4$

$x = -10$

61. $e^{1-x} = 10^x$

$1 - x = x \ln 10$

$x + x \ln 10 = 1$

$x(1 + \ln 10) = 1$

$x = \dfrac{1}{1 + \ln 10}$

$x \approx 0.30$

An alternative method uses common logarithms.

$(1 - x) \log e = x$

$\log e - x \log e = x$

$x + x \log e = \log e$

$x(1 + \log e) = \log e$

$x = \dfrac{\log e}{1 + \log e}$

$x \approx 0.30$

63. $\log_2(x^2 - 2x) - 1 = 2$

$\log_2(x^2 - 2x) = 3$

$x^2 - 2x = 2^3 = 8$

$x^2 - 2x - 8 = 0$

$(x + 2)(x - 4) = 0$

$x + 2 = 0$ or $x - 4 = 0$

$x = -2$ $x = 4$

© Houghton Mifflin Company. All rights reserved.

65. $\log_2(x-1) - \log_2(2x+5) = -2$

$$\log_2 \frac{x-1}{2x+5} = -2$$
$$\frac{x-1}{2x+5} = 2^{-2} = \frac{1}{4}$$
$$4(x-1) = 2x+5$$
$$4x-4 = 2x+5$$
$$2x = 9$$
$$x = \frac{9}{2}$$

67.
$$(\log_5 x)^2 + 3 = \log_5 x^4$$
$$(\log_5 x)^2 + 3 = 4\log_5 x$$
$$(\log_5 x)^2 - 4\log_5 x + 3 = 0$$
$$(\log_5 x - 1)(\log_5 x - 3) = 0$$

$\log_5 x - 1 = 0 \quad$ or $\quad \log_5 x - 3 = 0$

$\quad \log_5 x = 1 \qquad\qquad \log_5 x = 3$

$\qquad x = 5^1 \qquad\qquad\quad x = 5^3$

$\qquad x = 5 \qquad\qquad\quad\; x = 125$

69. We use the formula $A = A_0(2)^{-t/T}$ with
$A = 0.9A_0$ and $T = 5770$ and solve for t.

$$0.9A_0 = A_0(2)^{-t/5770}$$
$$0.9 = 2^{-t/5770}$$
$$\ln 0.9 = \ln(2^{-t/5770})$$
$$\ln 0.9 = -\frac{t}{5770}\ln 2$$
$$t = -5770\frac{\ln 0.9}{\ln 2}$$
$$t \approx 877$$

The skeleton was 877 years old.

71. a. The exponential regression equation is
$y = 778.59(0.83)^x$.

b. $778.59(0.83)^8 \approx 175$
There are 175 new cases predicted.

© Houghton Mifflin Company. All rights reserved.

Chapter 5: Systems of Equations and Inequalities

5.1 Exercises

1. Because the y-intercept of both lines is $(0,-3)$, the graphs intersect at $(0, -3)$. Thus $(0,-3)$ is the solution to the system.

3. F; $(4, -2)$

5. E; Every point of the line represents a solution of the system.

7. D; $(-6, -2)$

9. Replace y in the second equation with $4x - 13$.
$$3x = 29 - 2(4x - 13)$$
$$3x = 29 - 8x + 26$$
$$11x = 55$$
$$x = 5$$

   Replace x in the first equation with 5.
   $$y = 4(5) - 13$$
   $$y = 7$$
   The solution is $(5, 7)$.

11. Solve the first equation for x.
$$3 - x = 0$$
$$x = 3$$

   Replace x in the second equation with 3.
   $$3(3) + 5y = -1$$
   $$5y = -10$$
   $$y = -2$$

   The solution is $(3, -2)$.

13. Solve for y in the second equation.
$$y + 2x = 16$$
$$y = 16 - 2x$$

   Replace y in the first equation with $16 - 2x$.
   $$2x + 3(16 - 2x) + 8 = 0$$
   $$2x + 48 - 6x + 8 = 0$$
   $$-4x = -56$$
   $$x = 14$$
   Replace x in the second equation with 14.
   $$y + 2(14) = 16$$
   $$y = -12$$
   The solution is $(14, -12)$.

15. Replace y in the second equation with $-2x - 5$.
$$(-2x - 5) + 2x = 6$$
$$0 = 11$$
   The system is inconsistent. There is no solution.

17. The graphs are straight lines. If the lines do not coincide, they intersect at most once. So the maximum finite number of solutions of the system is 1.

19. Multiply the second equation by 3 and add the equations.
$$\begin{array}{rclcrcl} x - 3y &=& 9 & \to & x - 3y &=& 9 \\ 2x + y &=& 11 & \to & 6x + 3y &=& 33 \\ \hline & & & & 7x & =& 42 \\ & & & & x &=& 6 \end{array}$$
   Replace x in the second equation with 6.
   $$2(6) + y = 11$$
   $$y = -1$$
   The solution is $(6, -1)$.

21. We write both equations in standard form then multiply the second equation by 3 and add the equations.
$$\begin{array}{rclcrclcrcl} 4x + 3y + 2 &=& 0 & \to & 4x + 3y &=& -2 & \to & 4x + 3y &=& -2 \\ 2x &=& y - 1 & \to & 2x - y &=& -1 & \to & 6x - 3y &=& -3 \\ \hline & & & & & & & & 10x & =& -5 \\ & & & & & & & & x &=& -\dfrac{1}{2} \end{array}$$

   Replace x in the second equation with $-\dfrac{1}{2}$.
   $$2\left(-\frac{1}{2}\right) = y - 1$$
   $$y = 0$$
   The solution is $\left(-\dfrac{1}{2}, 0\right)$.

23. We write both equations in standard form then multiply the first equation by 2 and add the equations.
$$\begin{array}{rclcrclcrcl} 2x - 3y + 5 &=& 0 & \to & 2x - 3y &=& -5 & \to & 4x - 6y &=& -10 \\ 6y - 4x &=& 10 & \to & -4x + 6y &=& 10 & \to & -4x + 6y &=& 10 \\ \hline & & & & & & & & 0 &=& 0 \end{array}$$

   The system is dependent. There are infinitely many solutions of the form $\left\{\left(\dfrac{3}{2}y - \dfrac{5}{2}, y\right)\right\}$.

© Houghton Mifflin Company. All rights reserved.

25. Multiply the first equation by 2 and the second equation by –3 and add the equations.

$$6x + 5y = 11 \quad \rightarrow \quad 12x + 10y = 22$$
$$4x + 3y = 5 \quad \rightarrow \quad \underline{-12x - 9y = -15}$$
$$y = 7$$

Replace y in second equation with 7.

$$4x + 3(7) = 5$$
$$4x = -16$$
$$x = -4$$

The solution is (–4, 7).

27. We write both equations in standard form then multiply the first equation by –1 and add the equations.

$$5y = 2(3 - x) \quad \rightarrow \quad 2x + 5y = 6 \quad \rightarrow \quad -2x - 5y = -6$$
$$5y + 2x - 15 = 0 \quad \rightarrow \quad 2x + 5y = 15 \quad \rightarrow \quad \underline{2x + 5y = 15}$$
$$0 = 9$$

The system is inconsistent. There is no solution.

29. Multiply the first equation by –6 and the second equation by 20 and add the equations.

$$\frac{2}{3}x - \frac{1}{2}y = \frac{3}{2} \quad \rightarrow \quad -4x + 3y = -9$$
$$\frac{1}{5}x + y = \frac{8}{5} \quad \rightarrow \quad \underline{4x + 20y = 32}$$
$$23y = 23$$
$$y = 1$$

Replace y in the second equation with 1.

$$\frac{1}{5}x + 1 = \frac{8}{5}$$
$$\frac{1}{5}x = \frac{3}{5}$$
$$x = 3$$

The solution is (3,1).

31. Solve for y in the second equation.

$$y = -7$$

Replace y in the first equation with –7.

$$x + 2(-7) = 1$$
$$x = 15$$

The solution is (15, –7).
The system is consistent and independent.

33. Solve for y in the first equation.

$$y = \frac{2}{7}(4 - 3x)$$

Replace y in the second equation with $\frac{2}{7}(4 - 3x)$.

$$3x + 3.5\left[\frac{2}{7}(4 - 3x)\right] = -1$$
$$3x + 4 - 3x = -1$$
$$0 = -5$$

There is no solution. The system is inconsistent and independent.

© Houghton Mifflin Company. All rights reserved.

35. Multiply the first equation by 4 and add the equations.

$$\begin{aligned} 12x - 8y &= 4 \\ x + 8y &= -17 \\ \hline 13x \phantom{{}- 8y} &= -13 \\ x &= -1 \end{aligned}$$

Replace x in the first equation with -1.

$$\begin{aligned} 3(-1) - 2y &= 1 \\ -2y &= 4 \\ y &= -2 \end{aligned}$$

The solution is $(-1, -2)$. The system is consistent and independent.

37. Replace x in the second equation with $2y + 6$.

$$\begin{aligned} (2y + 6) + y &= -3 \\ 3y &= -9 \\ y &= -3 \end{aligned}$$

Replace y in the first equation with -3.

$$\begin{aligned} x &= 2(-3) + 6 \\ x &= 0 \end{aligned}$$

The solution is $(0, -3)$. The system is consistent and independent.

39. Multiply the first equation by 5, the second equation by -10 and add the equations.

$$\begin{array}{ccccc} x + 0.6y = 3.8 & \to & x + 0.6y = 3.8 & \to & 5x + 3y = 19 \\ 0.3y + 0.5x = 1.9 & \to & 0.5x + 0.3y = 1.9 & \to & \underline{-5x - 3y = -19} \\ & & & & 0 = 0 \end{array}$$

There are infinite solutions of the form $\{(3.8 - 0.6y, y)\}$. The system is consistent and dependent.

41. The only solution of (*i*) is $(0,0)$, whereas (*ii*) has infinitely many solutions.

43. Replace x with 7 and y with 11 in both equations.

$$\begin{array}{ccc} a(7) - b(11) = 2 & \to & 7a - 11b = 2 \\ b(7) - (a - b)(11) = -1 & \to & -11a + 18b = -1 \end{array}$$

Multiply the first equation by 11, the second equation by 7, and add the equations.

$$\begin{aligned} 77a - 121b &= 22 \\ -77a + 126b &= -7 \\ \hline 5b &= 15 \\ b &= 3 \end{aligned}$$

Replace b in the first equation with 3.

$$\begin{aligned} 7a - 11(3) &= 2 \\ 7a &= 35 \\ a &= 5 \end{aligned}$$

The solution is $a = 5$ and $b = 3$.

© Houghton Mifflin Company. All rights reserved.

45. Using the given substitutions,

$$\frac{4}{x}-\frac{1}{y}=7 \quad \rightarrow \quad 4u-v=7$$

$$-\frac{2}{x}+\frac{3}{y}=9 \quad \rightarrow \quad -2u+3v=9$$

Multiply the second equation by 2 and add the equations.

$$\begin{array}{r} 4u-v= 7 \\ -4u+6y=18 \\ \hline 5v=25 \\ v= 5 \end{array}$$

Replace v in the first equation with 5.

$$4u-(5)=7$$
$$u=3$$

Thus $u=\dfrac{1}{x}=3$, so $x=\dfrac{1}{3}$, and $v=\dfrac{1}{y}=5$, so

$y=\dfrac{1}{5}$.

The solution is $\left(\dfrac{1}{3},\dfrac{1}{5}\right)$.

47. Multiply the second equation by 3 and add the second and third equations.

$$\begin{array}{rcl} x-2y=-4 & \rightarrow & 3x-6y=-12 \\ -3x-y= 5 & \rightarrow & \underline{-3x-y= 5} \\ & & -7y=-7 \\ & & y= 1 \end{array}$$

Replace y in the third equation with 1.

$$-3x-1=5$$
$$x=-2$$

Since $(-2, 1)$ also satisfies the first equation, the solution is $(-2, 1)$.

49. Solve the first equation for y.

$$2x-y=5$$
$$y=2x-5$$

Replace y in the second equation with $2x-5$.

$$kx=3(2x-5)+15$$
$$kx=6x$$
$$(k-6)x=0$$

For the equations to be dependent, $k-6=0$ or $k=6$.

51. Let x = number of shoes at regular price
y = number of shoes at $20 off.
Then

$$80x+60y=3640 \quad (1)$$
$$x+ y= 50 \quad (2)$$

Multiply both sides of equation (2) by –60 and add the equations.

$$\begin{array}{r} 80x+60y=3640 \\ -60x-60y=-3000 \\ \hline 20x=640 \\ x=32 \end{array}$$

Replace x in equation (2) with 32.

$$32+y=50$$
$$y=18$$

18 coupons were redeemed.

53. Let x = milliliters of 20% antiseptic solution
y = milliliters of 60% antiseptic solution
Then

$$x+ y=50 \quad\quad (1)$$
$$0.20x+0.60y=0.36(50)=18 \quad (2)$$

Multiply equation (1) by –2 and equation (2) by 10 and add the equations.

$$\begin{array}{r} -2x-2y=-100 \\ 2x+6y= 180 \\ \hline 4y= 80 \\ y= 20 \end{array}$$

Replace y with 20 in equation (1).

$$x+(20)=50$$
$$x=30$$

Use 30 ml of the 20% solution and 20 ml of the 60% solution.

55. Let x = daily rate
y = mileage rate

Then

$$x+235y=39.45 \quad (1)$$
$$x+148y=33.36 \quad (2)$$

Subtract equation (2) from equation (1).

$$87y=6.09$$
$$y=0.07$$

Replace y with 0.07 in equation (1).

$$x+235(0.07)=39.45$$
$$x=23$$

The daily rate is $23 and the mileage rate is $0.07.

© Houghton Mifflin Company. All rights reserved.

57. Let x = volume of small box
y = volume of large box.
Then

$$10x + 6y = 576 \quad (1)$$
$$14x + 10y = 896 \quad (2)$$

Multiply equation (1) by 10, equation (2) by –6, and add the equations.

$$\begin{array}{rl} 100x + 60y = & 5760 \\ -84x - 60y = & -5376 \\ \hline 16x = & 384 \\ x = & 24 \end{array}$$

Replace x with 24 in equation (2).
$$14(24) + 10y = 896$$
$$10y = 560$$
$$y = 56$$
The volume of the small box is 24 cubic inches and the volume of the large box is 56 cubic inches.

59. Let x = daily wages of an electrician
y = daily wages of a sheet-metal worker
Then

$$x + 6y = 1152 \quad (1)$$
$$3x + 5y = 1350 \quad (2)$$
Multiply equation (1) by –3 and add the equations.

$$\begin{array}{rl} -3x - 18y = & -3456 \\ 3x + 5y = & 1350 \\ \hline -13y = & -2106 \\ y = & 162 \end{array}$$
Replace y with 162 in equation (1).
$$x + 6(162) = 1152$$
$$x = 180$$
An electrician earns \$180 per day, or \$22.50 per hour, and a sheet-metal worker earns \$162 per day, or \$20.25 per hour.

61. Let x = enrollment in Mathematics 101
y = enrollment in English 109

Then

$$x = y + 30 \quad (1)$$
$$54x + 58y = 11,700 \quad (2)$$

Replace x with y + 30 in equation (2).
$$54(y + 30) + 58y = 11,700$$
$$112y = 10,080$$
$$y = 90$$

Replace y with 90 in equation (1)
$$x = 90 + 30 = 120$$
120 students were enrolled Mathematics 101 and 90 students were enrolled in English 109.

63. Let x = number of customers who bought single frames
y = number of customers who bought two-for-\$20 frames

Then

$$x + 2y = 121 \quad (1)$$
$$11x + 20y = 1259 \quad (2)$$
Multiply equation (1) by –10 and add the equations.

$$\begin{array}{rl} -10x - 20y = & -1210 \\ 11x + 20y = & 1259 \\ \hline x = & 49 \end{array}$$
Replace x with 49 in equation (1)
$$49 + 2y = 121$$
$$2y = 72$$
$$y = 36$$
36 customers took advantage of the two-for-\$20 offer.

65. Let x = pounds of sunflower seeds
y = pounds of commercial mix

Then

$$x + y = 50 \quad (1)$$
$$1.00x + 0.15y = 0.32(50) = 16 \quad (2)$$
Multiply equation (2) by –1 and add the equations.

$$\begin{array}{rl} x + y = & 50 \\ -x - 0.15y = & -16 \\ \hline 0.85y = & 34 \\ y = & 40 \end{array}$$
Replace y with 40 in equation (1)
$$x + 40 = 50$$
$$x = 10$$
10 pounds of sunflower seeds were added to the commercial mix.

67. Let x = length of bedroom
y = width of bedroom
Then
$$x = y + 3 \quad \rightarrow \quad x = y + 3 \quad (1)$$
$$6 \cdot 9x + 9 \cdot 9y = 1782 \quad \rightarrow \quad 54x + 81y = 1782 \quad (2)$$
Replace x with y + 3 in equation (2).
$$54(y + 3) + 81y = 1782$$
$$135y = 1620$$
$$y = 12$$
Replace y with 12 in equation (1).
$$x = 12 + 13 = 15$$
The bedroom is 15 feet by 12 feet.

© Houghton Mifflin Company. All rights reserved.

69. Let x = number of hours the cyclist travels until they meet
y = number of hours her brother travels until they meet

Then, since distance = rate $\times$ time, and 1:12 is 1.2 hours past noon,

Then
$x = y + 1.2$ (1)
$20x = 50y$ (2)
Replace x in equation (2) with $y + 1.2$
$20(y + 1.2) = 50y$
$30y = 24$
$y = 0.8$
Thus, her brother has traveled 0.8 hour when they meet. They are at a distance of $0.8 \cdot 50 = 40$ miles from home.

71. Let x = number of faculty members
y = number of staff members

Then
$$x + y = 51 \qquad (1)$$
$$1500x + 500y = 67,500 \quad (2)$$

Solve for y in equation (1).
$y = 51 - x$
Replace y with $51 - x$ in equation (2)
$1500x + 500(51 - x) = 67,500$
$1000x = 42,000$
$x = 42$
Replace x with 42 in equation (1)
$42 + y = 51$
$y = 9$
There are 51 faculty members and 9 staff members.

73. Let x = number of hours to complete the job by electrician
y = number of hours to complete the job by apprentice

Then
$\dfrac{1}{x} + \dfrac{1}{y} = \dfrac{7}{24}$ (1)
$\dfrac{1}{x} + \dfrac{2}{y} = \dfrac{10}{24}$ (2)
Let $u = \dfrac{1}{x}$ and $v = \dfrac{1}{y}$.
We re-write the equations.
$u + v = \dfrac{7}{24}$ (3)
$u + 2v = \dfrac{10}{24}$ (4)
Subtract equation (3) from equation (4).
$v = \dfrac{3}{24} = \dfrac{1}{8}$
Replace v with $\dfrac{1}{8}$ in equation (3).
$u + \dfrac{1}{8} = \dfrac{7}{24}$
$u = \dfrac{4}{24} = \dfrac{1}{6}$
And, if $u = \dfrac{1}{x} = \dfrac{1}{6}$, then $x = 6$. It would take the electrician 6 hours working alone.

75. a. In 1982, the math score was lower than the verbal score, whereas in 1997, it was higher than the verbal score. Sometime between 1982 and 1997, the scores must have been the same.

b. We use the data points (2, 504) and (17, 505) to obtain the linear regression model $y = 0.07x + 503.87$ for the verbal scores. We use the data points (2, 493) and (17, 511) to obtain the linear regression model $y = 1.2x + 490.6$ for the math scores.

c. $y = 0.07x + 503.87$ (1)
$y = 1.2x + 490.6$ (2)
Subtract equation (1) from equation (2).
$0 = 1.13x - 13.27$
$x = 11.74$
The solution, approximately (12, 505), indicates that in 1992 both the math and verbal scores were about 505.

© Houghton Mifflin Company. All rights reserved.

5.2 Exercises

1. No. In the second equation, substitute 5 for z and solve for y.

3.
$$\begin{aligned}
2x + y - 3z &= 5 \quad (1) \\
y + 2z &= 15 \quad (2) \\
2x \qquad - z &= 6 \quad (3)
\end{aligned}$$

$$\begin{aligned}
x + \frac{1}{2}y - \frac{3}{2}z &= \frac{5}{2} \quad (4) \qquad \text{\textbf{Multiply equation (1) by} } \frac{1}{2}. \\
y + 2z &= 15 \quad (2) \\
2x \qquad - z &= 6 \quad (3)
\end{aligned}$$

$$\begin{aligned}
x + \frac{1}{2}y - \frac{3}{2}z &= \frac{5}{2} \quad (4) \\
y + 2z &= 15 \quad (2) \\
-y + 2z &= 1 \quad (5) \qquad \text{\textbf{Multiply equation (4) by} } -2 \text{ \textbf{and add the result to equation (3).}}
\end{aligned}$$

$$\begin{aligned}
x + \frac{1}{2}y - \frac{3}{2}z &= \frac{5}{2} \quad (4) \\
y + 2z &= 15 \quad (2) \\
4z &= 16 \quad (6) \qquad \text{\textbf{Add equation (2) to equation (5).}}
\end{aligned}$$

$$\begin{aligned}
x + \frac{1}{2}y - \frac{3}{2}z &= \frac{5}{2} \quad (4) \\
y + 2z &= 15 \quad (2) \\
z &= 4 \quad (7) \qquad \text{\textbf{Multiply equation (6) by} } \frac{1}{4}.
\end{aligned}$$

This is row-echelon form. From equation (7) we know $z = 4$. Substitute 4 for z in equation (2).
$$\begin{aligned}
y + 2(4) &= 15 \\
y &= 7
\end{aligned}$$
Substitute 4 for z and 7 for y in equation (4).
$$\begin{aligned}
x + \frac{1}{2}(7) - \frac{3}{2}(4) &= \frac{5}{2} \\
x &= 5
\end{aligned}$$
The solution is (5, 7, 4).

© Houghton Mifflin Company. All rights reserved.

5. $4x - y - z = 3$ (1)
  $2x - 2y + 3z = -5$ (2)
  $x + y + z = 2$ (3)

  $x + y + z = 2$ (3) **Interchange equations (1) and (3).**
  $2x - 2y + 3z = -5$ (2)
  $4x - y - z = 3$ (1)

  $x + y + z = 2$ (3)
  $-4y + z = -9$ (4) **Multiply equation (3) by -2 and add the result to equation (2).**
  $-5y - 5z = -5$ (5) **Multiply equation (3) by -4 and add the result to equation (1).**

  $x + y + z = 2$ (3)
  $y - \dfrac{1}{4}z = \dfrac{9}{4}$ (6) **Multiply equation (4) by $-\dfrac{1}{4}$.**
  $-5y - 5z = -5$ (5)

  $x + y + z = 2$ (3)
  $y - \dfrac{1}{4}z = \dfrac{9}{4}$ (6)
  $-\dfrac{25}{4}z = \dfrac{25}{4}$ (7)

  $x + y + z = 2$ (3)
  $y - \dfrac{1}{4}z = \dfrac{9}{4}$ (6)
  $z = -1$ (8) **Multiply equation (7) by $-\dfrac{4}{25}$.**

This is row-echelon form. From equation (8) we know $z = -1$.
Substitute -1 for z in equation (6).
$$y - \frac{1}{4}(-1) = \frac{9}{4}$$
$$y = 2$$
Substitute -1 for z and 2 for y in equation (3).
$$x + 2 - 1 = 2$$
$$x = 1$$
The solution is $(1, 2, -1)$.

7. $3x + y + z = 5$
  $x - 2y - 5z = -3$
  $5x - 3y - 9z = -1$

The row-echelon form is
$$x - \frac{3}{5}y - \frac{9}{5}z = -\frac{1}{5}$$
$$y + \frac{16}{7}z = 2$$
$$0 = 0$$

The system is dependent. There are infinite
solutions of the form $\left\{ \left(\dfrac{3}{7}z + 1, \ 2 - \dfrac{16}{7}z, \ z \right) \right\}$.

9. $x - y - z = -1$
  $x + y + z = 3$
  $2x + y - z = 6$

The row-echelon form is
$$x + \frac{1}{2}y - \frac{1}{2}z = 3$$
$$y + \frac{1}{3}z = \frac{8}{3}$$
$$z = -1$$
The solution is $(1, 3, -1)$.

© Houghton Mifflin Company. All rights reserved.

11.
$$5x + z = 5 + 4y \quad \rightarrow \quad 5x - 4y + z = 5$$
$$2z + x + 4 = y \quad \rightarrow \quad x - y + 2z = -4$$
$$5y - 13z = 8x + 29 \quad \rightarrow \quad 8x - 5y + 13z = -29$$
The row-echelon form is
$$x - \frac{5}{8}y + \frac{13}{8}z = -\frac{29}{8}$$
$$y + \frac{57}{7}z = -\frac{185}{7}$$
$$z = -3$$
The solution is (0, –2, –3).

13. The planes do not intersect, so the system has no solution.

15.
$$y - 3z = 1$$
$$x \quad - 2z = 5$$
$$x + y \quad = 1$$
The row-echelon form is
$$x \quad - 2z = 5$$
$$y - 3z = 1$$
$$z = -1$$
The solution is (3, –2, –1).

17.
$$2x - y + 3z = 4$$
$$2x + 7y - 11z = 0$$
$$4x + 2y - z = 5$$
The row-echelon form is
$$x + \frac{1}{2}y - \frac{1}{4}z = \frac{5}{4}$$
$$y - \frac{7}{4}z = -\frac{5}{12}$$
$$0 = 1$$
The system is inconsistent. There are no solutions.

19.
$$x - z = 2 \quad \rightarrow \quad x \quad - z = 2$$
$$3x + 4y + 3z = 10 \quad \rightarrow \quad 3x + 4y + 3z = 10$$
$$2y + 3z = 2 \quad \rightarrow \quad 2y + 3z = 2$$
The row-echelon form is
$$x + \frac{4}{3}y + z = \frac{10}{3}$$
$$y + \frac{3}{2}z = 1$$
$$0 = 0$$
The system is dependent. There are infinite solutions of the form $\left\{ \left(z + 2, 1 - \frac{3}{2}z, z \right) \right\}$.

21.
$$x = 6 + y + 3z \quad \rightarrow \quad x - y - 3z = 6$$
$$z = 2x + 4y \quad \rightarrow \quad 2x + 4y - z = 0$$
$$x = 14 + 5y - 2z \quad \rightarrow \quad x - 5y + 2z = 14$$
The row-echelon form is
$$x + 2y - \frac{1}{2}z = 0$$
$$y - \frac{5}{14}z = -2$$
$$z = 0$$
The solution is (4, –2, 0).

23. In the solution $y = 0$ and $x + z = 1$: Because $y = 0$, the solution cannot be written in terms of y. Because $x + z = 1$, we can solve for x and write $(1 - z, 0, z)$, or we can solve for z and write $(x, 0, 1 - x)$.

25.
$$f(x) = ax^2 + bx + c$$
$$f(-4) = 16a - 4b + c = 16$$
$$f(2) = 4a + 2b + c = -8$$
$$f(5) = 25a + 5b + c = 7$$
The row-echelon form is
$$a + \frac{1}{5}b + \frac{1}{25}c = \frac{7}{25}$$
$$b - \frac{1}{20}c = -\frac{8}{5}$$
$$c = -8$$
The solution is (1, –2, –8). Thus,
$$f(x) = x^2 - 2x - 8.$$

27.
$$f(x) = ax^2 + bx + c$$
$$f(-3) = 9a - 3b + c = -24$$
$$f(2) = 4a + 2b + c = -9$$
$$f(3) = 9a + 3b + c = -18$$
The row-echelon form is
$$a - \frac{1}{3}b + \frac{1}{9}c = -\frac{8}{3}$$
$$b = 1$$
$$c = -3$$
The solution is (–2, 1, –3). Thus,
$$f(x) = -2x^2 + x - 3.$$

29.
$$x - 2y + x = 5$$
$$2x + y - z = 16$$
The row-echelon form is
$$x + \frac{1}{2}y - \frac{1}{2}z = 8$$
$$y - \frac{3}{5}z = \frac{6}{5}$$
The system is dependent. There are infinite solutions of the form $\left\{ \left(\frac{1}{5}z + \frac{37}{5}, \frac{3}{5}z + \frac{6}{5}, z \right) \right\}$.

31. Let $u = \frac{1}{x}$, $v = \frac{1}{y}$, and $w = \frac{1}{z}$. The system becomes
$$u - v + 2w = -3$$
$$3u + v - w = 3$$
$$2u + 2v + w = -2$$
The row-echelon form is
$$u + \frac{1}{3}v - \frac{1}{3}w = 1$$
$$v - \frac{7}{4}w = 3$$
$$w = -2$$
The solution to this system is $\left(\frac{1}{2}, -\frac{1}{2}, -2 \right)$.

So $x = \frac{1}{u} = 2$, $y = \frac{1}{v} = -2$, and $z = \frac{1}{w} = -\frac{1}{2}$, or the solution is $\left(2, -2, -\frac{1}{2} \right)$.

© Houghton Mifflin Company. All rights reserved.

33.
$$x + y - 2z + w = -4 \quad (1)$$
$$2x - y + z - 3w = -13 \quad (2)$$
$$x - 2y - 3z - w = -15 \quad (3)$$
$$3x + y - z + 2w = -1 \quad (4)$$

$$x + y - 2z + w = -4 \quad (1)$$
$$-3y + 5z - 5w = -5 \quad (5) \quad \textbf{Multiply equation (1) by } -2 \textbf{ and add the result to equation (2).}$$
$$-3y - z - 2w = -11 \quad (6) \quad \textbf{Multiply equation (1) by } -1 \textbf{ and add the result to equation (3).}$$
$$-2y + 5z - w = 11 \quad (7) \quad \textbf{Multiply equation (1) by } -3 \textbf{ and add the result to equation (4).}$$

$$x + y - 2z + w = -4 \quad (1)$$
$$y - \frac{5}{3}z + \frac{5}{3}w = \frac{5}{3} \quad (8) \quad \textbf{Multiply equation (5) by } -\frac{1}{3}.$$
$$-3y - z - 2w = -11 \quad (6)$$
$$-2y + 5z - w = 11 \quad (7)$$

$$x + y - 2z + w = -4 \quad (1)$$
$$y - \frac{5}{3}z + \frac{5}{3}w = \frac{5}{3} \quad (8)$$
$$-6z + 3w = -6 \quad (9) \quad \textbf{Multiply equation (8) by 3 and add the result to equation (6).}$$
$$\frac{5}{3}z + \frac{7}{3}w = \frac{43}{3} \quad (10) \quad \textbf{Multiply equation (8) by 2 and add the result to equation (7).}$$

$$x + y - 2z + w = -4 \quad (1)$$
$$y - \frac{5}{3}z + \frac{5}{3}w = \frac{5}{3} \quad (8)$$
$$z - \frac{1}{2}w = 1 \quad (11) \quad \textbf{Multiply equation (9) by } -\frac{1}{6}.$$
$$\frac{5}{3}z + \frac{7}{3}w = \frac{43}{3} \quad (10)$$

$$x + y - 2z + w = -4 \quad (1)$$
$$y - \frac{5}{3}z + \frac{5}{3}w = \frac{5}{3} \quad (8)$$
$$z - \frac{1}{2}w = 1 \quad (11)$$
$$\frac{19}{6}w = \frac{38}{3} \quad (12) \quad \textbf{Multiply equation (11) by } -\frac{5}{3} \textbf{ and add the result to equation (10).}$$

$$x + y - 2z + w = -4 \quad (1)$$
$$y - \frac{5}{3}z + \frac{5}{3}w = \frac{5}{3} \quad (8)$$
$$z - \frac{1}{2}w = 1 \quad (11)$$
$$w = 4 \quad (13) \quad \textbf{Multiply equation (12) by } \frac{6}{19}.$$

This is the row-echelon form. The solution is $(-2, 0, 3, 4)$.

© Houghton Mifflin Company. All rights reserved.

35. Let $x = $ cost of a small envelope
$\quad\quad y = $ cost of a medium envelope
$\quad\quad z = $ cost of a large envelope

Then
$z = x + y$
$y - x = 0.60$
$2x + y + z = 7.20$

We write the system in standard form.
$x + y - z = 0$
$-x + y \quad\quad = 0.60$
$2x + y + z = 7.20$

The row-echelon form is
$x + \frac{1}{2}y + \frac{1}{2}z = \frac{18}{5}$
$\quad\quad y + \frac{1}{3}z = \frac{14}{5}$
$\quad\quad\quad\quad z = 3$

The solution is $\left(\frac{6}{5}, \frac{9}{5}, 3\right)$, or \$1.20 for a small envelope, \$1.80 for a medium envelope, and \$3.00 for a large envelope.

37. Let the four digits be x, y, z, and w. Then
$x + w = y + z$
$w = x + y$
$x + y + z + w = 14$
$z = 4$

We write the equations in standard form.
$x - y - z + w = 0$
$x + y \quad\quad - w = 0$
$x + y + z + w = 14$
$\quad\quad z \quad\quad = 4$

The row-echelon form is
$x - y \quad - z \quad + w = 0$
$\quad y + \frac{1}{2}z \quad - w = 0$
$\quad\quad\quad z + 2w = 14$
$\quad\quad\quad\quad w = 5$

The solution is $(2, 3, 4, 5)$. Thus the PIN number is 2345.

39. Let $x = $ number of 1-point shots
$\quad\quad y = $ number of 2-point shots
$\quad\quad z = $ number of 3-point shots

Then
$x + 2y + 3z = 68$
$2y = x + 3z$
$y = 2z - 1$

We write the equations in standard form.
$x + 2y + 3z = 68$
$x - 2y + 3z = 0$
$\quad\quad y - 2z = -1$

The row-echelon form is
$x + 2y + 3z = 68$
$\quad\quad y \quad\quad = 17$
$\quad\quad\quad\quad z = 9$

The solution is $(7, 17, 9)$.
Thus there were 7 one-point shots,
17 two-point shots, and 9 three-point shots.

41. Let $x = $ number of faculty lot permits
$\quad\quad y = $ number of central lot permits
$\quad\quad z = $ number of perimeter lot permits

Then
$x + y + z = 5740$
$120y = 150x + 70z + 4800$
$z - y = 2x$

We write the equations in standard form.
$x \quad + y \quad + z = 5740$
$150x - 120y + 70z = -4800$
$2x \quad + y \quad - z = 0$

The row-echelon form is
$x - \frac{4}{5}y + \frac{7}{15}z = -32$
$\quad\quad y - \frac{29}{39}z = \frac{320}{13}$
$\quad\quad\quad\quad z = 3060$

The solution is $(380, 2300, 3060)$. 380 faculty permits, 2300 central permits, and 3060 perimeter permits were sold.

© Houghton Mifflin Company. All rights reserved.

43. Let x = price of a 2x3 print
 y = price of a 3x5 print
 z = price of a 6x8 print

Then
$$8x + 2y + z = 19$$
$$10x + 4y + 2z = 32$$
$$12x + 6y + 4z = 51$$

The row-echelon form is
$$x + \frac{1}{2}y + \frac{1}{3}z = \frac{17}{4}$$
$$y + \frac{5}{6}z = \frac{15}{2}$$
$$z = 6$$

The solution is (1, 2.5, 6). Each 2x3 print cost $1, each 3x5 print cost $2.50, and each 6x8 print cost $6.00.

45. Let x = number of female households (in millions)
 y = number of male households (in millions)
 z = number of couple households (in millions)

Then
$$x + y + z = 69 \quad \rightarrow \quad x + y + z = 69$$
$$x = 4y \quad \rightarrow \quad x - 4y = 0$$
$$z = 4(x + y) - 6 \quad \rightarrow \quad 4x + 4y - z = 6$$

The row-echelon form is
$$x + y - \frac{1}{4}z = \frac{3}{2}$$
$$y - \frac{1}{20}z = \frac{3}{10}$$
$$z = 54$$

The solution is (12, 3, 54). There are 12 million female households, 3 million male households, and 54 million couple households.

47. Let x = number of children
 y = number of adults
 z = number of senior citizens
Then
$$x + y + z = 42$$
$$14y + 10z = 268$$
$$5x + 12y + 8z = 324$$

The row-echelon form is
$$x + \frac{12}{5}y + \frac{8}{5}z = \frac{324}{5}$$
$$y + \frac{5}{7}z = \frac{134}{7}$$
$$z = 10$$

The solution is (20, 12, 10). There were 20 children, 12 adults, and 10 senior citizens.

49. Let x = last year's enrollment at Chapman
 y = last year's enrollment at Bascomb
 z = last year's enrollment at Booth

Then
$$x + y + z = 1763 \quad \rightarrow \quad x + y + z = 1763$$
$$0.92x + 1.10y + \frac{9}{8}z = 1835 \quad \rightarrow \quad 0.92x + 1.10y + \frac{9}{8}z = 1835$$
$$1.10y = 0.92x + \frac{9}{8}z - 53 \quad \rightarrow \quad 0.92x - 1.10y + \frac{9}{8}z = 53$$

The row-echelon form is
$$x + y + z = 1763$$
$$y - \frac{41}{404}z = \frac{78,448}{101}$$
$$z = 328$$

The solution is (625, 810, 328). But those are last year's numbers. This year, there were $0.92 \cdot 625 = 575$ students at Chapman, $1.10 \cdot 810 = 891$ at Bascomb, and $\frac{9}{8} \cdot 328 = 369$ at Booth.

© Houghton Mifflin Company. All rights reserved.

51. $T(x) = ax^2 + bx + c$
$T(5) = 25a + 5b + c = 13$
$T(12) = 144a + 12b + c = 23$
$T(14) = 196a + 14b + c = 19$

The row-echelon form is

$a + \dfrac{1}{14}b + \dfrac{1}{196}c = \dfrac{19}{196}$
$b + \dfrac{19}{70}c = \dfrac{691}{210}$
$c = -17$

The solution is $\left(-\dfrac{8}{21}, \dfrac{166}{21}, -17\right)$. Thus the

function is $T(x) = -0.38x^2 + 7.90x - 17$.

52. $E(t) = at^2 + bt + c$
$E(6) = 36a + 6b + c = 30.5$
$E(11) = 121a + 11b + c = 23.0$
$E(13) = 169a + 13b + c = 26.6$

The row-echelon form is

$a + \dfrac{1}{13}b + \dfrac{1}{169}c = \dfrac{133}{845}$
$b + \dfrac{19}{78}c = \dfrac{41,969}{5460}$
$c = \dfrac{4943}{70}$

The solution is $\left(\dfrac{33}{70}, -\dfrac{333}{35}, \dfrac{4943}{70}\right)$. Thus the

function is $E(t) = 0.47t^2 - 9.51t + 70.61$.

5.3 Exercises

1. The coefficient matrix contains the coefficients of x and y.

3. 1×2

5. 2×3

7. $\begin{bmatrix} 1 & 3 & | & 7 \\ 2 & -5 & | & -8 \end{bmatrix}$

9. $\begin{bmatrix} 1 & 0 & 5 & | & 13 \\ 0 & 2 & 1 & | & 4 \\ 1 & 1 & -1 & | & 0 \end{bmatrix}$

11. $x - y = -2$
$3x + 2y = 14$

13. $-2x \qquad + z = 2$
$4x + y - 3z = -2$
$2x - 4y + z = -10$

15. For $c = 1$, the operations $c \cdot R_i + R_j \to R_j$
becomes $R_i + R_j \to R_j$.

17. a. 1 4 −3

b. 0 1 −2

c. 1 0 5

19. a. 0 1 −3 7

b. 0 −4 13 −31

c. 1 0 −5 16

d. 0 0 1 −3

e. 1 0 0 1

f. 0 1 0 −2

21. No, it is not in reduced row-echelon form.

$-6R_2 + R_1 \to \begin{bmatrix} 1 & 0 & | & 9 \\ 0 & 1 & | & -1 \end{bmatrix}$

23. Yes, it is in reduced row-echelon form.

25. The row corresponds to the equation $0 = 0$. The system has infinitely many solutions.

27. The augmented matrix is

$\begin{bmatrix} 1 & 3 & | & -3 \\ 2 & -1 & | & 8 \end{bmatrix}$

$-2R_1 + R_2 \to \begin{bmatrix} 1 & 3 & | & -3 \\ 0 & -7 & | & 14 \end{bmatrix}$

$R_3 \div -7 \to \begin{bmatrix} 1 & 3 & | & -3 \\ 0 & 1 & | & -2 \end{bmatrix}$

$-3R_2 + R_1 \to \begin{bmatrix} 1 & 0 & | & 3 \\ 0 & 1 & | & -2 \end{bmatrix}$

This is in reduced row-echelon form. The solution is $(3, -2)$.

29. The augmented matrix is

$\begin{bmatrix} 1 & 3 & | & 6 \\ -2 & -6 & | & 1 \end{bmatrix}$

The reduced row-echelon form is

$\begin{bmatrix} 1 & 3 & | & 0 \\ 0 & 0 & | & 1 \end{bmatrix}$

The system is inconsistent. There is no solution.

© Houghton Mifflin Company. All rights reserved.

31. We write the equations in standard form.

$4x = 1 + y \quad \rightarrow \quad 4x - y = 1$

$20x - 5y = 5 \quad \rightarrow \quad 20x - 5y = 5$

The augmented matrix is

$$\begin{bmatrix} 4 & -1 & | & 1 \\ 20 & -5 & | & 5 \end{bmatrix}$$

The reduced row-echelon form is

$$\begin{bmatrix} 1 & -\frac{1}{4} & | & \frac{1}{4} \\ 0 & 0 & | & 0 \end{bmatrix}$$

The system is dependent. The first row corresponds to the equation

$x - \frac{1}{4}y = \frac{1}{4}$ or $x = \frac{1}{4}y + \frac{1}{4}$. Thus there are infinite

solutions of the form $\left\{ \left(\frac{1}{4}y + \frac{1}{4}, y \right) \right\}$.

33. The augmented matrix is

$$\begin{bmatrix} 1 & 0 & -1 & | & 8 \\ 0 & 3 & -2 & | & 12 \\ 2 & 1 & 0 & | & 4 \end{bmatrix}$$

$$-2R_1 + R_3 \rightarrow \begin{bmatrix} 1 & 0 & -1 & | & 8 \\ 0 & 3 & -2 & | & 12 \\ 0 & 1 & 2 & | & -12 \end{bmatrix}$$

$$R_2 \leftrightarrow R_3 \begin{bmatrix} 1 & 0 & -1 & | & 8 \\ 0 & 1 & 2 & | & -12 \\ 0 & 3 & -2 & | & 12 \end{bmatrix}$$

$$-3R_2 + R_3 \rightarrow \begin{bmatrix} 1 & 0 & -1 & | & 8 \\ 0 & 1 & 2 & | & -12 \\ 0 & 0 & -8 & | & 48 \end{bmatrix}$$

$$R_3 \div -8 \rightarrow \begin{bmatrix} 1 & 0 & -1 & | & 8 \\ 0 & 1 & 2 & | & -12 \\ 0 & 0 & 1 & | & -6 \end{bmatrix}$$

$$R_3 + R_1 \rightarrow \begin{bmatrix} 1 & 0 & 0 & | & 2 \\ 0 & 1 & 2 & | & -12 \\ 0 & 0 & 1 & | & -6 \end{bmatrix}$$

$$-2R_3 + R_2 \rightarrow \begin{bmatrix} 1 & 0 & 0 & | & 2 \\ 0 & 1 & 0 & | & 0 \\ 0 & 0 & 1 & | & -6 \end{bmatrix}$$

This is in reduced row-echelon form.
The solution is (2, 0, –6).

35. The augmented matrix is

$$\begin{bmatrix} 4 & -2 & 1 & | & 1 \\ 1 & 3 & -5 & | & 16 \\ 1 & -1 & 1 & | & 0 \end{bmatrix}$$

The reduced row-echelon form is

$$\begin{bmatrix} 1 & 0 & -\frac{1}{2} & | & 0 \\ 0 & 1 & -\frac{3}{2} & | & 0 \\ 0 & 0 & 0 & | & 1 \end{bmatrix}$$

The system is inconsistent. There is no solution.

37. The augmented matrix is

$$\begin{bmatrix} 4 & 10 & | & 27 \\ 3 & 1 & | & 4 \end{bmatrix}$$

The reduced row-echelon form is

$$\begin{bmatrix} 1 & 0 & | & \frac{1}{2} \\ 0 & 1 & | & \frac{5}{2} \end{bmatrix}$$

The solution is $\left(\frac{1}{2}, \frac{5}{2} \right)$.

39. We write the equations in standard form.

$x - y = 6$
$2x - 2y = 12$

The augmented matrix is

$$\begin{bmatrix} 1 & -1 & | & 6 \\ 2 & -2 & | & 12 \end{bmatrix}$$

The reduced row-echelon form is

$$\begin{bmatrix} 1 & -1 & | & 6 \\ 0 & 0 & | & 0 \end{bmatrix}$$

The system is dependent. There are infinite solutions of the form $\{(y + 6, y)\}$.

41. We write the equations in standard form.

$x - y = 2$
$x + 2y = 14$

The augmented matrix is

$$\begin{bmatrix} 1 & -1 & | & 2 \\ 1 & 2 & | & 14 \end{bmatrix}$$

The reduced row-echelon form is

$$\begin{bmatrix} 1 & 0 & | & 6 \\ 0 & 1 & | & 4 \end{bmatrix}$$

The solution is (6,4).

© Houghton Mifflin Company. All rights reserved.

43. The augmented matrix is

$$\begin{bmatrix} 1 & -1 & 3 & | & 15 \\ 1 & 1 & 1 & | & 5 \\ 2 & -2 & 1 & | & 10 \end{bmatrix}$$

The reduced row-echelon form is

$$\begin{bmatrix} 1 & 0 & 0 & | & 2 \\ 0 & 1 & 0 & | & -1 \\ 0 & 0 & 1 & | & 4 \end{bmatrix}$$

The solution is $(2, -1, 4)$.

45. The augmented matrix is

$$\begin{bmatrix} 3 & 4 & 1 & | & 4 \\ 2 & -5 & 6 & | & 0 \\ 1 & -14 & 11 & | & 0 \end{bmatrix}$$

The reduced row-echelon form is

$$\begin{bmatrix} 1 & 0 & \frac{29}{23} & | & 0 \\ 0 & 1 & -\frac{16}{23} & | & 0 \\ 0 & 0 & 0 & | & 1 \end{bmatrix}$$

The system is inconsistent. There is no solution.

47. The augmented matrix is

$$\begin{bmatrix} 6 & -3 & 3 & | & 12 \\ -4 & 2 & -2 & | & -8 \\ 2 & -1 & 1 & | & 4 \end{bmatrix}$$

The reduced row-echelon form is

$$\begin{bmatrix} 1 & -\frac{1}{2} & \frac{1}{2} & | & 2 \\ 0 & 0 & 0 & | & 0 \\ 0 & 0 & 0 & | & 0 \end{bmatrix}$$

The system is dependent. The first row represents

the equation $x - \frac{1}{2}y + \frac{1}{2}z = 2$ or $x = \frac{1}{2}y - \frac{1}{3}z + z$.

Thus the solution is $\left\{ \left(\frac{1}{2}y - \frac{1}{2}z + 2, \ y, \ z \right) \right\}$.

Geometrically, the planes coincide and the
solution set is $\{(x, y, z) \mid 2x - y + z = 4\}$.

49. The augmented matrix is

$$\begin{bmatrix} 0 & 1 & -2 & | & -2 \\ 1 & -1 & 1 & | & 1 \\ 4 & 6 & 2 & | & 11 \end{bmatrix}$$

The reduced row-echelon form is

$$\begin{bmatrix} 1 & 0 & 0 & | & \frac{1}{2} \\ 0 & 1 & 0 & | & 1 \\ 0 & 0 & 1 & | & \frac{3}{2} \end{bmatrix}$$

The solution is $\left(\frac{1}{2}, 1, \frac{3}{2} \right)$.

51. We write the equations in standard form.

$$\begin{array}{rcl} x \quad\quad + 3z &=& 9 \\ 2y - \ z &=& 0 \\ 2x - 6y + 9z &=& 18 \end{array}$$

The augmented matrix is

$$\begin{bmatrix} 1 & 0 & 3 & | & 9 \\ 0 & 2 & -1 & | & 0 \\ 2 & -6 & 9 & | & 18 \end{bmatrix}$$

The reduced row-echelon form is

$$\begin{bmatrix} 1 & 0 & 3 & | & 9 \\ 0 & 1 & -\frac{1}{2} & | & 0 \\ 0 & 0 & 0 & | & 0 \end{bmatrix}$$

The system is dependent. There are infinite

solutions of the form $\left\{ \left(-3z + 9, \ \frac{1}{2}z, \ z \right) \right\}$.

53. We write the equations in standard form.

$$\begin{array}{rcl} 2x + \ y - \ 4z &=& -1 \\ 2x - 6y + 12z &=& 1 \\ 4x + 5y - \ 2z &=& 10 \end{array}$$

The augmented matrix is

$$\begin{bmatrix} 2 & 1 & -4 & | & -1 \\ 2 & -6 & 12 & | & 1 \\ 4 & 5 & -2 & | & 10 \end{bmatrix}$$

The reduced row-echelon form is

$$\begin{bmatrix} 1 & 0 & 0 & | & \frac{1}{2} \\ 0 & 1 & 0 & | & 2 \\ 0 & 0 & 1 & | & 1 \end{bmatrix}$$

The solution is $\left(\frac{1}{2}, 2, 1 \right)$.

© Houghton Mifflin Company. All rights reserved.

55. The augmented matrix is

$$\begin{bmatrix} 2 & 1 & -1 & 1 & | & 2 \\ 1 & -1 & 1 & -2 & | & -1 \\ 3 & 2 & 1 & -1 & | & 0 \\ 1 & 3 & -2 & 4 & | & 4 \end{bmatrix}$$

The reduced row-echelon form is

$$\begin{bmatrix} 1 & 0 & 0 & 0 & | & 1 \\ 0 & 1 & 0 & 0 & | & -1 \\ 0 & 0 & 1 & 0 & | & 1 \\ 0 & 0 & 0 & 1 & | & 2 \end{bmatrix}$$

The solution is $(1, -1, 1, 2)$.

57. The augmented matrix is

$$\begin{bmatrix} 1 & 1 & 0 & 0 & | & 3 \\ 0 & 1 & -1 & 0 & | & 2 \\ 0 & 0 & 1 & -1 & | & -4 \\ 1 & 1 & 0 & 1 & | & 6 \end{bmatrix}$$

The reduced row-echelon form is

$$\begin{bmatrix} 1 & 0 & 0 & 0 & | & 2 \\ 0 & 1 & 0 & 0 & | & 1 \\ 0 & 0 & 1 & 0 & | & -1 \\ 0 & 0 & 0 & 1 & | & 3 \end{bmatrix}$$

The solution is $(2, 1, -1, 3)$.

59. Let x = value of coffee (in \$billions)
y = value of sugar (in \$billions)
z = value of wheat (in \$billions)

Then

$x + y + z = 3.107 \rightarrow x + y + z = 3.107$
$x = y + z + 1.433 \rightarrow x - y - z = 1.433$
$y = 2z - 0.021 \qquad y - 2z = -0.021$

The augmented matrix is

$$\begin{bmatrix} 1 & 1 & 1 & | & 3.107 \\ 1 & -1 & -1 & | & 1.433 \\ 0 & 1 & -2 & | & -0.021 \end{bmatrix}$$

The reduced row-echelon form is

$$\begin{bmatrix} 1 & 0 & 0 & | & 2.27 \\ 0 & 1 & 0 & | & 0.551 \\ 0 & 0 & 1 & | & 0.286 \end{bmatrix}$$

The solution is $(2.27, 0.551, 0.286)$. Thus \$2.27 billion of coffee, \$0.551 billion of sugar, and \$0.286 billion of wheat were imported.

61. a. Midwest: Using the data points $(0, 24)$ ands $(5, 23.5)$, we obtain the linear regression model $y = -0.1x + 24$.
West: Using the data points $(0, 21.2)$ and $(5, 21.9)$, we obtain the linear regression model $y = 0.14x + 21.2$.

b. We write the equations in standard form
$0.1x + y = 24$
$0.14x - y = -21.2$

The augmented matrix is

$$\begin{bmatrix} 0.1 & 1 & | & 24 \\ 0.14 & -1 & | & -21.2 \end{bmatrix}$$

The reduced row-echelon form is

$$\begin{bmatrix} 1 & 0 & | & 11.67 \\ 0 & 1 & | & 22.83 \end{bmatrix}$$

Thus in the year $1990 + 11$ or 2001, the population in the two regions will be the same.

63. a. Army: Using the data points $(2, 668)$, $(4, 658)$, $(6, 603)$, $(8, 480)$, and $(10, 422)$, we obtain the linear regression model $y = -33.5x + 767.2$.
Navy: Using the data points $(2, 510)$, $(4, 506)$, $(6, 495)$, $(8, 439)$, and $(10, 372)$, we obtain the linear regression model $y = -17.15x + 567.3$.

b. The solution $(12.23, 357.62)$ indicates that in 1997 the number of enlisted personnel in the Army and in the Navy will be the same: 357.62 thousand.

© Houghton Mifflin Company. All rights reserved.

65. a. Let x = amount of 20% solution
y = amount of 30% solution
z = amount of 50% solution
Then
$$x + \quad y + \quad z = 50$$
$$0.2x + 0.3y + 0.5z = 0.32 \cdot 50 = 16$$
The augmented matrix is
$$\begin{bmatrix} 1 & 1 & 1 & | & 50 \\ 0.2 & 0.3 & 0.5 & | & 16 \end{bmatrix}$$
The reduced row-echelon form is
$$\begin{bmatrix} 1 & 0 & -2 & | & -10 \\ 0 & 1 & 3 & | & 60 \end{bmatrix}$$
The system is dependent. The solutions are of the form $\{(2z - 10, -3z + 60, z)\}$.

b. The least amount is 5 liters. The most is 20 liters.

c. We add the following equation to the system in part (a).
$$y = x + 20 \quad \rightarrow \quad -x + y = 20$$
The new augmented matrix is
$$\begin{bmatrix} 1 & 1 & 1 & 50 \\ 0.2 & 0.3 & 0.5 & 16 \\ -1 & 1 & 0 & 20 \end{bmatrix}$$
The new reduced row-echelon form is
$$\begin{bmatrix} 1 & 0 & 0 & | & 10 \\ 0 & 1 & 0 & | & 30 \\ 0 & 0 & 1 & | & 10 \end{bmatrix}$$
The solution is (10, 30, 10). Thus the worker should use 10 liters of the 20% solution, 30 liters of the 30%, and 10 liters of the 50%.

67. We write the equations in standard form.
$$i_1 - i_2 + i_3 = 0$$
$$4i_1 + 2i_2 \qquad = 8$$
$$\qquad 2i_2 + 5i_3 = 9$$
The augmented matrix is
$$\begin{bmatrix} 1 & -1 & 1 & | & 0 \\ 4 & 2 & 0 & | & 8 \\ 0 & 2 & 5 & | & 9 \end{bmatrix}$$
The reduced row-echelon form is
$$\begin{bmatrix} 1 & 0 & 0 & | & 1 \\ 0 & 1 & 0 & | & 2 \\ 0 & 0 & 1 & | & 1 \end{bmatrix}$$
The solution is (1, 2, 1). Thus, the first current is 1 amp, the second is 2 amps, and the third is 1 amp.

5.4 Exercises

1. The corresponding entries of A and B are equal.

3. $a + 3 = 0 \qquad 3b = 5$
$\qquad a = -3 \qquad b = \dfrac{5}{3}$

5. $2a = 10 \quad b - 5 = -2 \quad c = 0$
$\qquad a = 5 \qquad b = 3$

7. $A + B = \begin{bmatrix} 1+2 & 2+0 \\ -1-1 & 5+3 \end{bmatrix} = \begin{bmatrix} 3 & 2 \\ -2 & 8 \end{bmatrix}$

9. $4B = \begin{bmatrix} 4 \cdot 2 & 4 \cdot 0 \\ 4(-1) & 4 \cdot 3 \end{bmatrix} = \begin{bmatrix} 8 & 0 \\ -4 & 12 \end{bmatrix}$

11. $IA = A = \begin{bmatrix} 1 & 2 \\ -1 & 5 \end{bmatrix}$

13. $AB = \begin{bmatrix} 1 \cdot 2 + 2(-1) & 1 \cdot 0 + 2 \cdot 3 \\ -1 \cdot 2 + 5(-1) & -1 \cdot 0 + 5 \cdot 3 \end{bmatrix} = \begin{bmatrix} 0 & 6 \\ -7 & 15 \end{bmatrix}$

© Houghton Mifflin Company. All rights reserved.

15. $C - D = \begin{bmatrix} 1-4 & 0-1 & 2-(-2) \\ 3-1 & -1-2 & 1-5 \\ 1-(-1) & 2-2 & 0-(-1) \end{bmatrix} = \begin{bmatrix} -3 & -1 & 4 \\ 2 & -3 & -4 \\ 2 & 0 & 1 \end{bmatrix}$

17. $2C + D = \begin{bmatrix} 2+4 & 0+1 & 4+(-2) \\ 6+1 & -2+2 & 2+5 \\ 2+(-1) & 4+2 & 0+(-1) \end{bmatrix} = \begin{bmatrix} 6 & 1 & 2 \\ 7 & 0 & 7 \\ 1 & 6 & -1 \end{bmatrix}$

19. $2I + C = \begin{bmatrix} 2+1 & 1+1 & 0+2 \\ 0+3 & 2-1 & 0+1 \\ 0+1 & 0+2 & 2+0 \end{bmatrix} = \begin{bmatrix} 3 & 0 & 2 \\ 3 & 1 & 1 \\ 1 & 2 & 2 \end{bmatrix}$

21. $CD = \begin{bmatrix} 1\cdot4+0\cdot1+2\cdot(-1) & 1\cdot1+0\cdot2+2\cdot2 & 1\cdot(-2)+0\cdot5+2\cdot(-1) \\ 3\cdot4+(01)\cdot1+1\cdot(-1) & 3\cdot1+(-1)\cdot2+1\cdot2 & 3\cdot(-2)+(-2)\cdot5+1\cdot(-1) \\ 1\cdot4+2\cdot1+0(-1) & 1\cdot1+2\cdot2+0\cdot2 & 1(-2)+2\cdot5+0(-1) \end{bmatrix} = \begin{bmatrix} 2 & 5 & -4 \\ 10 & 3 & -12 \\ 6 & 5 & 8 \end{bmatrix}$

23. **a.** $A + B = \begin{bmatrix} 3+1 & -2+0 \\ 1+2 & 4+(-1) \end{bmatrix} = \begin{bmatrix} 4 & -2 \\ 3 & 3 \end{bmatrix}$

b. $A - 2B = \begin{bmatrix} 3-2 & -2+0 \\ 1-4 & 4+2 \end{bmatrix} = \begin{bmatrix} 1 & -2 \\ -3 & 6 \end{bmatrix}$

c. $5A = \begin{bmatrix} 5\cdot3 & 5\cdot(-2) \\ 5\cdot1 & 5\cdot4 \end{bmatrix} = \begin{bmatrix} 15 & -10 \\ 5 & 20 \end{bmatrix}$

d. $AB = \begin{bmatrix} 3\cdot1+(-2)\cdot2 & 3\cdot0+(-2)(-1) \\ 1\cdot1+4\cdot2 & 1\cdot0+4\cdot(-1) \end{bmatrix} = \begin{bmatrix} -1 & 2 \\ 9 & -4 \end{bmatrix}$

e. $BA = \begin{bmatrix} 1\cdot3+0\cdot1 & 1\cdot(-2)+0\cdot4 \\ 2\cdot3+(-1)\cdot1 & 2\cdot(-2)+(-1)\cdot4 \end{bmatrix} = \begin{bmatrix} 3 & -2 \\ 5 & -8 \end{bmatrix}$

25. **a.** $A + B$ is not possible

b. $A - 2B$ is not possible

c. $5A = \begin{bmatrix} 5\cdot1 \\ 5\cdot3 \end{bmatrix} = \begin{bmatrix} 5 \\ 15 \end{bmatrix}$

d. $AB = \begin{bmatrix} 1\cdot2 & 1\cdot(-6) \\ 3\cdot2 & 3\cdot(-6) \end{bmatrix} = \begin{bmatrix} 2 & -6 \\ 6 & -18 \end{bmatrix}$

e. $BA = [2\cdot1+(-6)\cdot3] = [-16]$

27. **a.** The matrices must have the same dimensions, so $m = p$ and $n = q$.

b. To determine AB, the number of columns of A must be the same as the number of rows of B, so $n = p$.

29. $[2 \quad 3 \quad -1] \begin{bmatrix} 1 & 2 \\ -1 & 0 \\ 4 & -1 \end{bmatrix} = [2\cdot1+3\cdot(-1)+(-1)\cdot4 \quad 2\cdot2+3\cdot0+(-1)(-1)] = [-5 \quad 5]$

31. $\begin{bmatrix} 4 & -1 \\ -3 & 1 \\ 7 & 2 \end{bmatrix} \begin{bmatrix} 3 & 1 \\ -2 & 7 \end{bmatrix} = \begin{bmatrix} 4\cdot3+(-1)\cdot(-2) & 4\cdot1+(-1)\cdot7 \\ -3\cdot3+1\cdot(-2) & -3\cdot1+1\cdot7 \\ 7\cdot3+2(-2) & 7\cdot1+2\cdot7 \end{bmatrix} = \begin{bmatrix} 14 & -3 \\ -11 & 4 \\ 17 & 21 \end{bmatrix}$

© Houghton Mifflin Company. All rights reserved.

33. $\begin{bmatrix} 2 \\ 4 \end{bmatrix} \begin{bmatrix} 1 & 3 \\ 4 & 2 \\ -1 & 1 \end{bmatrix}$ not possible

35. $\begin{bmatrix} 2 & -1 \\ 3 & -2 \\ 1 & 0 \end{bmatrix} \begin{bmatrix} 4 & 1 & -2 \\ 2 & 5 & -3 \end{bmatrix} = \begin{bmatrix} 2\cdot4+(-1)\cdot2 & 2\cdot1+(-1)\cdot5 & 2\cdot(-2)+(-1)(-3) \\ 3\cdot4+(-2)\cdot2 & 3\cdot1+(-2)\cdot5 & 3\cdot(-2)+(-2)(-3) \\ 1\cdot4+0\cdot2 & 1\cdot1+0\cdot5 & 1\cdot(-2)+0\cdot(-3) \end{bmatrix} = \begin{bmatrix} 6 & -3 & -1 \\ 8 & -7 & 0 \\ 4 & 1 & -2 \end{bmatrix}$

37. $x + 3y = 10$
$2x - y = -1$

39. $2x - 3y + z = -7$
$x + y - 4z = 6$
$3x + 2y - z = 5$

41. $\begin{bmatrix} 3 & -1 \\ 4 & 5 \end{bmatrix} \begin{bmatrix} x \\ y \end{bmatrix} = \begin{bmatrix} 10 \\ 26 \end{bmatrix}$

43. $\begin{bmatrix} 0 & 1 & 3 \\ 1 & 0 & 1 \\ 1 & -2 & 0 \end{bmatrix} \begin{bmatrix} x \\ y \\ z \end{bmatrix} = \begin{bmatrix} -4 \\ -1 \\ -3 \end{bmatrix}$

45. Matrix I is a square matrix with 1's on the main diagonal and 0's elsewhere.

47. $\quad 2A + X = B$
$\quad\quad X = B - 2A$
$\quad\quad X = \begin{bmatrix} 2 & 0 & -5 \\ 1 & -3 & 1 \end{bmatrix} - \begin{bmatrix} 2 & 4 & 0 \\ -2 & 6 & -4 \end{bmatrix}$
$\quad\quad X = \begin{bmatrix} 0 & -4 & -5 \\ 3 & -9 & 5 \end{bmatrix}$

49. Because $AB = \begin{bmatrix} 1 & 1 \\ 1 & 1 \end{bmatrix} \begin{bmatrix} 1 & 1 \\ -1 & -1 \end{bmatrix} = \begin{bmatrix} 0 & 0 \\ 0 & 0 \end{bmatrix} = O$,
$AB = O$ does not imply that $A = O$ or $B = O$.

51. a. $A^2 - B^2 = \begin{bmatrix} -27 & -18 & 13 \\ 16 & -9 & -32 \\ 9 & 8 & -44 \end{bmatrix}$

b. $(A+B)(A-B) = \begin{bmatrix} -24 & -4 & -1 \\ 43 & 10 & -31 \\ 25 & 25 & -66 \end{bmatrix}$

53. a. $(A+B)^2 = \begin{bmatrix} 26 & 34 & 11 \\ -15 & -26 & 5 \\ 35 & 25 & 4 \end{bmatrix}$

b. $A^2 + 2AB + B^2 = \begin{bmatrix} 23 & 20 & 25 \\ -42 & -45 & 4 \\ 19 & 8 & 26 \end{bmatrix}$

55. a. $P = \begin{bmatrix} 38 & 15 \\ 46 & 28 \\ 20 & 42 \end{bmatrix}$

The rows represent the number of items made at each location. The columns represent the production capabilities at each plant location.

b. $5P$ would represent the weekly production capabilities.

c. $S = 0.20P = \begin{bmatrix} 7.6 & 3 \\ 9.2 & 5.6 \\ 4 & 8.4 \end{bmatrix}$

d. $P + S = \begin{bmatrix} 45.6 & 18 \\ 55.2 & 33.6 \\ 24 & 50.4 \end{bmatrix}$

The matrix represents the daily production capabilities after the equipment upgrade.

57. a. $C = [67 \quad 22 \quad 58]$

b. $CP = [4718 \quad 4057]$
The matrix represents the total cost at each plant.

c. $R = [84 \quad 35 \quad 74]$

d. $RP = [6282 \quad 5348]$
The matrix represents the total revenue at each plant.

e. $RP - CP = [1564 \quad 1291]$
The matrix represents the total profit at each plant.

f. $R - C = [17 \quad 13 \quad 16]$
The matrix represents the profit per item.

g. $(R - C)P = [1564 \quad 1291]$
The matrix represents the total profit at each plant.

© Houghton Mifflin Company. All rights reserved.

59. a.
$$C = \begin{bmatrix} 64 & 42 \\ 75 & 52 \\ 93 & 28 \\ 51 & 12 \end{bmatrix}$$

b.
$$\frac{1}{2}C = \begin{bmatrix} 32 & 21 \\ 37.5 & 26 \\ 46.5 & 14 \\ 25.5 & 6 \end{bmatrix}$$

c. The matrix $24C$ represents the daily traffic count.

d. The matrix $M(24C)$ represents the weekly traffic count.

61. a. $D = \begin{bmatrix} 0.75 & 0.34 & 15 \\ 1.25 & 0.68 & 22 \end{bmatrix}$

b.
$$CD = \begin{bmatrix} 100.5 & 50.32 & 1884 \\ 121.25 & 60.86 & 2269 \\ 104.75 & 50.66 & 2011 \\ 53.25 & 25.5 & 1029 \end{bmatrix}$$

The matrix represents the hourly toll, the hourly number of out-of-town vehicles, and the hourly pollution emitted.

63. a. $E = \begin{bmatrix} 258 & 180 & 146 \\ 135 & 110 & 94 \end{bmatrix}$

b. $F = [0.45 \quad 0.54]$

c. $FE = [189 \quad 140.4 \quad 116.46]$.

The matrix represents the estimated number of females in each of the courses.

65. a. $M = [0.55 \quad 0.46]$

b. The elements of ME represent the number of males in each course.

c. $FE + ME = [393 \quad 290 \quad 240]$
The matrix represents the total enrollment in each course.

5.5 Exercises

1. The matrix A is not a square matrix.

3. $\begin{bmatrix} 3 & 0 \\ 0 & 2 \end{bmatrix}\begin{bmatrix} \frac{1}{3} & 0 \\ 0 & \frac{1}{2} \end{bmatrix} = \begin{bmatrix} 1 & 0 \\ 0 & 1 \end{bmatrix}$ and

$\begin{bmatrix} \frac{1}{3} & 0 \\ 0 & \frac{1}{2} \end{bmatrix}\begin{bmatrix} 3 & 0 \\ 0 & 2 \end{bmatrix} = \begin{bmatrix} 1 & 0 \\ 0 & 1 \end{bmatrix}$
Thus, the matrices are inverses.

5. $\begin{bmatrix} 2 & 3 \\ 3 & 4 \end{bmatrix}\begin{bmatrix} -1 & -3 \\ 1 & 2 \end{bmatrix} = \begin{bmatrix} 1 & 0 \\ 1 & 1 \end{bmatrix}$
Thus, the matrices are not inverses.

7.
$$\begin{bmatrix} 1 & 2 \\ -4 & 3 \\ 2 & 5 \end{bmatrix}, \begin{bmatrix} 1 & \frac{1}{2} \\ -\frac{1}{4} & \frac{1}{3} \\ \frac{1}{2} & \frac{1}{5} \end{bmatrix}$$

are not inverses because they aren't square matrices.

9.
$$\begin{bmatrix} 1 & 3 & 2 \\ 0 & 1 & 3 \\ 0 & 0 & 1 \end{bmatrix}\begin{bmatrix} 1 & -3 & 7 \\ 0 & 1 & -3 \\ 0 & 0 & 1 \end{bmatrix} = \begin{bmatrix} 1 & 0 & 0 \\ 0 & 1 & 0 \\ 0 & 0 & 1 \end{bmatrix}$$

and
$$\begin{bmatrix} 1 & -3 & 7 \\ 0 & 1 & -3 \\ 0 & 0 & 1 \end{bmatrix}\begin{bmatrix} 1 & 3 & 2 \\ 0 & 1 & 3 \\ 0 & 0 & 1 \end{bmatrix} = \begin{bmatrix} 1 & 0 & 0 \\ 0 & 1 & 0 \\ 0 & 0 & 1 \end{bmatrix}$$
Thus the matrices are inverses.

11.
$$\begin{bmatrix} 3 & -7 & | & 1 & 0 \\ 2 & -5 & | & 0 & 1 \end{bmatrix}$$
$$R_1 \div 3 \to \begin{bmatrix} 1 & -\frac{7}{3} & | & \frac{1}{3} & 0 \\ 2 & -5 & | & 0 & 1 \end{bmatrix}$$
$$-2R_1 + R_2 \to \begin{bmatrix} 1 & -\frac{7}{3} & | & \frac{1}{3} & 0 \\ 0 & -\frac{1}{3} & | & -\frac{2}{3} & 1 \end{bmatrix}$$
$$R_2 \cdot (-3) \to \begin{bmatrix} 1 & -\frac{7}{3} & | & \frac{1}{3} & 0 \\ 0 & 1 & | & 2 & -3 \end{bmatrix}$$
$$\frac{7}{3}R_2 + R_1 \to \begin{bmatrix} 1 & 0 & | & 5 & -7 \\ 0 & 1 & | & 2 & -3 \end{bmatrix}$$

The inverse is $\begin{bmatrix} 5 & -7 \\ 2 & -3 \end{bmatrix}$.

13. $\begin{bmatrix} 3 & 2 \\ 5 & 4 \end{bmatrix}^{-1} = \begin{bmatrix} 2 & -1 \\ -\frac{5}{2} & \frac{3}{2} \end{bmatrix}$

15.
$$\begin{bmatrix} 15 & -12 & | & 1 & 0 \\ 10 & -8 & | & 0 & 1 \end{bmatrix}$$
$$R_1 \div 15 \to \begin{bmatrix} 1 & -\frac{4}{5} & | & \frac{1}{15} & 0 \\ 10 & -8 & | & 0 & 1 \end{bmatrix}$$
$$-10R_1 + R_2 \to \begin{bmatrix} 1 & -\frac{4}{5} & | & \frac{1}{15} & 0 \\ 0 & 0 & | & \frac{2}{3} & 1 \end{bmatrix}$$

The inverse does not exist.

© Houghton Mifflin Company. All rights reserved.

17.

$$\left[\begin{array}{ccc|ccc} 2 & 5 & 6 & 1 & 0 & 0 \\ 1 & -1 & 2 & 0 & 1 & 0 \\ 1 & 2 & 3 & 0 & 0 & 1 \end{array}\right]$$

$$R_1 \leftrightarrow R_2 \left[\begin{array}{ccc|ccc} 1 & -1 & 2 & 0 & 1 & 0 \\ 2 & 5 & 6 & 1 & 0 & 0 \\ 1 & 2 & 3 & 0 & 0 & 1 \end{array}\right]$$

$$\begin{array}{c} -2R_1 + R_2 \to \\ -R_1 + R_3 \to \end{array} \left[\begin{array}{ccc|ccc} 1 & -1 & 2 & 0 & 1 & 0 \\ 0 & 7 & 2 & 1 & -2 & 0 \\ 0 & 3 & 1 & 0 & -1 & 1 \end{array}\right]$$

$$R_2 \div 7 \to \left[\begin{array}{ccc|ccc} 1 & -1 & 2 & 0 & 1 & 0 \\ 0 & 1 & \frac{2}{7} & \frac{1}{7} & -\frac{2}{7} & 0 \\ 0 & 3 & 1 & 0 & -1 & 1 \end{array}\right]$$

$$\begin{array}{c} R_2 + R_1 \to \\ \\ -3R_2 + R_3 \to \end{array} \left[\begin{array}{ccc|ccc} 1 & 0 & \frac{16}{7} & \frac{1}{7} & \frac{5}{7} & 0 \\ 0 & 1 & \frac{2}{7} & \frac{1}{7} & -\frac{2}{7} & 0 \\ 0 & 0 & \frac{1}{7} & -\frac{3}{7} & -\frac{1}{7} & 1 \end{array}\right]$$

$$R_3 \cdot 7 \to \left[\begin{array}{ccc|ccc} 1 & 0 & \frac{16}{7} & \frac{1}{7} & \frac{5}{7} & 0 \\ 0 & 1 & \frac{2}{7} & \frac{1}{7} & -\frac{2}{7} & 0 \\ 0 & 0 & 1 & -3 & -1 & 7 \end{array}\right]$$

$$\begin{array}{c} -\frac{16}{7}R_3 + R_1 \to \\ -\frac{2}{7}R_3 + R_2 \to \end{array} \left[\begin{array}{ccc|ccc} 1 & 0 & 0 & 7 & 3 & -16 \\ 0 & 1 & 0 & 1 & 0 & -2 \\ 0 & 0 & 1 & -3 & -1 & 7 \end{array}\right]$$

The inverse is $\begin{bmatrix} 7 & 3 & -16 \\ 1 & 0 & -2 \\ -3 & -1 & 7 \end{bmatrix}$.

19.
$$\begin{bmatrix} 1 & 2 & 1 \\ 0 & 2 & -1 \\ 1 & 0 & 1 \end{bmatrix}^{-1} = \begin{bmatrix} -1 & 1 & 2 \\ \frac{1}{2} & 0 & -\frac{1}{2} \\ 1 & -1 & -1 \end{bmatrix}$$

21.
$$\begin{bmatrix} 4 & -1 & -3 \\ -1 & 2 & 4 \\ -6 & 5 & 11 \end{bmatrix}$$

is singular; the inverse does not exist.

23.
$$\begin{bmatrix} -1 & 1 & -2 \\ 1 & -2 & 0 \\ 2 & -1 & 1 \end{bmatrix}^{-1} = \begin{bmatrix} \frac{2}{5} & -\frac{1}{5} & \frac{4}{5} \\ \frac{1}{5} & -\frac{3}{5} & \frac{2}{5} \\ -\frac{3}{5} & -\frac{1}{5} & -\frac{1}{5} \end{bmatrix}$$

25. $\begin{bmatrix} 10 & -4 \\ 6 & -2 \end{bmatrix}^{-1} = \begin{bmatrix} -\frac{1}{2} & 1 \\ -\frac{3}{2} & \frac{5}{2} \end{bmatrix}$

27.
$$\begin{bmatrix} 1 & 0 & 0 \\ -2 & 1 & 0 \\ 4 & -2 & 1 \end{bmatrix}^{-1} = \begin{bmatrix} 1 & 0 & 0 \\ 2 & 1 & 0 \\ 0 & 2 & 1 \end{bmatrix}$$

29.
$$\begin{bmatrix} 1 & 2 & 3 & 4 \\ 1 & 0 & 1 & 1 \\ -1 & 2 & -2 & 0 \\ 2 & -1 & 3 & 2 \end{bmatrix}^{-1} = \begin{bmatrix} 0 & 4 & -1 & -2 \\ 1 & 6 & -3 & -5 \\ 1 & 4 & -3 & -4 \\ -1 & -7 & 4 & 6 \end{bmatrix}$$

31.
$$\begin{bmatrix} 2 & 1 & -3 \\ 1 & 0 & 1 \\ 1 & -1 & 6 \end{bmatrix}$$

is singular; the inverse does not exist.

33. We convert $[A \mid I]$ into $[I \mid A^{-1}]$.

35.
$$AX = B$$
$$\begin{bmatrix} 3 & 2 \\ 4 & 3 \end{bmatrix}\begin{bmatrix} x \\ y \end{bmatrix} = \begin{bmatrix} -5 \\ -6 \end{bmatrix}$$
$$X = A^{-1}B$$
$$\begin{bmatrix} x \\ y \end{bmatrix} = \begin{bmatrix} 3 & -2 \\ -4 & 3 \end{bmatrix}\begin{bmatrix} -5 \\ -6 \end{bmatrix} = \begin{bmatrix} -3 \\ 2 \end{bmatrix}$$

Thus the solution is $(-3, 2)$.

37.
$$AX = B$$
$$\begin{bmatrix} 2 & 3 \\ -6 & -10 \end{bmatrix}\begin{bmatrix} x \\ y \end{bmatrix} = \begin{bmatrix} -1 \\ 4 \end{bmatrix}$$
$$X = A^{-1}B$$
$$\begin{bmatrix} x \\ y \end{bmatrix} = \begin{bmatrix} 5 & \frac{3}{2} \\ -3 & -1 \end{bmatrix}\begin{bmatrix} -1 \\ 4 \end{bmatrix} = \begin{bmatrix} 1 \\ -1 \end{bmatrix}$$

Thus the solution is $(1, -1)$.

39.
$$AX = B$$
$$\begin{bmatrix} 1 & -1 & 1 \\ 2 & -1 & 0 \\ 0 & 2 & -2 \end{bmatrix}\begin{bmatrix} x \\ y \\ z \end{bmatrix} = \begin{bmatrix} -5 \\ 0 \\ 12 \end{bmatrix}$$
$$X = A^{-1}B$$
$$\begin{bmatrix} x \\ y \\ z \end{bmatrix} = \begin{bmatrix} 1 & 0 & \frac{1}{2} \\ 2 & -1 & 1 \\ 2 & -1 & \frac{1}{2} \end{bmatrix}\begin{bmatrix} -5 \\ 0 \\ 12 \end{bmatrix} = \begin{bmatrix} 1 \\ 2 \\ -4 \end{bmatrix}$$

Thus the solution is $(1, 2, -4)$.

© Houghton Mifflin Company. All rights reserved.

41.
$$AX = B$$
$$\begin{bmatrix} 2 & -1 & 0 \\ 1 & 3 & 2 \\ -2 & 1 & -1 \end{bmatrix} \begin{bmatrix} x \\ y \\ z \end{bmatrix} = \begin{bmatrix} -1 \\ 11 \\ -3 \end{bmatrix}$$
$$X = A^{-1}B$$
$$\begin{bmatrix} x \\ y \\ z \end{bmatrix} = \begin{bmatrix} \frac{5}{7} & \frac{1}{7} & \frac{2}{7} \\ \frac{3}{7} & \frac{2}{7} & \frac{4}{7} \\ -1 & 0 & -1 \end{bmatrix} \begin{bmatrix} -1 \\ 11 \\ -3 \end{bmatrix} = \begin{bmatrix} 0 \\ 1 \\ 4 \end{bmatrix}$$

Thus the solution is (0, 1, 4).

43.
$$AX = B$$
$$\begin{bmatrix} 2 & 4 & 6 \\ -2 & 2 & -1 \\ 1 & 0 & 3 \end{bmatrix} \begin{bmatrix} x \\ y \\ z \end{bmatrix} = \begin{bmatrix} 10 \\ -1 \\ 7 \end{bmatrix}$$
$$X = A^{-1}B$$
$$\begin{bmatrix} x \\ y \\ z \end{bmatrix} = \begin{bmatrix} \frac{3}{10} & -\frac{3}{5} & -\frac{4}{5} \\ \frac{1}{4} & 0 & -\frac{1}{2} \\ -\frac{1}{10} & \frac{1}{5} & \frac{3}{5} \end{bmatrix} \begin{bmatrix} 10 \\ -1 \\ 7 \end{bmatrix} = \begin{bmatrix} -2 \\ -1 \\ 3 \end{bmatrix}$$

Thus the solution is (−2, −1, 3).

45.
$$AX = B$$
$$\begin{bmatrix} -3 & 3 & 7 & 5 \\ 1 & -2 & -5 & -1 \\ -1 & 1 & 2 & 1 \\ 3 & -2 & -4 & -1 \end{bmatrix} \begin{bmatrix} x \\ y \\ z \\ w \end{bmatrix} = \begin{bmatrix} 24 \\ -7 \\ 6 \\ -11 \end{bmatrix}$$
$$X = A^{-1}B$$
$$\begin{bmatrix} x \\ y \\ z \\ w \end{bmatrix} = \begin{bmatrix} -\frac{1}{4} & -\frac{1}{4} & \frac{7}{4} & \frac{3}{4} \\ -\frac{3}{2} & \frac{1}{2} & \frac{19}{2} & \frac{3}{2} \\ \frac{1}{2} & -\frac{1}{2} & -\frac{7}{2} & -\frac{1}{2} \\ \frac{1}{4} & \frac{1}{4} & \frac{1}{4} & \frac{1}{4} \end{bmatrix} \begin{bmatrix} 24 \\ -7 \\ 6 \\ -11 \end{bmatrix} = \begin{bmatrix} -2 \\ 1 \\ 0 \\ 3 \end{bmatrix}$$

Thus the solution is (−2, 1, 0 3).

47. Matrix multiplication is not commutative. The product $A^{-1}AX$ results in X whereas AXA^{-1} does not.

49.
$$\begin{bmatrix} 1 & 0 & | & 1 & 0 \\ c & 1 & | & 0 & 1 \end{bmatrix}$$
$$-cR_1 + R_2 \rightarrow \begin{bmatrix} 1 & 0 & | & 1 & 0 \\ 0 & 1 & | & -c & 1 \end{bmatrix}$$

Thus, the inverse is $\begin{bmatrix} 1 & 0 \\ -c & 1 \end{bmatrix}$.

51. First note $(AB)^{-1}AB = I$, since $(AB)^{-1}$ is the inverse of AB.

We also have
$$\begin{aligned} (B^{-1}A^{-1})(AB) &= B^{-1}(A^{-1}A)B \\ &= B^{-1}IB \\ &= B^{-1}B \\ &= I \end{aligned}$$

Therefore $B^{-1}A^{-1}$ must also be the inverse of AB. Thus $(AB)^{-1} = B^{-1}A^{-1}$.

© Houghton Mifflin Company. All rights reserved.

53. Given $AB = AC$. If A is invertible then
$$A^{-1}AB = A^{-1}AC$$
$$B = C$$

55. a.
$$A^2 = \begin{bmatrix} 1 & 0 & 0 \\ 2 & 1 & 0 \\ 2 & 0 & 1 \end{bmatrix}$$

b.
$$A^3 = \begin{bmatrix} 1 & 0 & 0 \\ 3 & 1 & 0 \\ 3 & 0 & 1 \end{bmatrix}$$

c.
$$A^n = \begin{bmatrix} 1 & 0 & 0 \\ n & 1 & 0 \\ n & 0 & 1 \end{bmatrix}$$

57
$$\begin{bmatrix} M & E & E \\ T & - & M \\ E & - & A \\ T & - & J \\ O & E & S \end{bmatrix}$$

We use a 5x3 matrix M to write the message
     "MEET ME AT JOES."

$$MC = \begin{bmatrix} 13 & 5 & 5 \\ 20 & 0 & 13 \\ 5 & 0 & 1 \\ 20 & 0 & 10 \\ 15 & 5 & 19 \end{bmatrix} \begin{bmatrix} 3 & 2 & 1 \\ -1 & 2 & 0 \\ 2 & 0 & 1 \end{bmatrix} = \begin{bmatrix} 44 & 36 & 18 \\ 86 & 40 & 33 \\ 17 & 10 & 6 \\ 80 & 40 & 30 \\ 78 & 40 & 34 \end{bmatrix}$$

59. We must first find C^{-1}.
$$C^{-1} = \begin{bmatrix} 1 & -1 & 0 & 0 \\ -2 & 1 & -1 & 0 \\ 1 & 0 & -1 & 1 \\ -1 & 4 & -2 & 3 \end{bmatrix}$$

Our message will be found by multiplication.
$$\begin{bmatrix} -33 & -20 & 10 & -2 \\ 108 & 1 & -76 & 30 \\ -49 & -16 & 20 & -2 \\ -213 & -38 & 114 & -38 \end{bmatrix} \begin{bmatrix} 1 & -1 & 0 & 0 \\ -2 & 1 & -1 & 0 \\ 1 & 0 & -1 & 1 \\ -1 & 4 & -2 & 3 \end{bmatrix} = \begin{bmatrix} 19 & 5 & 14 & 4 \\ 0 & 13 & 15 & 14 \\ 5 & 25 & 0 & 14 \\ 15 & 23 & 0 & 0 \end{bmatrix}$$

Converting the numerals to letters, our message reads "SEND MONEY NOW."

© Houghton Mifflin Company. All rights reserved.

5.6 Exercises

1. The graph is the half plane above the line.

3. B

5. D

7..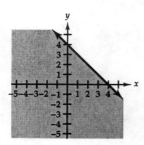

9.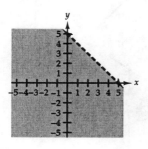

11.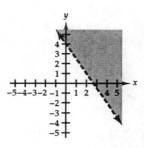

13.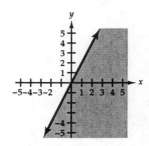

15.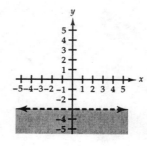

17.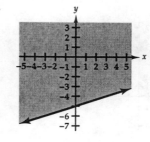

19.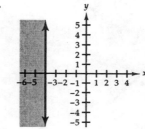

21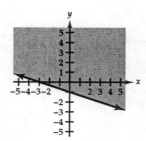

23. The graph of the one-variable inequality $x \leq -1$ is the interval $(-\infty, -1]$ on the number line. The graph of the two-variable inequality $x \leq -1$ is the line $x = -1$ and the half-plane to the left of the line.

25. D

27. B

© Houghton Mifflin Company. All rights reserved.

29.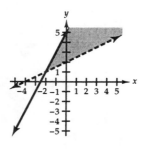

31.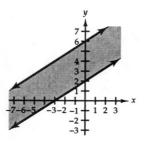

33.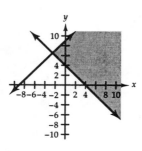

35.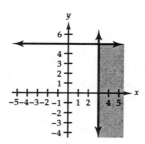

37.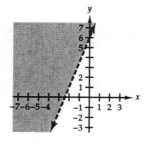

39.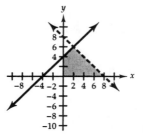

41.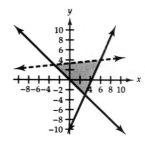

43.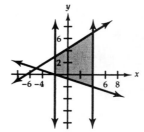

45. The feasible region is the graph of the system of inequalities.

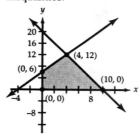

Vertex	$z = y - 2x$
$(0,0)$	0
$(0,6)$	6
$(4,12)$	4
$(10,0)$	-20

The maximum is 6. The minimum is –20.

© Houghton Mifflin Company. All rights reserved.

47. The feasible region is the graph of the system of inequalities.

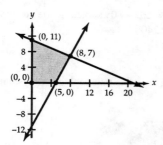

Vertex	$z = x + 4y$
$(0,0)$	0
$(0,11)$	44
$(8,7)$	36
$(5,0)$	5

The maximum is 44. The minimum is 0.

49. The feasible region is the graph of the system of inequalities.

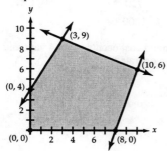

Vertex	$z = 3x + 4y$
$(0,0)$	0
$(0,4)$	16
$(3,9)$	45
$(10,6)$	54
$(8,0)$	24

The maximum is 54. The minimum is 0.

51. The feasible region is the graph of the system of inequalities.

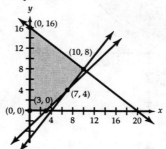

Vertex	$z = 2x + y$
$(0,0)$	0
$(0,16)$	16
$(10,8)$	28
$(7,4)$	18
$(3,0)$	6

The maximum is 28. The minimum is 0.

53.

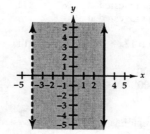

55.

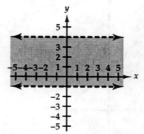

57.

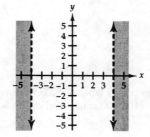

© Houghton Mifflin Company. All rights reserved.

59.

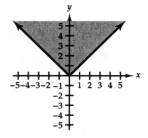

61.

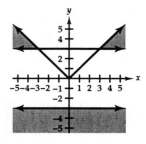

63. Let x = amount invested at 12%
$\quad\quad\ y$ = amount invested at 8%

Maximize $z = 0.12x + 0.08y$ subject to the constraints.
$$x + y \le 250{,}000$$
$$x \le 150{,}000$$
$$x \ge 0, y \ge 0.$$

The feasible region is the graph of the system of inequalities.

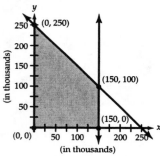

Vertex	$z = 0.12x + 0.08y$
$(0,0)$	0
$(0,250)$	20
$(150,100)$	26
$(150,0)$	18

The maximum occurs when $150,000 is invested in stocks and $100,000 is invested in bonds.

65. Let x = number of vans operated
$\quad\quad\ y$ = number of buses operated

Minimize $z = 60x + 80y$ subject to the constraints
$$12x + 32y \ge 240$$
$$x + y \ge 10$$
$$x \ge 0, y \ge 0.$$

The feasible region is the graph of the system of inequalities.

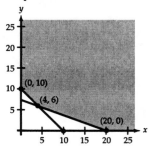

Vertex	$z = 60x + 80y$
$(0,10)$	800
$(4,6)$	720
$(20,0)$	1200

The minimum cost occurs when 4 vans and 6 buses are operated.

© Houghton Mifflin Company. All rights reserved.

67. Let x = number of birdhouses produced
y = number of mailboxes produced

Maximize $z = 18x + 30y$ subject to the constraints.
$$2x + y \leq 30$$
$$x + y \leq 18$$
$$x + 2y \leq 30$$
$$x \geq 0, \ y \geq 0$$

The feasible region is the graph of the system of inequalities.

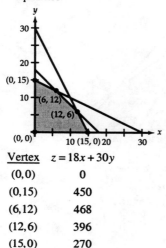

Vertex	$z = 18x + 30y$
$(0,0)$	0
$(0,15)$	450
$(6,12)$	468
$(12,6)$	396
$(15,0)$	270

The maximum profit occurs when 6 birdhouses and 12 mailboxes are produced.

69. Let x = number of acres of corn
y = number of acres of soybeans
Maximize $z = 230x + 210y$ subject to the constraints.
$$x + y \leq 100$$
$$80x + 50y \leq 7550$$
$$40x + 70y \leq 6250$$
$$x \geq 0, \ y \geq 0.$$
The feasible region is the graph of the system of inequalities.

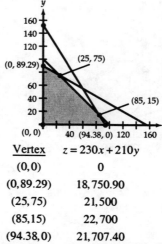

Vertex	$z = 230x + 210y$
$(0,0)$	0
$(0,89.29)$	18,750.90
$(25,75)$	21,500
$(85,15)$	22,700
$(94.38,0)$	21,707.40

The maximum profit is obtained when 85 acres of corn and 15 acres of soybeans are planted.

71. Let x = number of full-time employees
y = number of part-time employees

Minimize $z = 9x + 6y$ subject to the constraints.
$$x + y \geq 35$$
$$40x + 20y \geq 1000$$
$$y \leq 5x$$
$$x \geq 0, \ y \geq 0$$

The feasible region is the graph of the system of inequalities.

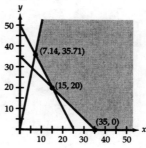

Vertex	$z = 9x + 6y$
$(7.14,35.71)$	278.52
$(15,20)$	255
$(35,0)$	315

The minimum labor cost is obtained when 15 full-time and 20 part-time employees are used.

© Houghton Mifflin Company. All rights reserved.

73. Let x = number of bleacher seats
y = number of reserved seats

Minimize $z = x + 1.5y$ subject to the constraints.
$$x + \quad y \geq 3000$$
$$\qquad y \leq 5x$$
$$5x + 15y \geq 21,000$$
$$x \geq 0, y \geq 0.$$

The feasible region is the graph of the system of inequalities.

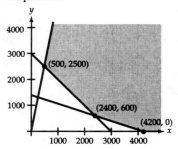

Vertex	$z = x + 1.5y$
$(500, 2500)$	4250
$(2400, 600)$	3300
$(4200, 0)$	4200

The minimum cost occurs when **2400 bleacher seats and 600 reserved seats** are used.

75. a. Let x = number of drivers processed
y = number of putters processed

Maximize $z = x + y$ subject to the constraints.
$$x + \quad y \geq 20$$
$$45x + 90y \leq 45 \cdot 60 = 2700$$
$$30x + 40y \leq 25 \cdot 60 = 1500$$
$$x \geq 0, y \geq 0$$

The feasible region is the graph of the system of inequalities.

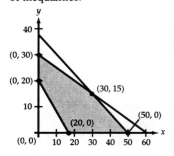

Vertex	$z = x + y$
$(0, 20)$	20
$(0, 30)$	30
$(30, 15)$	45
$(50, 0)$	50
$(20, 0)$	20

The maximum number of clubs that can be processed is 50.

b. Maximize $z = 5x + 15y$ subject to the constraints given in part (a). We obtain the same feasible region and the same vertices as before.

Vertex	$z = 5x + 15y$
$(0, 20)$	300
$(0, 30)$	450
$(30, 15)$	375
$(50, 0)$	250
$(20, 0)$	100

The maximum profit results when no drivers and 30 putters are processed.

© Houghton Mifflin Company. All rights reserved.

Review Exercises

1. The lines have the same slope but different *y*-intercepts, so they do not intersect. Thus the system has no solution.

3.

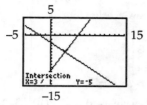

Replace *y* in the second equation with $2x - 11$.
$$x + y = -2$$
$$x + (2x - 11) = -2$$
$$3x = 9$$
$$x = 3$$

Replace *x* in the first equation with 3.
$$y = 2x - 11$$
$$y = 2(3) - 11 = -5$$

The solution is $(3, -5)$.

5. Multiply the second equation by 2 and add the equations.
$$\begin{array}{r} 3x + 2y = 0 \\ 8x - 2y = 44 \\ \hline 11x = 44 \\ x = 4 \end{array}$$

Replace *x* with 4 in the second equation and solve for *y*.
$$4(4) - y = 22$$
$$y = -6$$
The solution is $(4, -6)$.

7. Solve the first equation for *x*.
$$x + 2y = -6$$
$$x = -2y - 6$$

Replace *x* with $-2y - 6$ in the second equation.
$$\frac{y}{2} + \frac{-2y - 6}{4} = 1$$
$$2y - 2y - 6 = 4$$
$$-6 = 4$$
The system is inconsistent. There are no solutions.

9. Replace *x* with $\frac{y+1}{3}$ in the second equation.
$$3\left(\frac{y+1}{3}\right) - y - 1 = 0$$
$$0 = 0$$
The system is dependent. There are infinite solutions of the form $\left\{\left(\frac{1}{3}y + \frac{1}{3}, y\right)\right\}$.

11. Let *x* = amount of 15% cough suppressant
    y = amount of 40% cough suppressant

Then
$$x + y = 500$$
$$0.15x + 0.40y = 0.30 \cdot 500 = 150$$

Solve the first equation for *y*.
$$x + y = 500$$
$$y = 500 - x$$

Replace *y* with $500 - x$ in the second equation.
$$0.15x + 0.40(500 - x) = 150$$
$$-0.25x = -50$$
$$x = 200$$

Replace *x* with 200 in the first equation.
$$x + y = 500$$
$$200 + y = 500$$
$$y = 300$$

Use 200 milliliters of the 15% cough syrup and 300 milliliters of the 40% syrup.

13. The ordered triple corresponds to the point of intersection of three planes.

15. The row-echelon form is
$$x + 2y - z = 6$$
$$y - \frac{1}{2}z = 2$$
$$z = -4$$
The solution is $(2, 0, -4)$.

17. The row-echelon form is
$$x + \frac{1}{2}y - 2z = \frac{15}{2}$$
$$y - \frac{6}{5}z = \frac{17}{5}$$
$$0 = 0$$
The system is dependent. There are infinite solutions of the form $\left\{\left(\frac{7}{5}z + \frac{29}{5}, \frac{6}{5}z + \frac{17}{5}, z\right)\right\}$

19. We write the system in standard form.
$$\begin{array}{rcl} 11 + y = x + 3z & \rightarrow & x - y + 3z = 11 \\ 5z + 2x = 16 & \rightarrow & 2x + 5z = 16 \\ y = 4x - 14 & \rightarrow & 4x - y = 14 \end{array}$$

The row-echelon form is
$$x - \frac{1}{4}y = \frac{7}{2}$$
$$y - 4z = -10$$
$$z = 2$$
The solution is $(3, -2, 2)$.

21. Solve the second equation for *y*. Then substitute the value of *y* into the third equation and solve for *z*. Finally substitute the values of *y* and *z* into the first equation and solve for *x*.

© Houghton Mifflin Company. All rights reserved.

23. Let x = percentage of schools with enrollments of 0-299

y = percentage of schools with enrollments of 300-699

z = percentage of schools with enrollments over 699

Then

$$x + y + z = 100 \quad \rightarrow \quad x + y + z = 100$$
$$y = x + z - 4 \quad \rightarrow \quad x - y + z = 4$$
$$y + z = 2x + 10 \quad \rightarrow \quad -2x + y + z = 10$$

The row-echelon form is

$$x - \frac{1}{2}y - \frac{1}{2}z = -5$$
$$y + z = 70$$
$$z = 22$$

The solution is (30, 48, 22). 30% of schools fall in the 0-299 category, 48% fall in the 300-699 category, and 22% fall in the over 699 category.

25. a. The coefficient matrix is 2×2.

b. The augmented matrix is 2×3.

27. $2x - 3y = -7$
$-5x + y = -15$

29.
$$\begin{bmatrix} 1 & 3 & | & 1 \\ 2 & -1 & | & 9 \end{bmatrix}$$
$$-2R_1 + R_2 \rightarrow \begin{bmatrix} 1 & 3 & | & 1 \\ 0 & -7 & | & 7 \end{bmatrix}$$
$$R_2 \div (-7) \rightarrow \begin{bmatrix} 1 & 3 & | & 1 \\ 0 & 1 & | & -1 \end{bmatrix}$$
$$-3R_2 + R_1 \rightarrow \begin{bmatrix} 1 & 0 & | & 4 \\ 0 & 1 & | & -1 \end{bmatrix}$$

The solution is (4, −1).

31. The augmented matrix is

$$\begin{bmatrix} 2 & 7 & | & -4 \\ 5 & -6 & | & 37 \end{bmatrix}$$

The reduced row-echelon form is

$$\begin{bmatrix} 1 & 0 & | & 5 \\ 0 & 1 & | & -2 \end{bmatrix}$$

The solution is (5, −2).

33. The augmented matrix is

$$\begin{bmatrix} 3 & -1 & 6 & | & -6 \\ 2 & 3 & -5 & | & 16 \\ 4 & -2 & 1 & | & -3 \end{bmatrix}$$

The reduced row-echelon form is

$$\begin{bmatrix} 1 & 0 & 0 & | & 1 \\ 0 & 1 & 0 & | & 3 \\ 0 & 0 & 1 & | & -1 \end{bmatrix}$$

The solution is (1, 3, −1).

35. We use the data points (0, 30.5), (10, 23.0), and (34, 26.6).

$$E(t) = at^2 + bt + c$$
$$E(0) = c = 30.5$$
$$E(10) = 100a + 10b + c = 23.0$$
$$E(34) = 1156a + 34b + c = 26.6$$

The augmented matrix is

$$\begin{bmatrix} 0 & 0 & 1 & | & 30.5 \\ 100 & 10 & 1 & | & 23.0 \\ 1156 & 34 & 1 & | & 26.6 \end{bmatrix}$$

The reduced row-echelon form is

$$\begin{bmatrix} 1 & 0 & 0 & | & 0.026 \\ 0 & 1 & 0 & | & -1.015 \\ 0 & 0 & 1 & | & 30.5 \end{bmatrix}$$

The solution is (0.026, −1.015, 30.5). Thus the function is $E(t) = 0.02t^2 - 1.015t + 30.5$.

37. $a + 3 = 5 \qquad 9 = 1 - 2b \qquad 4c = -2 \qquad d = 0$
$\qquad a = 2 \qquad 2b = -8 \qquad c = -\frac{1}{2}$
$\qquad\qquad\qquad b = -4$

39. a.
$$A - B = \begin{bmatrix} 2-1 & -1-3 & 3-1 \\ 0-0 & 2-2 & 1-4 \\ 1-0 & -1-0 & 4-(-1) \end{bmatrix} = \begin{bmatrix} 1 & -4 & 2 \\ 0 & 0 & -3 \\ 1 & -1 & 5 \end{bmatrix}$$

b.
$$3B + 2I = \begin{bmatrix} 3+2 & 9+0 & 3+0 \\ 0+0 & 6+2 & 12+0 \\ 0+0 & 0+0 & -3+2 \end{bmatrix} = \begin{bmatrix} 5 & 9 & 3 \\ 0 & 8 & 12 \\ 0 & 0 & -1 \end{bmatrix}$$

© Houghton Mifflin Company. All rights reserved.

41. a. $A + B$ is not possible

b. $2A - B$ is not possible

c.
$$3A = \begin{bmatrix} 3 & -6 \\ 9 & -3 \\ 12 & 9 \end{bmatrix}$$

d.
$$AB = \begin{bmatrix} -3 & -2 & -1 \\ 1 & -1 & -3 \\ 10 & 3 & -4 \end{bmatrix}$$

e.
$$BA = \begin{bmatrix} -3 & -5 \\ 5 & -5 \end{bmatrix}$$

43. a.
$$A + B = \begin{bmatrix} 4 & 4 & -1 \\ 8 & -7 & 13 \end{bmatrix}$$

b.
$$2A - B = \begin{bmatrix} -1 & 2 & -2 \\ 4 & -8 & 5 \end{bmatrix}$$

c.
$$3A = \begin{bmatrix} 3 & 6 & -3 \\ 12 & -15 & 18 \end{bmatrix}$$

d. AB is not possible

e. BA is not possible

45. $2x + y = 1$
$3x + 4y = -1$

47.
$$P = \begin{bmatrix} 4 & 8 & 6 \\ 6 & 4 & 4 \\ 6 & 10 & 8 \\ 10 & 12 & 8 \end{bmatrix}$$

49. $\begin{bmatrix} 3 & 4 \\ -4 & -5 \end{bmatrix}\begin{bmatrix} -5 & -4 \\ 4 & 3 \end{bmatrix} = \begin{bmatrix} 1 & 0 \\ 0 & 1 \end{bmatrix}$ and

$\begin{bmatrix} -5 & -4 \\ 4 & 3 \end{bmatrix}\begin{bmatrix} 3 & 4 \\ -4 & -5 \end{bmatrix} = \begin{bmatrix} 1 & 0 \\ 0 & 1 \end{bmatrix}$

Thus the matrices are inverses.

51.
$$\begin{bmatrix} 3 & 2 & | & 1 & 0 \\ 5 & 3 & | & 0 & 1 \end{bmatrix}$$
$$R_1 \div 3 \rightarrow \begin{bmatrix} 1 & \frac{2}{3} & | & \frac{1}{3} & 0 \\ 5 & 3 & | & 0 & 1 \end{bmatrix}$$
$$-5R_1 + R_2 \rightarrow \begin{bmatrix} 1 & \frac{2}{3} & | & \frac{1}{3} & 0 \\ 0 & -\frac{1}{3} & | & -\frac{5}{3} & 1 \end{bmatrix}$$
$$R_2 \cdot (-3) \rightarrow \begin{bmatrix} 1 & \frac{2}{3} & | & \frac{1}{3} & 0 \\ 0 & 1 & | & 5 & -3 \end{bmatrix}$$
$$-\frac{2}{3}R_2 + R_1 \begin{bmatrix} 1 & 0 & | & -3 & 2 \\ 0 & 1 & | & 5 & -3 \end{bmatrix}$$

Thus $\begin{bmatrix} 3 & 2 \\ 5 & 3 \end{bmatrix}^{-1} = \begin{bmatrix} -3 & 2 \\ 5 & -3 \end{bmatrix}$.

53.
$$\begin{bmatrix} -1 & 2 & -1 \\ -2 & 2 & -1 \\ 1 & -1 & 1 \end{bmatrix}^{-1} = \begin{bmatrix} 1 & -1 & 0 \\ 1 & 0 & 1 \\ 0 & 1 & 2 \end{bmatrix}$$

55.
$$AX = B$$
$$\begin{bmatrix} 6 & 8 \\ 4 & 6 \end{bmatrix}\begin{bmatrix} x \\ y \end{bmatrix} = \begin{bmatrix} 5 \\ 3 \end{bmatrix}$$

$$X = A^{-1}B$$
$$\begin{bmatrix} x \\ y \end{bmatrix} = \begin{bmatrix} \frac{3}{2} & -2 \\ -1 & \frac{3}{2} \end{bmatrix}\begin{bmatrix} 5 \\ 3 \end{bmatrix} = \begin{bmatrix} \frac{3}{2} \\ -\frac{1}{2} \end{bmatrix}$$

Thus the solutions $\left(\dfrac{3}{2}, -\dfrac{1}{2}\right)$.

57.
$$AX = B$$
$$\begin{bmatrix} 3 & 2 & 1 \\ 1 & 1 & 2 \\ 1 & 2 & 1 \end{bmatrix}\begin{bmatrix} x \\ y \\ z \end{bmatrix} = \begin{bmatrix} 1 \\ 7 \\ 3 \end{bmatrix}$$

$$X = A^{-1}B$$
$$\begin{bmatrix} x \\ y \\ z \end{bmatrix} = \begin{bmatrix} \frac{1}{2} & 0 & -\frac{1}{2} \\ -\frac{1}{6} & -\frac{1}{3} & \frac{5}{6} \\ -\frac{1}{6} & \frac{2}{3} & -\frac{1}{6} \end{bmatrix}\begin{bmatrix} 1 \\ 7 \\ 3 \end{bmatrix} = \begin{bmatrix} -1 \\ 0 \\ 4 \end{bmatrix}$$

Thus the solution is $(-1, 0, 4)$.

© Houghton Mifflin Company. All rights reserved.

59.
$$AX = B$$

$$\begin{bmatrix} 2 & -3 & 4 & -5 \\ 1 & 0 & -3 & 3 \\ 7 & 1 & -2 & 0 \\ -1 & 2 & 0 & -1 \end{bmatrix} \begin{bmatrix} x \\ y \\ z \\ w \end{bmatrix} = \begin{bmatrix} 27 \\ -14 \\ 0 \\ -1 \end{bmatrix}$$

$$X = A^{-1}B$$

$$\begin{bmatrix} x \\ y \\ z \\ w \end{bmatrix} = \begin{bmatrix} -\frac{9}{86} & -\frac{11}{43} & \frac{15}{86} & -\frac{21}{86} \\ -\frac{25}{86} & -\frac{21}{43} & \frac{13}{86} & -\frac{1}{86} \\ -\frac{22}{43} & -\frac{49}{43} & \frac{8}{43} & -\frac{37}{43} \\ -\frac{41}{86} & -\frac{31}{43} & \frac{11}{86} & -\frac{67}{86} \end{bmatrix} \begin{bmatrix} 27 \\ -14 \\ 0 \\ -1 \end{bmatrix} = \begin{bmatrix} 1 \\ -1 \\ 3 \\ -2 \end{bmatrix}$$

61.
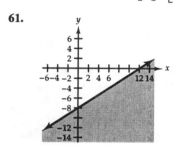

63.
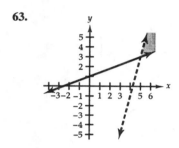

65.
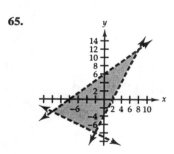

67. In (*ii*) the points of the line $y = x + 2$ correspond to solutions of the inequalities. In (*i*), the graphs of the two inequalities have no points in common.

69. The feasible region is the graph of the system of inequalities.

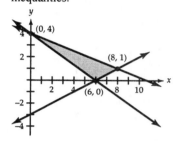

Vertex	$z = 3x + 7y$
$(0,4)$	28
$(8,1)$	31
$(6,0)$	18

The maximum is 31. The minimum is 18.

71. Let x = number of coach seats
y = number of first class seats
Maximize $z = 20x + 50y$ subject to the constraints.
$$x + y \le 250$$
$$3x + 8y \le 1040$$
$$x \ge 0, y \ge 0.$$
The feasible region is the graph of the system of inequalities.

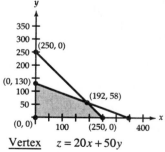

Vertex	$z = 20x + 50y$
$(0,0)$	0
$(0,130)$	6500
$(192,58)$	6740
$(250,0)$	5000

The maximum profit is obtained when the configuration is 192 coach and 58 first class seats.

© Houghton Mifflin Company. All rights reserved.

Chapter 6: Sequences, Series, Probability

6.1 Exercises

1. The absolute values of the terms are the same. The signs of some alternating terms differ.

3. The next two terms appear to be 20 and 22. Because the terms are even numbers, the general term should include the expression $2n$. Noting that each term is 10 more than $2n$, the pattern suggests that $a_n = 2n + 10$.

n	1	2	3	4	...
$2n$	2	4	6	8	...
$2n + 10$	12	14	16	18	...

5. The signs of the terms alternate and the pattern suggests the next two terms are 7 and −8. The absolute value of each term appears to be 2 greater than n. The pattern suggests that $a_n = (-1)^{n+1}(n+2)$.

n	1	2	3	4	...
$n + 2$	3	4	5	6	...
$(-1)^{n+1}(n+2)$	3	−4	5	−6	...

7. For each term, the denominator is n while the numerator is one greater. This suggests the next two terms are $\frac{6}{5}$ and $\frac{7}{6}$. Thus, $a_n = \frac{n+1}{n}$.

n	1	2	3	4
$n + 1$	2	3	4	5
$\frac{n+1}{n}$	2	$\frac{3}{2}$	$\frac{4}{3}$	$\frac{5}{4}$

9. Each term is of the form $1 + x$, where x is a fraction. The numerator and denominator of the fraction increase by one from term to term. The pattern suggests the next two terms are $1 + \frac{6}{7}$ and $1 + \frac{7}{8}$. The numerators are n greater than 1 and the denominators are $n + 1$ greater than 1. This pattern suggests $a_n = 1 + \frac{n+1}{n+2}$.

n	1	2	3	4	...
$\frac{n+1}{n+2}$	$\frac{2}{3}$	$\frac{3}{4}$	$\frac{4}{5}$	$\frac{5}{6}$	...
$1 + \frac{n+1}{n+2}$	$1 + \frac{2}{3}$	$1 + \frac{3}{4}$	$1 + \frac{4}{5}$	$1 + \frac{5}{6}$	...

11. a. $b_1 = 2^1 = 2 \qquad b_2 = 2^2 = 4$
$b_3 = 2^3 = 8 \qquad b_4 = 2^4 = 16$

b. $b_7 = 2^7 = 128$

c. $b_{12} = 2^{12} = 4096$

13. a. $c_1 = 1 - 3(1) = -2 \qquad c_2 = 1 - 3(2) = -5$
$c_3 = 1 - 3(3) = -8 \qquad c_4 = 1 - 3(4) = -11$

b. $c_7 = 1 - 3(7) = -20$

c. $c_{12} = 1 - 3(12) = -35$

15. a. $a_1 = \frac{(-1)^1 - 1}{1} = -2 \qquad a_2 = \frac{(-1)^2 - 1}{2} = 0$
$a_3 = \frac{(-1)^3 - 1}{3} = -\frac{2}{3} \qquad a_4 = \frac{(-1)^4 - 1}{4} = 0$

b. $a_7 = \frac{(-1)^7 - 1}{7} = -\frac{2}{7}$

c. $a_{12} = \frac{(-1)^{12} - 1}{12} = 0$

17. a. $b_1 = \frac{1^2}{1 + 4} = \frac{1}{5} \qquad b_2 = \frac{2^2}{2 + 4} = \frac{2}{3}$
$b_3 = \frac{3^2}{3 + 4} = \frac{9}{7} \qquad b_4 = \frac{4^2}{4 + 4} = 2$

b. $b_7 = \frac{7^2}{7 + 4} = \frac{49}{11}$

c. $b_{12} = \frac{12^2}{12 + 4} = 9$

19. a. $a_1 = \frac{1}{2} \qquad a_2 = \frac{2}{3}$
$a_3 = \frac{3}{4} \qquad a_4 = \frac{4}{5}$
$a_5 = \frac{5}{6}$

21. $b_1 = \frac{1 + (-1)^2}{1} = 2 \qquad b_2 = \frac{1 + (-1)^3}{2} = 0$
$b_3 = \frac{1 + (-1)^4}{3} = \frac{2}{3} \qquad b_4 = \frac{1 + (-1)^5}{4} = 0$
$b_5 = \frac{1 + (-1)^6}{5} = \frac{2}{5} \qquad b_6 = \frac{1 + (-1)^7}{6} = 0$
$b_7 = \frac{1 + (-1)^8}{7} = \frac{2}{7} \qquad b_8 = \frac{1 + (-1)^9}{8} = 0$

23. $c_1 = 2$, $c_2 = 2.25$, $c_3 = 2.3704$, $c_4 = 2.4414$

© Houghton Mifflin Company. All rights reserved.

25. $a_1 = 1$, $a_2 = 1.5874$, $a_3 = 2.0801$
$a_4 = 2.5198$

27. $b_1 = 0.4142$, $b_2 = 0.3178$, $b_3 = 0.2680$,
$b_4 = 0.2361$

29. The recursive definition includes the first term and a formula describing the nth term in relation to the preceding term.

31. a. $c_1 = 3$
$c_2 = c_1 + 5 = 3 + 5 = 8$
$c_3 = c_2 + 5 = 8 + 5 = 13$
$c_4 = c_3 + 5 = 13 + 5 = 18$

b.

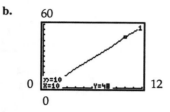

The graph indicates that $c_{10} = 48$.

33. a. $b_1 = 1$
$b_2 = 3$
$b_3 = b_2 - b_1 = 3 - 1 = 2$
$b_4 = b_3 - b_2 = 2 - 3 = -1$

b.

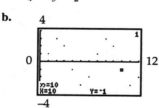

The graph indicates $b_{10} = -1$.

35. $a_1 = 1$, $a_2 = 1.4142$, $a_3 = 1.5538$,
$a_4 = 1.5981$

37. $a_1 = 2$, $a_2 = 3$, $a_3 = -0.1667$, $a_4 = -6.3333$

39. The fourth partial sum of a sequence is the sum of the first four terms of the sequence.

41. $S_1 = 2$
$S_2 = 2 + 4 = 6$
$S_3 = 6 + 6 = 12$
$S_4 = 12 + 8 = 20$
$S_5 = 20 + 10 = 30$
$S_6 = 30 + 12 = 42$
$S_7 = 42 + 14 = 56$
$S_8 = 56 + 16 = 72$
$S_9 = 72 + 18 = 90$

Thus, $S_4 = 20$ and $S_9 = 90$.

43. $S_1 = -1$
$S_2 = -1 + 4 = 3$
$S_3 = 3 + (-7) = -4$
$S_4 = -4 + 10 = 6$
$S_5 = 6 + (-13) = -7$
$S_6 = -7 + 16 = 9$
$S_7 = 9 + (-19) = -10$
$S_8 = -10 + 22 = 12$

Thus, $S_5 = -7$ and $S_8 = 12$

45. Given $S_n = \displaystyle\sum_{i=1}^{n} (1 + i^2)$, using the calculator we obtain $S_6 = 97$ and $S_{10} = 395$.

47. Given $S_n = \displaystyle\sum_{i=1}^{n} \frac{-i}{i+1}$, using the calculator we obtain $S_5 = -3.55$ and $S_9 = -7.07$.

49. First enter the sequence as $u(n) = 2u(n-1) + 1$ and enter -5 as $u(n\text{Min})$. Then, enter the command $\text{sum(seq}(u(n), n, 1, i, 1))$ for each partial sum S_i. Using this method, we obtain $S_5 = -129$ and $S_8 = -1028$.

51. Enter $u(n) = (n-1) / u(n-1)$ and $u(n\text{Min}) = 2$. Using the command $\text{sum (seq}(u(n), n, 1, i, 1))$ for each partial sum, we obtain $S_6 = 13.52$ and $S_9 = 28.33$.

53. Enter $u(n) = (u(n-1) \cdot u(n-2))$ and $u(n\text{Min}) = \{2, 1\}$. Using the command $\text{sum(seq}(u(n), n\ 1, i, 1))$ for each partial sum, we obtain $S_7 = 305$ and $S_{10} = 1.72 \cdot 10^{10}$.

55. $\displaystyle\sum_{i=1}^{5} (2i - 5) = (-3) + (-1) + 1 + 3 + 5 = 5$

57. $\displaystyle\sum_{i=1}^{4} (-2)^i = (-2) + 4 + (-8) + 16 = 10$

59. $\displaystyle\sum_{i=1}^{7} 3i = 3 + 6 + 9 + 12 + 15 + 18 + 21 = 84$

61. $\displaystyle\sum_{i=1}^{4} \left(1 - \frac{2}{i}\right) = (-1) + 0 + \frac{1}{3} + \frac{1}{2} = -\frac{1}{6}$

© Houghton Mifflin Company. All rights reserved.

63. The absolute values of the sequence terms begin with 3 and end with 8, increasing by one from term to term. The terms alternate in sign but are positive for odd terms and negative for even. Therefore,

$$3 - 4 + 5 - 6 + 7 - 8 = \sum_{i=1}^{6} (-1)^{i+1}(i+2).$$

65. The numerator is a constant 2, the denominator of the form $i + 2$, for $1 \le i \le 15$. Therefore

$$\frac{2}{1+2} + \frac{2}{2+2} + \frac{2}{3+2} + \frac{2}{4+2} + \ldots + \frac{2}{15+2}$$

$$= \sum_{i=1}^{15} \frac{2}{i+2}.$$

67. Each term is 3 greater than the previous one and is of the form $3i - 1$ for $1 \le i \le 10$. Thus,

$$2 + 5 + 8 + 11 + \ldots + 29 = \sum_{i=1}^{10} (3i - 1).$$

69. a. Each term is 2 less than the previous term. Since the sequence begins with 5, we must have $(-2)(1) + k = 5$, or $k = 7$. Therefore, $a_n = -2n + 7$.

b. From part (a), we know $a_1 = 5$ and $a_n = a_{n-1} - 2$.

71. a. Each term is ± 7; the odd terms are positive and even terms are negative. Thus,
$$a_n = (-1)^{n+1} \cdot 7.$$

b. From part (a), $a_1 = 7$ and $a_n = -1 \cdot a_{n-1}$.

73. Let the sequence of odd whole numbers be $a_1 = 1, 3, 5, 7, \ldots, 2n-1, \ldots$. Therefore, the sequence of partial sums S_n is

$S_1 = 1$
$S_2 = 1 + 3 = 4$
$S_3 = 4 + 5 = 9$
$S_4 = 9 + 7 = 16$
$S_5 = 16 + 9 = 25$

It appears that $S_n = n^2$.

75. $a_1 = 1$
$a_2 = 1$
$a_3 = 1 + 1 = 2$
$a_4 = 2 + 1 = 3$
$a_5 = 3 + 2 = 5$
Therefore, $a_{n+2} = a_{n+1} + a_n$.

77.

Number of Months	Start	1	2	3	4	5	6	7	8	9	10	11
Number of Pairs	1	1	2	3	5	8	13	21	34	55	89	144

79. FMFFMFMFFMFFM

81. The sequence in Exercise 80 is the Fibonacci sequence.

83. $a_{n+2} = a_{n+1} + a_n$. This sequence is the same as the Fibonacci sequence except for the first term.

6.2 Exercises

1. To find the common difference, determine the difference between any term and the preceding term.

3. a. $a_2 - a_1 = -2 - (-5) = 3$
$a_3 - a_2 = 1 - (-2) = 3$
$a_4 - a_3 = 4 - 1 = 3$
Thus the sequence is an arithmetic sequence with $a_1 = -5$ and $d = 3$.

b. $a_n = a_1 + (n-1)d$
$a_n = -5 + (n-1)(3)$
$\quad = -5 + 3n - 3 = -8 + 3n$
Thus $a_n = 3n - 8$.

5. a. $a_2 - a_1 = 1 - 3 = -2$
$a_3 - a_2 = -1 - 1 = -2$
$a_4 - a_3 = -3 - (-1) = -2$
Thus the sequence is an arithmetic sequence with $a_1 = 3$ and $d = -2$.

b. $a_n = a_1 + (n-1)d$
$a_n = 3 + (n-1)(-2)$
$\quad = 3 - 2n + 2 = 5 - 2n$
Thus, $a_n = -2n + 5$.

7. $a_2 - a_1 = 2 - 0 = 2$
$a_3 - a_2 = 6 - 2 = 4$
The sequence is not an arithmetic sequence.

9. a. $a_2 - a_1 = 0.6 - 0.2 = 0.4$
$a_3 - a_2 = 1.0 - 0.6 = 0.4$
$a_4 - a_3 = 1.4 - 1.0 = 0.4$
Thus the sequence is an arithmetic sequence with $a_1 = 0.2$ and $d = 0.4$.

b. $a_n = a_1 + (n-1)d$
$a_n = 0.2 + (n-1)(0.4)$
$\quad = 0.2 + 0.4n - 0.4 = -0.2 + 0.4n$
Thus $a_n = 0.4n - 0.2$.

11. a. $c_1 = -3$
$c_2 = c_1 + d = -3 + 7 = 4$
$c_3 = c_2 + d = 4 + 7 = 11$
$c_4 = c_3 + d = 11 + 7 = 18$

b $c_n = c_1 + (n-1)d$
$c_n = -3 + (n-1)(7)$
$\quad = -3 + 7n - 7 = 7n - 10$

c. $c_{10} = 7(10) - 10 = 70 - 10 = 60$

© Houghton Mifflin Company. All rights reserved.

13. a. $a_1 = 18$

$a_2 = a_1 + d = 18 + (-6) = 12$
$a_3 = a_2 + d = 12 + (-6) = 6$
$a_4 = a_3 + d = 6 + (-6) = 0$

b. $a_n = a_1 + (n-1)d$
$a_n = 18 + (n-1)(-6)$
$\quad = 18 - 6n + 6 = -6n + 24$

c. $a_{10} = -6(10) + 24 = -36$

15. a. $k_n = k_1 + (n-1)d$

$k_5 = k_1 + 4d = k_1 + 4(-4) = k_1 - 16$
$24 = k_1 - 16$
$k_1 = 40$
$k_2 = k_1 + d = 40 + (-4) = 36$
$k_3 = k_2 + d = 36 + (-4) = 32$
$k_4 = k_3 + d = 32 + (-4) = 28$

b. $k_n = k_1 + (n-1)d$
$k_n = 40 + (n-1)(-4)$
$\quad = 40 - 4n + 4 = -4n + 44$

c. $k_{10} = -4(10) + 44 = 4$

17. a. $c_n = c_1 + (n-1)d$

$c_4 = c_1 + 3d = c_1 + 3(0.3) = c_1 + 0.9$
$-3 = c_1 + 0.9$
$c_1 = -3.9$
$c_2 = c_1 + d = -3.9 + 0.3 = -3.6$
$c_3 = c_2 + d = -3.6 + 0.3 = -3.3$
$c_4 = c_3 + d = -3.3 + 0.3 = -3$

b. $c_n = c_1 + (n-1)d$
$c_n = -3.9 + (n-1)(0.3)$
$\quad = -3.9 + 0.3n - 0.3$
$\quad = 0.3n - 4.2$

c. $c_{10} = 0.3(10) - 4.2 = -1.2$

19. The graph of the equation is a solid line. The graph of the sequence contains only those points with a positive integer as a first coordinate.

21. $c_n = c_1 + (n-1)d$

$c_6 = c_1 + 5d$
$-11 = 4 + 5d$
$5d = -15$
$d = -3$
$c_n = 4 + (n-1)(-3) = 4 - 3n + 3 = -3n + 7$

23. $a_n = a_1 + (n-1)d$

$a_3 = a_1 + 2d$ and $a_{11} = a_1 + 10d$
$1 = a_1 + 2d$ and $3 = a_1 + 10d$

Multiply the first equation by -1 and add to the second equation.

$-1 = -a_1 - 2d$
$\underline{\ 3 = a_1 + 10d\ }$
$\ 2 = \quad\ 8d$
$d = \dfrac{1}{4}$

Now solve for a_1.

$a_3 = a_1 + 2d$
$1 = a_1 + 2\left(\dfrac{1}{4}\right)$
$a_1 = \dfrac{1}{2}$

Thus, $a_n = \dfrac{1}{2} + (n-1)\left(\dfrac{1}{4}\right)$
$\qquad = \dfrac{1}{2} + \dfrac{1}{4}n - \dfrac{1}{4} = \dfrac{1}{4}n + \dfrac{1}{4}$

25. $t_n = t_1 + (n-1)d$

$t_4 = t_1 + 3d$ and $t_8 = t_1 + 7d$
$6 = t_1 + 3d$ and $10.8 = t_1 + 7d$

Multiply the first equation by -1 and add to the second equation.

$-6 = t_1 - 3d$
$\underline{\ 10.8 = t_1 + 7d\ }$
$4.8 = \quad\ 4d$
$d = 1.2$

Now solve for t_1.

$t_4 = t_1 + 3d$
$6 = t_1 + 3(1.2)$
$t_1 = 2.4$
Thus, $t_n = 2.4 + (n-1)(1.2) = 1.2n + 1.2$.

27. $d = a_2 - a_1 = -3 - (-10) = 7$
$d = a_3 - a_2 = 4 - (-3) = 7$
$a_n = a_1 + (n-1)d = -10 + (n-1)(7) = 7n - 17$
$158 = 7n - 17$
$7n = 175$
$n = 25$

29. $d = a_2 - a_1 = \dfrac{7}{3} - \dfrac{5}{3} = \dfrac{2}{3}$
$d = a_3 - a_2 = 3 - \dfrac{7}{3} = \dfrac{2}{3}$
$a_n = a_1 + (n-1)d = \dfrac{5}{3} + (n-1)\left(\dfrac{2}{3}\right) = \dfrac{2}{3}n + 1$
$99 = \dfrac{2}{3}n + 1$
$\dfrac{2}{3}n = 98$
$n = 147$

© Houghton Mifflin Company. All rights reserved.

31. $a_1 = 9; d = a_2 - a_1 = 13 - 9 = 4$

To find the sum, we must find a_{24}.
$$a_n = a_1 + (n-1)d$$
$$a_{24} = 9 + (23)(4) = 101$$

Now determine the sum.
$$S_n = \frac{n(a_1 + a_2)}{2}$$
$$S_{24} = \frac{24(9 + 101)}{2} = 1320$$

33. $a_1 = 9; d = a_2 - a_1 = 2 - 9 = -7$

To find the sum, we must find a_{30}.
$$a_n = a_1 + (n-1)d$$
$$a_{30} = 9 + (29)(-7) = -194$$

Now determine the sum.
$$S_n = \frac{n(a_1 + a_n)}{2}$$
$$S_{30} = \frac{30(9 - 194)}{2} = -2775$$

35. $a_1 = \frac{3}{2}; d = a_2 - a_1 = 2 - \frac{3}{2} = \frac{1}{2}$

To find the sum, we must find a_{41}.
$$a_n = a_1 + (n-1)d$$
$$a_{41} = \frac{3}{2} + (40)\left(\frac{1}{2}\right) = \frac{43}{2}$$

Now determine the sum.
$$S_n = \frac{n(a_1 + a_n)}{2}$$
$$S_{41} = \frac{41\left(\frac{3}{2} + \frac{43}{2}\right)}{2} = 471.5$$

37. Part (iii) does not contain enough information to determine a_9, so we cannot determine the ninth partial sum.

39. $a_1 = -1; d = 1 - (-1) = 2$
To find the sum, we must find n.
$$a_n = a_1 + (n-1)d$$
$$45 = -1 + (n-1)(2) = 2n - 3$$
$$2n = 48$$
$$n = 24$$

Now determine the sum.
$$S_n = \frac{n(a_1 + a_n)}{2}$$
$$S_{24} = \frac{24(-1) + 45)}{2} = 528$$

41. $a_1 = \frac{1}{6}; d = a_2 - a_1 = -\frac{1}{6} - \frac{1}{6} = -\frac{1}{3}$

To find the sum, we must find n.
$$a_n = a_1 + (n-1)d$$
$$-\frac{13}{2} = \frac{1}{6} + (n-1)\left(-\frac{1}{3}\right)$$
$$= -\frac{1}{3}n + \frac{1}{2}$$
$$-\frac{1}{3}n = -7$$
$$n = 21$$

Now determine the sum.
$$S_n = \frac{n(a_1 + a_n)}{2}$$
$$S_{21} = \frac{21\left(\frac{1}{6} - \frac{13}{2}\right)}{2} = -66.5$$

43. $\displaystyle\sum_{i=1}^{24} (8 - 5i)$

The series has $n = 24$ terms with $a_1 = 8 - 5(1) = 3$
and $a_{24} = 8 - 5(24) = -112$.
$$S_n = \frac{n(a_1 + a_n)}{2} = \frac{24(3 - 112)}{2} = -1308$$

45. $\displaystyle\sum_{k=1}^{30} \left(\frac{1}{2}k + 4\right)$

The series has $n = 30$ terms with
$a_1 = \frac{1}{2}(1) + 4 = \frac{9}{2}$ and $a_{30} = \frac{1}{2}(30) + 4 = 19$.
$$S_n = \frac{n(a_1 + a_n)}{2} = \frac{30\left(\frac{9}{2} + 19\right)}{2} = 352.5$$

47. $\displaystyle\sum_{i=1}^{20} \frac{3(1 + 2i)}{5}$

The series has $n = 20$ terms with
$a_1 = \frac{3(1 + 2(1))}{5} = \frac{9}{5}$ and $a_{20} = \frac{3(1 + 2(20))}{5} = \frac{123}{5}$.
$$S_n = \frac{n(a_1 + a_n)}{2} = \frac{20\left(\frac{9}{5} + \frac{123}{5}\right)}{2} = 264$$

49. $\displaystyle\sum_{n=5}^{30} (n - 4)$

The series has $n = 30 - 5 + 1 = 26$ terms with
$a_1 = 5 - 4 = 1$ and $a_{26} = 30 - 4 = 26$.
$$S_n = \frac{n(a_1 + a_n)}{2} = \frac{26(1 + 26)}{2} = 351$$

© Houghton Mifflin Company. All rights reserved.

51. $\displaystyle\sum_{i=12}^{26} 4i$

The series has $n = 26 - 12 + 1 = 15$ terms with $a_1 = 4(12) = 48$ and $a_{15} = 4(26) = 104$.

$$S_n = \frac{n(a_1 + a_n)}{2} = \frac{15(48 + 104)}{2} = 1140$$

53. $k_1 = 14; k_2 = k_1 + 6$, so $d = 6$
$k_n = k_1 + (n-1)d = 14 + (n-1)(6) = 6n + 8$
$k_{50} = 6(50) + 8 = 308$

55. $c_1 = -7; c_2 = -7 - 3.2$, so $d = -3.2$
$\begin{aligned} c_n &= c_1 + (n-1)d = -7 + (n-1)(-3.2) \\ &= -3.2n - 3.8 \end{aligned}$
$c_{50} = -3.2(50) - 3.8 = -163.8$

57. $2 + 4 + 6 + \ldots + 2n$
There are n terms with $a_1 = 2$ and $a_n = 2n$

$$\begin{aligned} S_n &= \frac{n(a_1 + a_n)}{2} = \frac{n(2 + 2n)}{2} \\ &= \frac{n \cdot 2(1+n)}{2} = n(1+n) \end{aligned}$$

59. The sequence $-2, 1, 4, 7, \ldots$ has $a_1 = -2$ and $d = 3$. Therefore,
$a_n = a_1 + (n-1)d = -2 + (n-1)(3) = 3n - 5$.
If 2400 is a term of the sequence then, for some n, we have:
$$\begin{aligned} 2400 &= 3n - 5 \\ 3n &= 2405 \\ n &= \frac{2405}{3} \end{aligned}$$

But n must be an integer and $\dfrac{2405}{3} = 801.\overline{6}$. Thus, 2400 is not a term of the sequence. The first three terms of the sequence which exceed 2400 are
$a_{802} = 3(802) - 5 = 2401$,
$a_{803} = 3(803) - 5 = 2404$, and
$a_{804} = 3(804) - 5 = 2407$.

61. a. We have the arithmetic sequence with $a_1 = 12$ and $d = 6$.
$\begin{aligned} a_n &= a_1 + (n-1)d \\ &= 12 + (n-1)(6) = 6n + 6 \end{aligned}$
Thus, $a_n = 6n + 6, 1 \le n \le 18$.

b. We need to determine the sum S_{18}. To do so, we must find a_{18}.
$a_{18} = 12 + 17(6) = 114$
$\begin{aligned} S_{18} &= \frac{n(a_1 + a_n)}{2} \\ &= \frac{18(12 + 114)}{2} = 1134 \end{aligned}$
Each tray holds 6 plants so the gardener should purchase $\dfrac{1134}{6}$ or 189 trays.

63. From 12:15 P.M. to 1:15 P.M. on Thursday, the clock will strike 2 times – once at 12:30 and once at 1. The next hour, it will strike 3 times, and so on, until it will strike 13 times between 11:15 P.M. on Thursday and 12:15 A.M. on Friday.

This defines an arithmetic sequence with $a_1 = 2$, and $d = 1$, or
$a_n = 2 + (n-1)(1) = n + 1, 1 \le n \le 12$. The sum is
$S_{12} = \dfrac{12(2 + 13)}{2} = 90$. From 12:15 A.M. on Friday to 12:15 P.M. Friday, the clock will strike another 90 times for a total of 180.

65. a. We have the arithmetic sequence with $a_1 = 78$ and $d = -4$.
$a_n = a_1 + (n - 1)d = 78 + (n-1)(-4) = -4n + 82$

b. We need to determine S_{15}. To do so, we must find a_{15}.
$a_{15} = -4(15) + 82 = 22$
$S_{15} = \dfrac{n(a_1 + a_n)}{2} = \dfrac{15(78 + 22)}{2} = 750$

If each student volunteer provided 3 hours of tutoring per week, there were $3 \cdot 750$ or 2250 student-hours of tutoring.

67. a. We have the arithmetic sequence with $a_1 = 80.05$ (1 minute, 20.05 seconds) and $d = -0.08$.

$\begin{aligned} a_n &= a_1 + (n-1)d \\ &= 80.05 + (n-1)(-0.08) = -0.08n + 80.13 \end{aligned}$

b. We need to determine S_8. To do so, we must find a_8.
$a_8 = -0.08(8) + 80.13 = 79.49$
$\begin{aligned} S_8 &= \frac{n(a_1 + a_n)}{2} \\ &= \frac{8(80.05 + 79.49)}{2} = 638.16 \end{aligned}$

Her total time was 638.16 seconds or 10 minutes, 38.16 seconds.

© Houghton Mifflin Company. All rights reserved.

69. a. We use the data points (1, 72,74), (2, 68.68), (3, 61.48), (4, 56.21), and (5, 51.00) and obtain the linear regression model $b_n = -5.6n + 78.8$. b_{11} represents the average bill in the year 1990 + 11 or 2001.

b. For this sequence, $b_1 = -5.6(1) + 78.8 = 73.2$ and $b_5 = -5.6(5) + 78.8 = 50.8$. These are average monthly bills so we multiply each by 12 for an average yearly rate and find S_5.

$$S_5 = \frac{n(12b_1 + 12b_n)}{2}$$
$$= \frac{5 \cdot 12(73.2 + 50.8)}{2} = 3720$$

The total expenditure for five years is $3720.

c. For the years 1994–2000, we will have
$$12 \cdot \sum_{i=4}^{10}(-5.6i + 78.8)$$
$$= 12(177.20) = 3721.32.$$

6.3 Exercises

1. Determine the common ratio by dividing a term by the preceding term.

3. a. Geometric? $\frac{a_2}{a_1} = \frac{-6}{3} = -2$
$\frac{a_3}{a_2} = \frac{12}{-6} = -2$
$\frac{a_4}{a_3} = \frac{-24}{12} = -2$
The sequence is geometric.

b. Geometric? $\frac{a_2}{a_1} = \frac{2}{1} = 2$
$\frac{a_3}{a_2} = \frac{6}{2} = 3 \neq 2$
The sequence is not geometric.

Arithmetic? $a_2 - a_1 = 2 - 1 = 1$
$a_3 - a_2 = 6 - 2 = 4 \neq 1$
The sequence is not arithmetic.
It is neither.

c. Geometric? $\frac{a_2}{a_1} = \frac{-5}{-9} = \frac{5}{9}$
$\frac{a_3}{a_2} = \frac{-1}{-5} = \frac{1}{5} \neq \frac{5}{9}$
The sequence is not geometric.

Arithmetic? $a_2 - a_1 = -5 - (-9) = 4$
$a_3 - a_2 = -1 - (-5) = 4$
$a_4 = a_3 = 3 - (-1) = 4$
The sequence is arithmetic.

5. a. Write the first few terms of the sequence: $-1, -3, -7, -15, \dots$.

Geometric? $\frac{a_2}{a_1} = \frac{-3}{-1} = 3$
$\frac{a_3}{a_2} = \frac{-7}{-3} = \frac{7}{3} \neq 3$
The sequence is not geometric.
Arithmetic? $a_2 - a_1 = -3 - (-1) = -2$
$a_3 - a_2 = -7 - (-3) = -4 \neq -2$
The sequence is not arithmetic.
It is neither.

b. Write the first few terms of the sequence: $-\frac{2}{5}, 2, -10, 50, \dots$.

Geometric? $\frac{c_2}{c_1} = \frac{2}{-\frac{2}{5}} = -5$
$\frac{c_3}{c_2} = \frac{-10}{2} = -5$
$\frac{c_4}{c_3} = \frac{50}{-10} = -5$
The sequence is geometric.

c. Write the first few terms of the sequence: $\frac{2}{5}, \frac{1}{5}, 0, -\frac{1}{5}, \dots$.

Geometric? $\frac{k_2}{k_1} = \frac{\frac{1}{5}}{\frac{2}{5}} = \frac{1}{2}$
$\frac{k_3}{k_2} = \frac{0}{\frac{1}{5}} = 0 \neq \frac{1}{2}$
The sequence is not geometric.
Arithmetic? $k_2 - k_1 = \frac{1}{5} - \frac{2}{5} = -\frac{1}{5}$
$k_3 - k_2 = 0 - \frac{1}{5} = -\frac{1}{5}$
$k_4 - k_3 = -\frac{1}{5} - 0 = -\frac{1}{5}$
The sequence is arithmetic.

7. a. $a_1 = -2$
$a_2 = ra_1 = -\frac{3}{2}(-2) = 3$
$a_3 = ra_2 = -\frac{3}{2}(3) = -\frac{9}{2}$
$a_4 = ra_3 = -\frac{3}{2}\left(-\frac{9}{2}\right) = \frac{27}{4}$

b. $a_n = a_1 r^{n-1}$
$a_9 = -2\left(-\frac{3}{2}\right)^8 = -\frac{6561}{128} \approx -51.26$

9. a. $a_1 = (-2)^4 = 16$
$a_2 = (-2)^5 = -32$
$a_3 = (-2)^6 = 64$
$a_4 = (-2)^7 = -128$

b. $a_9 = (-2)^{12} = 4096$

© Houghton Mifflin Company. All rights reserved.

11. **a.** $c_1 = 0.004(0.2)^{-5} = 12.5$

$c_2 = 0.004(0.2)^{-4} = 2.5$

$c_3 = 0.004(0.2)^{-3} = 0.5$

$c_4 = 0.004(0.2)^{-2} = 0.1$

b. $c_9 = 0.004(0.2)^3 = 3.2 \cdot 10^{-5}$

13. We must first find b_1.

$b_4 = b_1 r^3$

$\dfrac{9}{7} = b_1 \left(\dfrac{3}{7}\right)^3$

$b_1 = \dfrac{\frac{9}{7}}{\left(\frac{3}{7}\right)^3} = \dfrac{49}{3}$

Now we can write b_n.

$b_n = b_1 r^{n-1} = \dfrac{49}{3}\left(\dfrac{3}{7}\right)^{n-1}$

$= 7 \cdot \dfrac{7}{3}\left(\dfrac{3}{7}\right)^{n-1} = 7\left(\dfrac{3}{7}\right)^{n-2}$

15. We must first find a_1.

$a_n = a_1 r^{n-1}$

$-\dfrac{125}{2} = a_1(-5)^2$

$a_1 = \dfrac{-\frac{125}{2}}{25} = -\dfrac{5}{2}$

Now we can write a_n.

$a_n = a_1 r^{n-1} = -\dfrac{5}{2}(-5)^{n-1} = \dfrac{(-5)^n}{2}$

17. First find r.

$c_4 = c_1 r^3$

$-\dfrac{8}{9} = 3 \cdot r^3$

$r^3 = -\dfrac{8}{27}$

$r = -\dfrac{2}{3}$

Thus, $c_n = c_1 r^{n-1} = 3\left(-\dfrac{2}{3}\right)^{n-1}$.

19. First find r.

$\dfrac{a_5}{a_2} = \dfrac{a_1 r^4}{a_1 r} = r^3$

$r^3 = \dfrac{-\frac{1215}{16}}{\frac{45}{2}} = -\dfrac{27}{8}$

$r = -\dfrac{3}{2}$

Now find a_1.

$a_2 = a_1 r$

$\dfrac{45}{2} = a_1\left(-\dfrac{3}{2}\right)$

$a_1 = -15$

Thus, $a_n = a_1 r^{n-1} = -15\left(-\dfrac{3}{2}\right)^{n-1}$.

21. First find r.

$\dfrac{c_6}{c_3} = \dfrac{c_1 r^5}{c_1 r^2} = r^3$

$r^3 = \dfrac{12}{\frac{3}{2}} = 8$

$r = 2$

Now find c_1.

$c_3 = c_1 r^2$

$\dfrac{3}{2} = c_1(2)^2$

$c_1 = \dfrac{3}{8}$

Thus, $c_n = c_1 r^{n-1} = \dfrac{3}{8}(2)^{n-1} = 3 \cdot 2^{n-4}$.

23. Use the two terms to determine r. Then determine the eighth partial sum.

25. The series has $n = 24$ terms with $a_1 = 0.75$ and

$r = \dfrac{1.5}{0.75} = 2$.

$S_n = \dfrac{a_1(1 - r^n)}{1 - r} = \dfrac{0.75(1 - 2^{24})}{1 - 2}$

$= 12{,}582{,}911.25$

27. The series has $n = 14$ terms with $a_1 = \dfrac{\sqrt{3}}{3}$ and

$r = \dfrac{1}{-\frac{\sqrt{3}}{3}} = -\sqrt{3}$.

$S_n = \dfrac{a_1(1 - r^n)}{1 - r}$

$= \dfrac{-\frac{\sqrt{3}}{3}\left[1 - (-\sqrt{3})^{14}\right]}{1 - (-\sqrt{3})} = 461.96$

© Houghton Mifflin Company. All rights reserved.

29. We have $a_1 = \dfrac{3}{4}$ and $r = \dfrac{-3/2}{3/4} = -2$. We must find n.

$$a_n = a_1 r^{n-1}$$
$$3072 = \frac{3}{4}(-2)^{n-1}$$
$$(-2)^{n-1} = 4096 = 2^{12} = (-2)^{12}$$
$$n - 1 = 12$$
$$n = 13$$

Therefore

$$S_n = \frac{a_1(1 - r^n)}{1 - r}$$
$$= \frac{\left(\frac{3}{4}\right)\left[1 - (-2)^{13}\right]}{1 - (-2)} = 2048.25.$$

31. We have $a_1 = 8$ and $r = \dfrac{6}{8} = \dfrac{3}{4}$. We must find n.

$$a_n = a_1 r^{n-1}$$
$$\frac{6561}{8192} = 8\left(\frac{3}{4}\right)^{n-1}$$
$$\left(\frac{3}{4}\right)^{n-1} = \frac{6561}{65,536} = \left(\frac{3}{4}\right)^8$$
$$n - 1 = 8$$
$$n = 9$$

Therefore

$$S_n = \frac{a_1(1 - r^n)}{1 - r}$$
$$= \frac{8\left[1 - \left(\frac{3}{4}\right)^9\right]}{1 - \frac{3}{4}}$$
$$= \frac{242,461}{8192} = 29.60$$

33. The series has $n = 15$ terms with

$$a_1 = \frac{1}{8}(2^0) = \frac{1}{8} \text{ and } r = 2.$$

$$S_n = \frac{a_1(1 - r^n)}{1 - r}$$
$$= \frac{\left(\frac{1}{8}\right)(1 - 2^{15})}{1 - 2}$$
$$= \frac{32,767}{8} = 4095.875$$

35. Be careful: the initial index is 0. So the series has $n = 8 + 1 = 9$ terms with $a_1 = 7(0.1)^0 = 7$ and $r = 0.1$.

$$S_n = \frac{a_1(1 - r^n)}{1 - r} = \frac{7(1 - 0.1^9)}{1 - 0.1} = 7.78$$

37. The series has $n = 9$ terms with

$$a_1 = 25\left(\frac{1}{5}\right)^{-2} = 625 \text{ and } r = \frac{1}{5}.$$

$$S_n = \frac{a_1(1 - r^n)}{1 - r} = \frac{625\left[1 - \left(\frac{1}{5}\right)^9\right]}{1 - \frac{1}{5}} = 781.25$$

39. The series has $n = 10$ terms with

$$a_1 = 5\left(-\frac{5}{3}\right)^{-3} = -\frac{27}{25} \text{ and } r = -\frac{5}{3}$$

$$S_n = \frac{a_1(1 - r^n)}{1 - r}$$
$$= \frac{\left(-\frac{27}{25}\right)\left[1 - \left(-\frac{5}{3}\right)^{10}\right]}{1 - \left(-\frac{5}{3}\right)} = 66.57$$

41. The common ratio $\left(r = \dfrac{-1}{0.5} = -2\right)$ is less than -1.

43. Use $a_1 = \dfrac{9}{4}\left(\dfrac{2}{3}\right)^1 = \dfrac{3}{2}$ and $r = \dfrac{2}{3}$ $(-1 < r < 1)$ to determine the sum.

$$S = \frac{a_1}{1 - r} = \frac{\frac{3}{2}}{1 - \frac{2}{3}} = \frac{9}{2} = 4.5$$

45. Rewrite the series in standard form with initial index $j = 1$.

$$\sum_{i=0}^{\infty} 7^{1-i} = 7 + 1 + \frac{1}{7} + \frac{1}{49} + \ldots$$
$$= \left(\frac{1}{7}\right)^{-1} + \left(\frac{1}{7}\right)^{0} + \left(\frac{1}{7}\right)^{1} + \left(\frac{1}{7}\right)^{2} + \ldots$$
$$= \sum_{j=1}^{\infty} \left(\frac{1}{7}\right)^{j-2}$$

Use $a_1 = 7$ and $r = \dfrac{1}{7}$ $(-1 < r < 1)$ to determine the sum.

$$S = \frac{a_1}{1 - r} = \frac{7}{1 - \frac{1}{7}} = \frac{49}{6}$$

© Houghton Mifflin Company. All rights reserved.

47. Rewrite the series in standard form.

$$\sum_{i=1}^{\infty} \frac{4}{(-3)^{i-1}} = \sum_{i=1}^{\infty} 4\left(-\frac{1}{3}\right)^{i-1}$$

$$= \sum_{i=1}^{\infty} 4\left(-\frac{1}{3}\right)^{i}\left(-\frac{1}{3}\right)^{-1}$$

$$= \sum_{i=1}^{\infty} (-12)\left(-\frac{1}{3}\right)^{i}$$

Use $a_1 = (-12)\left(-\frac{1}{3}\right)^1 = 4$ and $r = -\frac{1}{3}$

$(-1 < r < 1)$ to determine the sum.

$$S = \frac{a_1}{1-r} = \frac{4}{1-\left(-\frac{1}{3}\right)} = 3$$

49. Rewrite the series in standard form.

$$\sum_{i=1}^{\infty} 2(\sqrt{7})^{1-i} = \sum_{i=1}^{\infty} 2(\sqrt{7})^{1}(\sqrt{7})^{-i}$$

$$= \sum_{i=1}^{\infty} 2\sqrt{7}\left(\frac{1}{\sqrt{7}}\right)^{i}$$

Use $a_1 = 2\sqrt{7}\left(\frac{1}{\sqrt{7}}\right) = 2$ and $r = \frac{1}{\sqrt{7}}$

$(-1 < r < 1)$ to determine the sum.

$$S = \frac{a_1}{1-r} = \frac{2}{1-\frac{1}{\sqrt{7}}} = 3.22$$

51. This is a geometric series with $a_1 = 4$ and

$$r = \frac{2}{4} = \frac{1}{2} \ (-1 < r < 1).$$

$$S = \frac{a_1}{1-r} = \frac{4}{1-\frac{1}{2}} = 8$$

53. This is a geometric series with $a_1 = 5$ and $r = \frac{-1}{5}$

$(-1 < r < 1).$

$$S = \frac{a_1}{1-r} = \frac{5}{1-\left(-\frac{1}{5}\right)} = \frac{25}{6}$$

55.

n	1	2	3	4	5	6	7	8
0.7^{n-1}	1	0.7	0.49	0.343	0.2401	0.16807	0.117649	0.0823543
0.5^{n-1}	1	0.5	0.25	0.125	0.0625	0.03125	0.015625	0.0078125
0.1^{n-1}	1	0.1	0.01	0.001	0.0001	0.00001	0.000001	0.0000001

In each case r^n approaches 0 as n becomes larger and larger.

57. $\displaystyle\sum_{i=1}^{N} 4(-1)^{i-1} = 4 - 4 + 4 - 4 + \ldots + 4(-1)^{N-1}$.

a. For any even integer N, the 4's cancel in pairs, so the sum is 0.

b. For any odd integer N, the 4's cancel in pairs leaving one term of 4, so the sum is 4.

© Houghton Mifflin Company. All rights reserved.

59. $0.\overline{8} = 0.8888\ldots$

$\quad = 0.8 + 0.08 + 0.008 + 0.0008 + \ldots$

$\quad = \dfrac{8}{10} + \dfrac{8}{100} + \dfrac{8}{1000} + \dfrac{8}{10,000} + \ldots$

$\quad = \displaystyle\sum_{i=1}^{\infty} \dfrac{8}{10^i}$

$\quad = \displaystyle\sum_{i=1}^{\infty} 8\left(\dfrac{1}{10}\right)^i$

This is a geometric series with $a_1 = \dfrac{8}{10}$

and $r = \dfrac{1}{10}$.

$S = \dfrac{a_1}{1-r} = \dfrac{\frac{8}{10}}{1-\frac{1}{10}} = \dfrac{8}{9}$

61. $5.8\overline{3} = 5.83333\ldots$

$\quad = 5.8 + 0.03 + 0.003 + 0.0003 + 0.00003 + \ldots$

$\quad = 5.8 + \dfrac{3}{100} + \dfrac{3}{1000} + \dfrac{3}{10,000} + \dfrac{3}{100,000} + \ldots$

$\quad = 5.8 + \displaystyle\sum_{i=1}^{\infty} \dfrac{3}{10^{i+1}}$

Ignoring the 5.8, this is a geometric series with $a_1 = \dfrac{3}{100}$ and $r = \dfrac{1}{10}$.

$S = \dfrac{a_1}{1-r} = \dfrac{\frac{3}{100}}{1-\frac{1}{10}} = \dfrac{1}{30}$.

Thus, $5.8\overline{3} = 5.8 + \dfrac{1}{30} = \dfrac{35}{6}$.

63. If $b_1{}^2, b_2{}^2, \ldots$ is a geometric sequence, then $\dfrac{b_2}{b_1} = r$ for some value of r.

a. Is $b_1, b_2, \ldots$ geometric? Since $\dfrac{b_2{}^2}{b_1{}^2} = \left(\dfrac{b_2}{b_1}\right)^2 = r^2$, there is a common ratio so the sequence is geometric.

b. Is $\ln b_1, \ln b_2, \ldots$ a geometric sequence? Since $\dfrac{\ln b_2}{\ln b_1} \neq \ln \dfrac{b_2}{b_1} = \ln r$, the sequence is not geometric.

Is $\ln b_1, \ln b_2, \ldots$ an arithmetic sequence? Since $\ln b_2 - \ln b_1 = \ln \dfrac{b_2}{b_1} = \ln r$, the sequence is arithmetic.

© Houghton Mifflin Company. All rights reserved.

65. a. $\ln 2, \ln 4, \ln 6, \ln 8, \ldots$

Geometric? $\quad \dfrac{a_2}{a_1} = \dfrac{\ln 4}{\ln 2} = \dfrac{\ln 2^2}{\ln 2} = \dfrac{2\ln 2}{\ln 2} = 2$

$\qquad\qquad \dfrac{a_3}{a_2} = \dfrac{\ln 6}{\ln 4} = \dfrac{\ln(3 \cdot 2)}{\ln(2 \cdot 2)} = \dfrac{\ln 3 + \ln 2}{\ln 2 + \ln 2} \neq 2$

The sequence is not geometric.

Arithmetic? $\quad a_2 - a_1 = \ln 4 - \ln 2 = \ln(2 \cdot 2)\ln 2 = (\ln 2 + \ln 2) - \ln 2 = \ln 2$

$\qquad\qquad a_3 - a_2 = \ln 6 - \ln 4 = \ln(2 \cdot 3) - \ln 2 = (\ln 2 + \ln 3) - \ln 2 = \ln 3 \neq \ln 2$

The sequence is not arithmetic.

The sequence is neither.

b. $\ln 2, \ln 4, \ln 8, \ln 16, \ldots$

Geometric? $\quad \dfrac{a_2}{a_1} = \dfrac{\ln 4}{\ln 2} = \dfrac{\ln 2^2}{\ln 2} = \dfrac{2\ln 2}{\ln 2} = 2$

$\qquad\qquad \dfrac{a_3}{a_1} = \dfrac{\ln 8}{\ln 2} = \dfrac{\ln 2^3}{\ln 2} - \dfrac{3\ln 2}{\ln 2} = 3 \neq 2$

The sequence is not geometric.

Arithmetic? $\quad a_2 - a_1 = \ln 4 - \ln 2 = \ln \dfrac{4}{2} = \ln 2$

$\qquad\qquad a_3 - a_2 = \ln 8 - \ln 4 = \ln \dfrac{8}{4} = \ln 2$

$\qquad\qquad a_4 - a_3 = \ln 16 - \ln 8 = \ln \dfrac{16}{8} = \ln 2$

The sequence is arithmetic.

c. $\ln 2, \ln 4, \ln 16, \ln 25 \, b,$

Geometric? $\quad \dfrac{a_2}{a_1} = \dfrac{\ln 4}{\ln 2} = \dfrac{\ln 2^2}{\ln 2} = \dfrac{2\ln 2}{\ln 2} = 2$

$\qquad\qquad \dfrac{a_3}{a_2} = \dfrac{\ln 16}{\ln 4} = \dfrac{\ln 4^2}{\ln 4} = \dfrac{2\ln 4}{\ln 4} = 2$

$\qquad\qquad \dfrac{a_4}{a_3} = \dfrac{\ln 256}{\ln 16} = \dfrac{\ln 16^2}{\ln 16} = \dfrac{2\ln 16}{\ln 16} = 2$

The sequence is geometric.

67. a. $a_n = 200,000(0.7)^n$

b. $a_6 = 200,000(0.7)^6 = 23,529.8$

$23,529.80

c. $\displaystyle\sum_{i=0}^{12} 200,000(0.7)^i$

d. The series has $n = 12 + 1 = 13$ terms with $a_1 = 200,000$ and $r = 0.7$.

$S_n = \dfrac{a_1(1 - r^n)}{1 - r} = \dfrac{200,000(1 - 0.7^{13})}{1 - 0.7} = 660,207.40$

The sum is $660,207.40.

© Houghton Mifflin Company. All rights reserved.

69. a. $a_n = 2(0.8)^{n-1}$, where saleμs are in millions of dollars.

b. We solve the inequality $a_n < 0.5$ for n.

$$2(0.8)^{n-1} < 0.5$$
$$(0.8)^{n-1} < 0.25$$

Using a calculator, we see $0.8^7 = 0.262144$ and $0.8^8 = 0.2097152$.
Therefore, after 8 months sales will drop below $0.5 million.

c. $\displaystyle\sum_{i=1}^{12} 2(0.8)^{i-1}$

This series has $n = 12$ terms with $a_1 = 2$ and $r = 0.8$.

$$S_n = \frac{a_1(1-r^n)}{1-r} = \frac{2(1-0.8^{12})}{1-0.8} = 9.31$$

Total sales for the first year are $9.31 million.

d. $S = \dfrac{a_1}{1-r} = \dfrac{2}{1-0.8} = 10$

Since the sum of the infinite series is 10, sales can never be greater than $10 million.

71. a. Use Example 9 as a guide to obtain $a_n = 20\left(\dfrac{4}{5}\right)^{n-1}$

b. $b_n = 20\left(\dfrac{4}{5}\right)^{n}$

c. We solve the inequality $b_n < 1 \cdot \dfrac{1}{12} = \dfrac{1}{12}$. (Remember to convert to inches.)

$$20\left(\frac{4}{5}\right)^{n} < \frac{1}{12}$$
$$\left(\frac{4}{5}\right)^{n} < \frac{1}{20 \cdot 12} = \frac{1}{240} \approx 0.0042$$

sing a calculator, we see $\left(\dfrac{4}{5}\right)^{24} \approx 0.0047$ and $\left(\dfrac{4}{5}\right)^{25} \approx 0.0038$.
The ball will hit the ground 25 times.

d. $\displaystyle\sum_{i=1}^{\infty} 20\left(\frac{4}{5}\right)^{i-1} + \sum_{i=1}^{\infty} 20\left(\frac{4}{5}\right)^{i} = 20 + 2\sum_{i=1}^{\infty} 20\left(\frac{4}{5}\right)^{i}$

The series has $a_1 = 20\left(\dfrac{4}{5}\right)^{1} = 16$ and $r = \dfrac{4}{5}$

$$S = \frac{a_1}{1-r} = \frac{16}{1-\frac{4}{5}} = 80$$

Thus, the ball will travel $20 + 2(80)$ or 180 feet.

© Houghton Mifflin Company. All rights reserved.

73. a. $a_n = 30,271(1.05)^{n-1}$

b. $a_{20} = 30,271(1.05)^{19} = 76,493.31$
The person can expect \$76,493.31.

c. $\displaystyle\sum_{i=1}^{20} 30,271(1.05)^{i-1}$

The series has 20 terms with $a_1 = 30,271$ and $r = 1.05$.

$$S_n = \frac{a_1(1-r^n)}{1-r} = \frac{30,271(1-1.05^{20})}{1-1.05} = 1,000,939.50$$
The total earnings are \$1,000,939.50.

d. We need to solve $S_n > 2,000,000$.

$$S_n = \sum_{i=1}^{n} 30,271(1.05)^{i-1} > 2,000,000$$

Using the calculator, we find $S_{29} = 1,886,570.81$ and $S_{30} = 2,011,170.35$. Thus, it will take 30 years.

6.4 Exercises

1. The notation 4! means $4 \cdot 3 \cdot 2 \cdot 1$, whereas 4^4 means $4 \cdot 4 \cdot 4 \cdot 4$.

3. $7! = 7 \cdot 6 \cdot 5 \cdot 4 \cdot 3 \cdot 2 \cdot 1 = 5040$

5. $\dfrac{12!}{8!} = \dfrac{12 \cdot 11 \cdot 10 \cdot 9 \cdot 8 \cdot 7 \cdot 6 \cdot 5 \cdot 4 \cdot 3 \cdot 2 \cdot 1}{8 \cdot 7 \cdot 6 \cdot 5 \cdot 4 \cdot 3 \cdot 2 \cdot 1}$
$= 12 \cdot 11 \cdot 10 \cdot 9 = 11,880$

7. $\dfrac{16!}{10!6!} = \dfrac{16 \cdot 15 \cdot 14 \cdot 13 \cdot 12 \cdot 11 \cdot 10!}{10!6!}$
$= \dfrac{16 \cdot 15 \cdot 14 \cdot 13 \cdot 12 \cdot 11}{6 \cdot 5 \cdot 4 \cdot 3 \cdot 2 \cdot 1} = 8008$

9. $_6P_3 = \dfrac{6!}{(6-3)!} = \dfrac{6!}{3!} = 6 \cdot 5 \cdot 4 = 120$

11. $_{10}P_4 = \dfrac{10!}{(10-4)!} = \dfrac{10!}{4!} = 10 \cdot 9 \cdot 8 \cdot 7 = 5040$

13. $_7C_5 = \dfrac{7!}{(7-5)!5!} = \dfrac{7!}{2!5!} = \dfrac{7 \cdot 6}{2 \cdot 1} = 21$

15. $_{15}C_7 = \dfrac{15!}{(15-7)!7!}$
$= \dfrac{15!}{8!7!}$
$= \dfrac{15 \cdot 14 \cdot 13 \cdot 12 \cdot 11 \cdot 10 \cdot 9}{7 \cdot 6 \cdot 5 \cdot 4 \cdot 3 \cdot 2 \cdot 1} = 6435$

17. $\displaystyle\sum_{i=0}^{3} {}_3C_i = {}_3C_0 + {}_3C_1 + {}_3C_2 + {}_3C_3$
$= 1 + 3 + 3 + 1 = 8$

19. $\dfrac{_nP_r}{_nC_r} = \dfrac{\frac{n!}{(n-r)!}}{\frac{n!}{(n-r)!r!}} = \dfrac{n!}{(n-r)!} \cdot \dfrac{(n-r)!r!}{n!} = r!$

21. There are $24 \cdot 15$ or 360 ways to fill the positions.

23. A student can complete the requirements in $5 \cdot 4 \cdot 6 \cdot 2$ or 240 ways.

25. a. There are 10^7 possible numbers.

b. There are $8 \cdot 10^6$ possible numbers.

27. The test can be answered in 4^{10} or 1,048,576 ways.

29. a. There are 4 possible ways to select a girl for the head of the line. There are then 3 girls and 6 boys left. So the students can line up in $4 \cdot 3! \cdot 6! = 17,280$ ways.

b. There are 6 possible ways to select a boy for the head of the line. There are then 4 girls and 5 boys left. So the students can line up in $6 \cdot 4! \cdot 5! = 17,280$ ways.

c. There are 10 students so they can line up in $10! = 3,628,800$ ways.

31. There are $3 \cdot 4 \cdot 7 = 84$ different sandwiches possible.

33. For a permutation, the order of the items is important, whereas, for a combination, the order of the items is not important.

35. The final order can be drawn in $_6P_6 = 6! = 720$ ways.

37. The parade can be lined up in $18! \approx 6.40 \cdot 10^{15}$ ways.

39. The program can be arranged in $_8P_8 = 8! = 40,320$ ways.

41. There are seven colors and three are chosen. Thus, there are $_7P_3 = 210$ flags possible.

© Houghton Mifflin Company. All rights reserved.

43. There are 10 professors from which to select the 7 teachers. Thus, there are $_{10}P_7 = 604,800$ assignments possible.

45. a. There are $_{26}P_5 = 7,893,600$ different PINs possible.

b. There would be $26^5 = 11,881,376$ PINs possible.

47. There are 1 M, 4 I's, 4 S's, and 2 P's and $\dfrac{11!}{1!4!4!2!} = 34,650$ distinguishable permutations.

49. There are $\dfrac{9!}{4!3!2!} = 1260$ distinguishable assignments.

51. There are $\dfrac{15!}{6!5!4!} = 630,630$ distinguishable arrangements.

53. Because the value of $_5C_3$ and $_5C_2$ are both $\dfrac{5!}{3!2!}$, the number of combinations is the same.

55. There are $_{32}C_4 = 35,960$ ways the teams can reach the semifinal round.

57. There are $_{24}C_{12} = 2,704,156$ possible juries that can be chosen.

59. There are $_{36}C_3 = 7140$ cones possible.

61. $4 \cdot _{13}C_5 = 5148$ different hands are possible.

63. $_{100}C_{20} = 5.36 \cdot 10^{20}$ tests can be generated.

65. The Democratic members can be chosen in $_{54}C_7$ ways and the Republican members can be chosen in $_{46}C_5$ ways. Thus, the committee can be chosen in $_{54}C_7 \cdot _{46}C_5 = 2.43 \cdot 10^{14}$ ways.

67. The men can be chosen in $_{42}C_5$ ways and the women can be chosen in $_{52}C_5$ ways. Thus, the advisory group can be formed in $_{42}C_5 \cdot _{52}C_5 = 2.21 \cdot 10^{12}$.

6.5 Exercises

1. The coin has no memory, so the probability is $\dfrac{1}{2}$.

3. Sample space: {1, 2, 3, 4, 5, 6}

a. Event: {5}
$$P(E) = \dfrac{1}{6}$$

b. Event: {1, 2, 3}
$$P(E) = \dfrac{1}{2}$$

c. Event: {2, 6}
$$P(E) = \dfrac{1}{3}$$

d. Event: {2, 4, 6}
$$P(E) = \dfrac{1}{2}$$

5. Sample space: {HHHH, HHHT, …, TTTT}. There are $2^4 = 16$ outcomes in the sample space.

a. Event: "exactly two are heads." There are $_4C_2 = 6$ outcomes corresponding to the event.
$$P(E) = \dfrac{6}{16} = \dfrac{3}{8}$$

b. Event: "at least two are heads" is equivalent to "exactly two or exactly three or exactly four are heads." There are $_4C_2 + _4C_3 + _4C_4 = 6+4+1 = 11$ outcomes corresponding to the event.
$$P(E) = \dfrac{11}{16}$$

c. Event: {HHHH, TTTT}
$$P(E) = \dfrac{2}{16} = \dfrac{1}{8}$$

d. Event: "one is heads and three are tails." There are $_4C_1 = 4$ outcomes corresponding to the event.
$$P(E) = \dfrac{4}{16} = \dfrac{1}{4}$$

© Houghton Mifflin Company. All rights reserved.

7. The sample space is given in Example 4.

   a. Event: {(3, 6), (4, 5), (5, 4), (6, 3)}

      $$P(E) = \frac{4}{36} = \frac{1}{9}$$

   b. Event: "the sum is at least five" is the complement of "the sum is exactly two, three, or four." There are 1, 2, and 3 outcomes, respectively, corresponding to these three results.

      $$1 - P(E) = 1 - \left(\frac{6}{36}\right) = \frac{30}{36} = \frac{5}{6}$$

   c. Event: {(1,2), (2,1), (5,6), (6,5)}

      $$P(E) = \frac{4}{36} = \frac{1}{9}$$

   d. Event: {(1,1), (2,2), (3,3), (4,4), (5,5), (6,6)}

      $$P(E) = \frac{6}{36} = \frac{1}{6}$$

9. The sample space is the set of 52 cards in a standard deck.

   a. Event: There are 13 clubs in the deck.

      $$P(E) = \frac{13}{52} = \frac{1}{4}$$

   b. Event: There are 4 aces in the deck.

      $$P(E) = \frac{4}{52} = \frac{1}{13}$$

   c. Event: There are three face cards in each of the four suits for a total of 12 face cards.

      $$P(E) = \frac{12}{52} = \frac{3}{13}$$

   d. Event: There are 26 hearts or diamonds and 6 of these are face cards. Thus

      $$P(E) = \frac{26}{52} - \frac{6}{52} = \frac{20}{52} = \frac{3}{13}$$

   e. Event: There are 13 clubs, 12 face cards, and 3 clubs which are face cards. These are non-mutually exclusive events. Thus,

      $$P(E) = \frac{13}{52} + \frac{12}{52} - \frac{3}{52} = \frac{22}{52} = \frac{11}{26}$$

11. The sample space is the set of $_{52}C_2 = 1326$ outcomes possible with a standard deck.

    a. Event: There are $_4C_2 = 6$ outcomes corresponding to "both are aces."

       $$P(E) = \frac{6}{1326} = \frac{1}{221}$$

    b. Event: There are $_{40}C_2 = 780$ outcomes corresponding to "neither card is a face card."

       $$P(E) = \frac{780}{1326} = \frac{10}{17}$$

    c. Event: The complementary event is "neither card is a face card." Using the results from part (b),

       $$1 - P(E) = 1 - \frac{10}{17} = \frac{7}{17}$$

    d. Event: The complementary event is "the cards are the same color." There are $_{26}C_2$ outcomes corresponding to two black cards and $_{26}C_2$ outcomes corresponding to two red cards, or $2 \cdot {}_{26}C_2 = 2 \cdot 325 = 650$ outcomes all together.

       $$1 - P(E) = 1 - \frac{650}{1326} = \frac{676}{1326} = \frac{26}{51}$$

    e. Event: Let E_1 be "both cards are face cards" and E_2 be "both cards are red cards." The event E_1 and E_2, "both cards are red face cards," has $_6C_2 = 15$ outcomes. Using our previous results,

       $$P(E_1 \text{ or } E_2)$$
       $$= P(E_1) + P(E_2) - P(E_1 \text{ and } E_2)$$
       $$= \frac{66}{1326} + \frac{325}{1326} - \frac{15}{1326}$$
       $$= \frac{376}{1326} = \frac{188}{663}$$

13. There are 14 marbles total.

    a. $P(\text{white}) = \frac{4}{14} = \frac{2}{7}$

    b. $P(\text{not red}) = 1 - P(\text{red}) = 1 - \frac{6}{14} = \frac{8}{14} = \frac{4}{7}$

    c. $P(\text{blue or red}) = P(\text{blue}) + P(\text{red})$

       $$= \frac{3}{14} + \frac{6}{14} = \frac{9}{14}$$

© Houghton Mifflin Company. All rights reserved.

15. There are 16 keys total. There are $_{16}C_2 = 120$ outcomes in the sample space.

    a. There are $_6C_2 = 15$ outcomes corresponding to the event E_1 "both keys open door C."
$$P(E) = \frac{15}{120} = \frac{1}{8}$$

    b. Let E be the event "both keys open the same door." There are $_7C_2 = 21$ outcomes corresponding to "both keys open door A," $_3C_2 = 3$ outcomes corresponding to "both keys open door B," and $_6C_2 = 15$ outcomes corresponding to "both keys open door C." These are usually exclusive outcomes, so
$$P(E) = \frac{21}{120} + \frac{3}{120} + \frac{15}{120} = \frac{39}{120} = \frac{13}{40}$$

    c. There are $7 + 3 = 10$ keys that open door A or door B so there are $_{10}C_2 = 45$ outcomes corresponding to the event E, "both keys open either door A or door B."
$$P(E) = \frac{45}{120} = \frac{3}{8}$$

17. Because a van could be blue, the events in (*i*) are not mutually exclusive. The events in (*ii*) cannot occur simultaneously, so they are mutually exclusive.

19. Let E be the event "a student receives a room."
$$P(E) = \frac{300}{420} = \frac{5}{7}$$

21. There are $_4P_4 = 24$ different arrangements of the four cards and only one of them is in correct alphabetical order: $P(E) = \frac{1}{24}$.

23. The problem simplifies to finding the probability the first digit is 1 or 2 out of the five possible digits, so $P(E) = \frac{2}{5}$.

25. $P(E) = \dfrac{_{16}C_6 \cdot {}_{20}C_6}{_{36}C_{12}} = \dfrac{8008 \cdot 38,760}{1,251,677,700} \approx 0.25$

27. **a.** $P(E_1) = \dfrac{_6C_6}{_{25}C_6} = \dfrac{1}{177,100}$

    b. "At least 4 numbers" is equivalent to "exactly 4, 5, or 6 numbers."
$$P(E_2) = \frac{_6C_4 \cdot {}_{19}C_2 + {}_6C_5 \cdot {}_{19}C_1 + {}_6C_6 \cdot {}_{19}C_0}{_{25}C_6}$$
$$= \frac{15 \cdot 171 + 6 \cdot 19 + 1 \cdot 1}{177,100}$$
$$= \frac{2680}{177,100} = \frac{134}{8855}$$

29. There is only one way to select 3 balls if each is less than 4.
$$P(E) = \frac{1}{_8C_3} = \frac{1}{56}$$

31. $P(E) = \dfrac{3}{8} \cdot \dfrac{3}{8} \cdot \dfrac{3}{8} = \dfrac{27}{512}$

33. **a.** $P(E_1) = \dfrac{1}{_5P_5} = \dfrac{1}{120}$

    b. $P(E_2) = \dfrac{1}{5}$

35. **a.** $P(E) = \dfrac{_3C_0 \cdot {}_{17}C_4}{_{20}C_4} = \dfrac{28}{57}$

    b The complementary event is "none are spoiled" whose probability was found in part (a).
$$1 = P(E) = 1 - \frac{28}{57} = \frac{29}{57}$$

37. The total number of workers is 89 (million).

    a. $P(E_1) = \dfrac{6+4}{89} = \dfrac{10}{89}$

    b. $P(E_2) = \dfrac{6+45}{89} = \dfrac{51}{89}$

    c. $P(E_3) = \dfrac{45+34}{89} = \dfrac{79}{89}$

    d. $P(E_1) = \dfrac{4}{89}$

39. The total number of people is 171 (million).

    a. $P(E_1) = \dfrac{31+58}{171} = \dfrac{89}{171}$

    b. $P(E_2) = \dfrac{42+40}{171} = \dfrac{82}{171}$

    c. $P(E_3) = \dfrac{40}{171}$

41. We use the following formula where E_1 is "high blood pressure" and E_2 is "overweight."
$$P(E_1 \text{ or } E_2) = P(E_1) + P(E_2) - P(E_1 \text{ and } E_2)$$
$$P(E_1 \text{ or } E_2) = 0.25 + 0.35 - 0.10$$
$$P(E_1 \text{ or } E_2) = 0.50$$

43. We use the following formula where E_1 is "female" and E_2 is "lives more than 10 miles from campus."
$$P(E_1 \text{ or } E_2) = P(E_1) + P(E_2) - P(E_1 \text{ and } E_2) \text{Th}$$
$$0.70 = 0.60 + 0.30 - P(E_1 \text{ and } E_2)$$
$$P(E_1 \text{ and } E_2) = 0.20$$
e probability is 20%.

45. The odds in favor of E is the reciprocal of the odds in favor of $\overline{E}$.

© Houghton Mifflin Company. All rights reserved.

47. There are 36 possible outcomes of which 4 correspond to rolling a 5 and 32 correspond to rolling something other than a 5. Thus, the odds in favor of rolling a 5 are $\dfrac{P(E)}{P(\overline{E})} = \dfrac{\frac{4}{36}}{\frac{32}{36}} = \dfrac{1}{8}$, or 1 to 8. The odds against rolling a 5 are 8 to 1.

49. We have $P(E) = 0.15$ so that
$P(\overline{E}) = 1 - P(E) = 0.85$.
Thus, the odds in favor of completing a call are
$\dfrac{P(\overline{E})}{P(E)} = \dfrac{0.85}{0.15} = \dfrac{17}{3}$ or 17 to 3.

51. If the odds in favor are 5 to 2, the probability that the Packers will win is $\dfrac{5}{5+2} = \dfrac{5}{7}$.

53. There are 22 socks all together.

a. Let E_1 be "a matching pair." Then
$$P(E_1) = \frac{_{10}C_2 + _8C_2 + _4C_2}{_{22}C_2}$$
$$= \frac{45 + 28 + 6}{231} = \frac{79}{231}$$
$$P(E_1) = 1 - \frac{79}{231} = \frac{152}{231}$$
Thus, the odds in favor of E_1 are
$\dfrac{P(E_1)}{P(\overline{E}_1)} = \dfrac{\frac{79}{231}}{\frac{152}{231}} = \dfrac{79}{152}$, or 79 to 152.

b. Let E_2 be "a pair of black socks." Then
$$P(E_2) = \frac{_8C_2}{_{22}C_2} = \frac{28}{231}$$
$$P(\overline{E}_2) = 1 - \frac{28}{231} = \frac{203}{231}$$
Thus, the odds in favor of E_2 are
$\dfrac{P(E_2)}{P(\overline{E}_2)} = \dfrac{\frac{28}{231}}{\frac{203}{231}} = \dfrac{28}{203} = \dfrac{4}{29}$
or 4 to 29.

c. Let E_3 be "one orange and one white sock." Then
$$P(E_3) = \frac{_4C_1 \cdot _{10}C_1}{_{22}C_2} = \frac{40}{231}$$
$$P(\overline{E}_3) = 1 - \frac{40}{231} = \frac{191}{231}$$
Thus, the odds in favor of E_3 are
$\dfrac{P(E_3)}{P(\overline{E}_3)} = \dfrac{\frac{40}{231}}{\frac{191}{231}} = \dfrac{40}{191}$, or 40 to 191.

55. Let E be "no one with previous experience." Then
$$P(E) = \frac{_{16}C_{12}}{_{24}C_{12}} = \frac{1820}{2,704,156} = \frac{5}{7429}$$
$$P(\overline{E}) = 1 - \frac{5}{7429} = \frac{7424}{7429}$$
Thus, the odds in favor of E are
$\dfrac{P(E)}{P(\overline{E})} = \dfrac{\frac{5}{7429}}{\frac{7424}{7429}} = \dfrac{5}{7424}$, or 5 to 7424.

6.6 Exercises

1. No, line 13 contains the coefficients.

3. From Pascal's triangle, the coefficients of the expansion are 1 4 6 4 1.
Thus the model for expansion is $(a+b)^4 = a^4 + 4a^3b + 6a^2b^2 + 4ab^3 + b^4$

Replace a with $2x$ and b with $-y$.
$(2x-y)^4 = (2x)^4 + 4(2x)^3(-y) + 6(2x)^2(-y)^2 + 4(2x)(-y)^3 + (-y)^4$
$(2x-y)^4 = 16x^4 - 32x^3y + 24x^2y^2 - 8xy^3 + y^4$

© Houghton Mifflin Company. All rights reserved.

5. The model for expansion is $(a+b)^5 = a^5 + 5a^4b + 10a^3b^2 + 10a^2b^3 + 5ab^4 + b^5$

Replace a with $3y$ and b with 2.

$(3y+2)^5 = (3y)^5 + 5(3y)^4(2) + 10(3y)^3(2)^2 + 10(3y)^3(2)^3 + 5(3y)(2)^4 + (2)^5$

$(3y+2)^5 = 243y^5 + 810y^4 + 1080y^3 + 720y^2 + 240y + 32$

7. The binomial coefficients are

$_6C_0 = 1 \quad _6C_1 = 6 \quad _6C_2 = 15 \quad _6C_3 = 20 \quad _6C_4 = 15 \quad _6C_5 = 6 \quad _6C_6 = 1$

Thus the model for expansion is

$(a+b)^6 = a^6 + 6a^5b + 15a^4b^2 + 20a^3b^3 + 15a^2b^4 + 6ab^5 + b^5$

Replace a with y and b with -2.

$(y-2)^6 = y^6 + 6y^5(-2) + 15y^4(-2)^2 + 20y^3(-2)^3 + 15y^2(-2)^4 + 6y(-2)^5 + (-2)^6$

$(y-2)^6 = y^6 - 12y^5 + 60y^4 - 160y^3 + 240y^2 - 192y + 64$

9. The binomial coefficients are

$_5C_0 = 1 \quad _5C_1 = 5 \quad _5C_2 = 10 \quad _5C_3 = 10 \quad _5C_4 = 5 \quad _5C_5 = 1$

Thus the model for expansion is

$(a+b)^5 = a^5 + 5a^4b + 10a^3b^2 + 10a^2b^3 + 5ab^4 + b^5$

Replace a with 1 and b with $2x$.

$(1+2x)^5 = (1)^5 + 5(1)^4(2x) + 10(1)^3(2x)^2 + 10(1)^2(2x)^3 + 5(1)(2x)^4 + (2x)^5$

$(1+2x)^5 = 1 + 10x + 40x^2 + 80x^3 + 80x^4 + 32x^5$

11. The binomial coefficients are

$_4C_0 = 1 \quad _4C_1 = 4 \quad _4C_2 = 6 \quad _4C_3 = 4 \quad _4C_4 = 1$

Thus the model for expansion is

$(a+b)^4 = a^4 + 4a^3b + 6a^2b^2 + 4ab^3 + 6^4$

Replace a with $2x$ and b with $-3\sqrt{y}$.

$(2x-3\sqrt{y})^4 = (2x)^4 + 4(2x)^3(-3\sqrt{y}) + 6(2x)^2(-3\sqrt{y})^2 + 4(2x)(-3\sqrt{y})^3 + (-3\sqrt{y})^4$

$(2x-3\sqrt{y})^4 = 16x^4 - 96x^3\sqrt{y} + 216x^2y - 216xy\sqrt{y} + 81y^2$

13. The binomial coefficients are

$_5C_0 = 1 \quad _5C_1 = 5 \quad _5C_2 = 10 \quad _5C_3 = 10 \quad _5C_4 = 5 \quad _5C_5 = 1$

Thus the model for expansion is

$(a+b)^5 = a^5 + 5a^4b + 10a^3b^2 + 10a^2b^3 + 5ab^4 + b^5$

Replace a with t^2 and b with $4w$.

$(t^2+4w)^5 = (t^2)^5 + 5(t^2)^4(4w) + 10(t^2)^3(4w)^2 + 10(t^2)^2(4w)^3 + 5(t^2)(4w)^4 + (4w)^5$

$(t^2+4w)^5 = t^{10} + 20t^8w + 160t^6w^2 + 640t^4w^3 + 1280t^2w^4 + 1024w^5$

15. The binomial coefficients are

$_6C_0 = 1 \quad _6C_1 = 6 \quad _6C_2 = 15 \quad _6C_3 = 20 \quad _6C_4 = 15 \quad _6C_5 = 6 \quad _6C_6 = 1$

Thus the model for expansion is

$(a+b)^6 = a^6 + 6a^5b + 15a^4b^2 + 20a^3b^3 + 15a^2b^4 + 6ab^5 + b^5$

Replace a with $\dfrac{x}{y}$ and b with $-\dfrac{z}{2}$.

$$\left(\frac{x}{y}-\frac{z}{2}\right)^6 = \left(\frac{x}{y}\right)^6 + 6\left(\frac{x}{y}\right)^5\left(-\frac{z}{2}\right) + 15\left(\frac{x}{y}\right)^4\left(-\frac{z}{2}\right)^2 + 20\left(\frac{x}{y}\right)^3\left(-\frac{z}{2}\right)^3$$

$$+ 15\left(\frac{x}{y}\right)^2\left(-\frac{z}{2}\right)^4 + 6\left(\frac{x}{y}\right)\left(-\frac{z}{2}\right)^5 + \left(-\frac{z}{2}\right)^6$$

$$\left(\frac{x}{y}-\frac{z}{2}\right)^6 = \frac{x^6}{y^6} - \frac{3x^5z}{y^5} + \frac{15x^4z^2}{4y^2} - \frac{5x^3z^3}{2y^3} + \frac{15x^2z^4}{16y^4} - \frac{3xz^5}{16y^5} + \frac{z^6}{64}$$

© Houghton Mifflin Company. All rights reserved.

17. The first four binomial coefficients are
$$_{10}C_0 = 1 \quad _{10}C_1 = 10 \quad _{10}C_2 = 45 \quad _{10}C_3 = 120$$
$$(a+b)^{10} = a^{10} + 10a^9 b + 45a^8 b^2 + 120a^7 b^3 + \ldots$$
Replace a with x and b with $4y$.
$$(x+4y)^{10} = x^{10} + 10x^9 (4y) + 45x^8 (4y)^2 + 120x^7 (4y)^3 + \ldots$$
$$(x+4y)^{10} = x^{10} + 40x^9 y + 720x^8 y^2 + 7680x^7 y^3 + \ldots$$

19. The first four binomial coefficients are
$$_{9}C_8 = 1 \quad _{9}C_1 = 9 \quad _{9}C_2 = 36 \quad _{9}C_3 = 84$$
Thus the model for expansion is
$$(m+n)^9 = m^9 + 9m^8 n + 36m^7 n^2 + 84m^6 n^3 + \ldots$$
Replace m with a^2 and n with $-2b^4$
$$(a^2 - 2b^4)^9 = (a^2)^9 + 9(a^2)^8(-2b^4) + 36(a^2)^7(-2b^4)^2 + 84(a^2)^6(-2b^4)^3 + \ldots$$
$$(a^2 - 2b^4)^9 = a^{18} - 18a^{16}b^4 + 144a^{14}b^8 - 672a^{12}b^{12} + \ldots$$

21. The first four binomial coefficients are
$$_{8}C_0 = 1 \quad _{8}C_1 = 8 \quad _{8}C_2 = 28 \quad _{8}C_3 = 56$$
Thus the model for expansion is
$$(a+b)^8 = a^8 + 8a^7 b + 28a^6 b^2 + 56a^5 b^3 + \ldots$$
Replace a with $4y$ and b with $5z$.
$$(4y+5z)^8 = (4y)^8 + 8(4y)^7(5z) + 28(4y)^6(5z)^2 + 56(4y)^5(5z)^3 + \ldots$$
$$(4y+5z)^8 = 65,536y^8 + 655,360y^7 z + 2,867,200y^6 z^2 + 7,168,000y^5 z^3 + \ldots$$

23. The first four binomial coefficients are
$$_{12}C_0 = 1 \quad _{12}C_1 = 12 \quad _{12}C_2 = 66 \quad _{12}C_3 = 220$$
Thus the model for expansion is
$$(a+b)^{12} = a^{12} + 12a^{11}b + 66a^{10}b^2 + 220a^9 b^3 + \ldots$$
Replace a with $\dfrac{1}{y}$ and b with $-\dfrac{z}{2}$

$$\left(\frac{1}{y} - \frac{z}{2}\right)^{12} = \left(\frac{1}{y}\right)^{12} + 12\left(\frac{1}{y}\right)^{11}\left(-\frac{z}{2}\right) + 66\left(\frac{1}{y}\right)^{10}\left(-\frac{z}{2}\right)^2 + 220\left(\frac{1}{y}\right)^9\left(-\frac{z}{2}\right)^3 + \ldots$$

$$\left(\frac{1}{y} - \frac{z}{2}\right)^{12} = \frac{1}{y^{12}} - \frac{6z}{y^{11}} + \frac{33z^2}{2y^{10}} - \frac{55z^3}{2y^9} + \ldots$$

25. No, several terms of the expansion are like terms and can be combined.

27. $(x+2y)^{50} = {_{50}C_0} \cdot (x)^{50} + {_{50}C_1} \cdot (x)^{49}(2y) + \ldots = x^{50} + 100x^{49}y + \ldots$

29. $(z-3)^{200} = {_{200}C_0} \cdot z^{200} + {_{200}C_1} \cdot z^{199}(-3) + \ldots = z^{200} - 600z^{199} + \ldots$

31. For the second term, $k = 1$, $n = 12$, $a = z$, and $b = 4w$
$$_{n}C_k a^{n-k} b^k = {_{12}C_1}(z)^{11}(4w)^1 = 48wz^{11}$$

33. For the last term, $k = 9$, $n = 9$, $a = x$, and $b = -3y$.
$$_{n}C_k a^{n-k} b^k = {_{9}C_9}(x)^0(-3y)^9 = -19,683y^9$$

35. For the seventh term, $k = 6$, $n = 10$, $a = 2t^2$, and $b = s$.
$$_{n}C_k a^{n-k} b^k = {_{10}C_6}(2t^2)^4(s)^6 = 3360t^8 s^6$$

37. For the third term, $k = 2$, $n = 8$, $a = 3$, and $b = -x^2$.
$$_{n}C_k a^{n-k} b^k = {_{8}C_2}(3)^6(-x^2)^2 = 20,412x^4$$

© Houghton Mifflin Company. All rights reserved.

39. $_nC_j = \dfrac{n!}{(n-j)!\,j!}$

$_nC_{n-j} = \dfrac{n!}{[n-(n-j)]!(n-j)!} = \dfrac{n!}{j!(n-j)!}$

Thus $_nC_j = {}_nC_{n-j}$.

41. $_nC_0 = \dfrac{n!}{(n-0)!\,0!} = \dfrac{n!}{n!} = 1$

43. y^3 appears in the term where $n-k=3$. Since $n=10$, $k=7$; $a=y$ and $b=4$.

$_nC_k a^{n-k}b^k = {}_{10}C_7 y^3(4)^7 = 1,966,080y^3$

The coefficient is 1,966,080.

45. x^7y^4 appears in the term where $k=4$, $n=11$, $a=2x$ and $b=5y$.

$_nC_k a^{n-k}b^k = {}_{11}C_4 (2x)^7(5y)^4 = 26,400,000x^7y^4$

The coefficient is 26,400,000.

47. x^2y^6 appears in the term where $k=6$, $n=8$, $a=x$, and $b=-2y$.

$_nC_k a^{n-k}b^k = {}_8C_6 (x)^2(-2y)^6 = 1792x^2y^6$

The coefficient is 1792.

49. a^4b^3 appears in the term where $2k=4$, so $k=2$; $n=5$, $x=b$, and $y=-3a^2$.

$_nC_k x^{n-k}y^k = {}_5C_2 b^3(-3a^2)^2 = 90a^4b^3$

The coefficient is 90.

51. $(1-i)^{10} = {}_{10}C_0(1)^{10} + {}_{10}C_1(1)^9(-i) + {}_{10}C_2(1)^8(-i)^2 + {}_{10}C_3(1)^7(-i)^3$
$\qquad + {}_{10}C_4(1)^6(-i)^4 + {}_{10}C_5(1)^5(-i)^5 + {}_{10}C_6(1)^4(-i)^6 + {}_{10}C_7(1)^3(-i)^7$
$\qquad + {}_{10}C_8(1)^2(-i)^8 + {}_{10}C_9(1)(-i)^9 + {}_{10}C_{10}(-i)^{10}$
$\qquad = 1 + -10i - 45 + 120i + 210 - 252i - 210 + 120i + 45 - 10i - 1$
$\qquad = -32i$

53. Diagonal 1: 1
Diagonal 2: 1
Diagonal 3: 2
Diagonal 4: 3
Diagonal 5: 5
Diagonal 6: 8
The sequence 1, 1, 2, 3, 5, 8, ... is the Fibonacci sequence.

55. $(2-y^{-3})^6 = {}_6C_0 2^6 + {}_6C_1 2^5(-y^{-3}) + {}_6C_2 2^4(-y^{-3})^2 + {}_6C_3 2^3(-y^{-3})^3$
$\qquad + {}_6C_4 2^2(-y^{-3})^4 + {}_6C_5 2(-y^{-3})^5 + {}_6C_6(-y^{-3})^6$
$\qquad = 64 - 192y^{-3} + 240y^{-6} - 160y^{-9} + 60y^{-12} - 12y^{-15} + y^{-18}$
$\qquad = 64 - \dfrac{192}{y^3} + \dfrac{240}{y^6} - \dfrac{160}{y^9} + \dfrac{60}{y^{12}} - \dfrac{12}{y^{15}} + \dfrac{1}{y^{18}}$

57. There are seven options from which to choose so there are $2^7 = 128$ different option packages available.

59. a. For the eighth term, $k=7$, $n=10$, $a=\dfrac{1}{5}$, and $b=\dfrac{4}{5}$.

$_nC_k a^{n-k}b^k = {}_{10}C_7\left(\dfrac{1}{5}\right)^3\left(\dfrac{4}{5}\right)^7 = 0.20$

b. For at least 7 men, $k=7,8,9,10$, $n=10$, $a=\dfrac{1}{5}$, and $b=\dfrac{4}{5}$.

$_{10}C_7\left(\dfrac{1}{5}\right)^3\left(\dfrac{4}{5}\right)^7 + {}_{10}C_8\left(\dfrac{1}{5}\right)^2\left(\dfrac{4}{5}\right)^8 + {}_{10}C_9\left(\dfrac{1}{5}\right)\left(\dfrac{4}{5}\right)^9 + {}_{10}C_{10}\left(\dfrac{4}{5}\right)^{10}$
$= 0.20 + 0.30 + 0.27 + 0.11$
$= 0.88$

© Houghton Mifflin Company. All rights reserved.

Review Exercises

1. a. $10, -11$

   $$a_n = (-1)^n(n+4)$$

   b. $\dfrac{13}{12}, \dfrac{15}{14}$

   $$a_n = \dfrac{2n+1}{2n}$$

3. a. $b_1 = \dfrac{2(1)-1}{1+3} = \dfrac{1}{4}$

   $b_2 = \dfrac{2(2)-1}{2+3} = \dfrac{3}{5}$

   $b_3 = \dfrac{2(3)-1}{3+3} = \dfrac{5}{6}$

   $b_4 = \dfrac{2(4)-1}{4+3} = \dfrac{7}{7} = 1$

   b. $b_6 = \dfrac{2(6)-1}{6+3} = \dfrac{11}{9}$

   c. $b_{10} = \dfrac{2(10)-1}{10+3} = \dfrac{19}{13}$

5. $b_1 = -1$

   $b_2 = 2$

   $b_3 = b_2 + 2b_1 = 2 + 2(-1) = 0$

   $b_4 = b_3 + 2b_2 = 0 + 2(2) = 4$

   $b_5 = b_4 + 2b_3 = 4 + 2(0) = 4$

7. $S_1 = 1$

   $S_2 = 1 + 2 = 3$

   $S_3 = 3 + 0 = 3$

   $S_4 = 3 + 3 = 6$

   $S_5 = 6 + (-1) = 5$

   $S_6 = 5 + 4 = 9$

   $S_7 = 9 - 2 = 7$ since $a_7 = -2$

9. $\displaystyle\sum_{i=1}^{5}(1-2i) = -1-3-5-7-9 = -25$

11. a. $p_n = 250$

    b. $I_n = 26.25 - 3.75n$

    c. $P_6 = 250$

       $I_6 = 26.25 - 3.75(6) = 3.75$

    d. $p_n + I_n$ represents the total monthly payment.

13. $a_2 - a_1 = 4 - (-1) = 5$

    $a_3 - a_2 = -7 - 4 = -11$

    No, the sequence is not arithmetic.

15. $3 - \dfrac{5}{2} = \dfrac{1}{2}, \ \dfrac{7}{2} - 3 = \dfrac{1}{2}, \ 4 - \dfrac{7}{2} = \dfrac{1}{2}$

    $a_n = \dfrac{1}{2}n + 2$

    $a_7 = \dfrac{1}{2}(7) + 2 = \dfrac{11}{2}$

17. $k_3 = k_1 + 2d$

    $8 = k_1 + 2(-2)$

    $k_1 = 12$

    $k_n = 12 + (n-1)(-2) = -2n + 14$

    $k_7 = -2(7) + 14 = 0$

19. No, you need to know at least one term.

21. $42 + 33 + 24 + \ldots -192$

    The terms are from the arithmetic sequence with

    $a_1 = 42$ and $d = -9$. Thus

    $a_n = a_1 + (n-1)d = 42 + (n-1)(-9) = -9n + 51$ We

    have $a_n = -9n + 51 = -192$ so that $n = 27$. Thus

    $42 + 33 + 24 + \ldots -192 = S_{27} = \dfrac{27(42-192)}{2}$

    $= -2025$

23. a. $t_n = 4n + 8$

    b. We need to solve for n where

       $4n + 8 = 40$

       $4n = 32$

       $n = 8$

       They will add 40 teachers in year 8.

25. a. $a_2 - a_1 = 1 - \dfrac{3}{2} = -\dfrac{1}{2}$,

       $a_3 - a_2 = \dfrac{1}{2} - 1 = -\dfrac{1}{2}$,

       $a_4 - a_3 = 0 - \dfrac{1}{2} = -\dfrac{1}{2}$

       The sequence is arithmetic.

    b. The sequence is neither.

    c. $\dfrac{a_2}{a_1} = \dfrac{-\frac{1}{4}}{\frac{1}{8}} = -2$,

       $\dfrac{a_3}{a_2} = \dfrac{\frac{1}{2}}{-\frac{1}{4}} = -2$,

       $\dfrac{a^4}{a_3} = \dfrac{-1}{\frac{1}{2}} = -2$

       The sequence is geometric.

27. $c_n = c_1 r^{n-1} = 25\left(\dfrac{1}{5}\right)^{n-1}$

© Houghton Mifflin Company. All rights reserved.

29. $b_3 = b_1 r^2 = -54$

$$b_8 = b_1 r^7 = -\frac{6561}{16}$$

$$\frac{b_8}{b_3} = \frac{b_1 r^7}{b_1 r^2} = \frac{-\frac{6561}{16}}{-54}$$

$$r^5 = \frac{243}{32}$$

$$r = \frac{3}{2}$$

$$b_3 = b_1 \left(\frac{3}{2}\right)^2 = -54$$

$$b_1 = -54 \left(\frac{2}{3}\right)^2 = -24$$

$$b_n = -24 \left(\frac{3}{2}\right)^{n-1}$$

31. For the geometric series, $a_1 = -32$ and

$$r = \frac{a_2}{a_1} = \frac{16}{-32} = -\frac{1}{2}.$$

$$S = \frac{a_1}{1-r} = \frac{-32}{1-\left(-\frac{1}{2}\right)} = -\frac{64}{3}$$

33. The series has $a_1 = 6$ and $r = \frac{1}{3}$.

$$S = \frac{a_1}{1-r} = \frac{6}{1-\frac{1}{3}} = 9$$

35. Series (i) is finite and can be evaluated. Series (ii) is an infinite series with $|r| > 1$, so it cannot be evaluated.

37. $\dfrac{18!}{14!} = \dfrac{18 \cdot 17 \cdot 16 \cdot 15 \cdot 14!}{14!}$
$$= 18 \cdot 17 \cdot 16 \cdot 15 = 73{,}440$$

39. $_{12}C_4 = \dfrac{12!}{(12-4)!4!}$
$$= \dfrac{12!}{8!4!}$$
$$= \dfrac{12 \cdot 11 \cdot 10 \cdot 9}{4 \cdot 3 \cdot 2 \cdot 1} = 495$$

41. $8 \cdot 5 \cdot 3 = 120$
They can sell a package in 120 different ways.

43. $12! = 479{,}001{,}600$
The deliveries can be arranged 479,001,600 ways.

45. $\dfrac{20!}{4! \, 10! \, 6!} = 38{,}798{,}760$
The jobs can be assigned in 38,798,760 ways.

47. $\dfrac{15!}{6! \, 9!} = 5005$
The finalists can be chosen in 5005 ways.

49. There are 15.5 (million) total students.

a. $P(E_1) = \dfrac{4.7}{15.5} = 0.30$

b. $P(E_2) = \dfrac{2.5}{15.5} = 0.16$

51. The sample space is given in Section 6.5, Example 4.

a. Event: {(3,6), (4,5), (5,4), (6.3)}
$$P(E_1) = \frac{4}{36} = \frac{1}{9}$$

b. Event: The complementary event is "the sum is at least 9" or, equivalently, "the sum is 9, 10, 11, or 12."
{(3,6), (6,3), (4,5), (5,4), (4,6), (6,4), (5,5), (5,6), (6,5), (6,6)}
$$P(E_2) = \frac{10}{36} = \frac{5}{18}$$

c. Event: {(2,2), (2,4), (2,6), (4,2), (4,4), (4,6), (6,2), (6,4), (6,6)}
$$P(E_3) = \frac{9}{36} = \frac{1}{4}$$

53. The sample space consists of the set of $_{52}C_2 = 1326$ outcomes possible with a standard deck.

a. There are $_{13}C_2 = 78$ outcomes corresponding to "both are clubs."
$$P(E) = \frac{78}{1326} = \frac{1}{17}$$

b. There are 16 cards which are either face cards or aces, so there are $_{16}C_2 = 120$ outcomes corresponding to "both are either face cards or aces."
$$P(E) = \frac{120}{1326} = \frac{20}{221}$$

c. There are 26 red cards so there are $_{26}C_2 = 325$ outcomes corresponding to "both are red cards."
$$P(E) = \frac{325}{1326} = \frac{25}{102}$$

55. There are 18 socks total

a. $P(\text{both white}) = \dfrac{_8C_2}{_{18}C_2} = \dfrac{28}{153}$

b. $P(\text{socks match}) = P(\text{both white}) + P(\text{both blue}) + P(\text{both brown})$
$$= \frac{_8C_2}{_{18}C_2} + \frac{_6C_2}{_{18}C_2} + \frac{_4C_2}{_{18}C_2}$$
$$= \frac{49}{153}$$

© Houghton Mifflin Company. All rights reserved.

57. a. $P(\text{both defective}) = \dfrac{_3C_2}{_{15}C_2} = \dfrac{1}{35}$

b. $P(\text{both good}) = \dfrac{_{12}C_2}{_{15}C_2} = \dfrac{22}{35}$

c. $P(\text{exactly one good})$

$= P(\text{one good and one bad}) = \dfrac{_{12}C_1 \cdot {_3}C_1}{_{15}C_2} = \dfrac{12}{35}$

59. The probability of heads is $\dfrac{1}{2}$. The odds in favor of heads is the probability of heads divided by the probability of tails: 1 to 1.

61. Each entry in the triangle is the sum of the two numbers above it.

63. The binomial coefficients are
$_4C_0 = 1 \quad _4C_1 = 4 \quad _4C_2 = 6 \quad _4C_3 = 4 \quad _4C_4 = 1$

Thus the model for expansion is
$(a+b)^4 = a^4 + 4a^3b + 6a^2b^2 + 4ab^3 + b^4$

Replace a with x and b with 5.
$(x+5)^4 = x^4 + 4x^3(5) + 6x^2(5)^2 + 4x(5)^3 + (5)^4$
$(x+5)^4 = x^4 + 20x^3 + 150x^2 + 500x + 625$

65. The binomial coefficients are
$_5C_0 = 1 \quad _5C_1 = 5 \quad _5C_2 = 10 \quad _5C_3 = 10 \quad _5C_4 = 5 \quad _5C_5 = 1$

Thus the model for expansion is
$(a+b)^5 = a^5 + 5a^4b + 10a^3b^2 + 10a^2b^3 + 5ab^4 + b^5$

Replace a with $\sqrt{3}$ and b with $-y$.
$(\sqrt{3}-y)^5 = (\sqrt{3})^5 + 5(\sqrt{3})^4(-y) + 10(\sqrt{3})^3(-y)^2 + 10(\sqrt{3})^2(-y)^3 + 5(\sqrt{3})(-y)^4 + (-y)^5$
$(\sqrt{3}-y)^5 = 9\sqrt{3} - 45y + 30\sqrt{3}y^2 - 30y^3 + 5\sqrt{3}y^4 - y^5$

67. b^3 appears in the term with $k = 3$. We have $n = 9$, $x = a^2$, and $y = -2b$.
$_nC_k x^{n-k} y^k = {_9}C_3(a^2)^6(-2b)^3 = -672a^{12}b^3$

The coefficient is -672.

69. $(1-3i)^5 = 1^5 + 5(1)^4(-3i) + 10(1)^3(-3i)^2 + 10(1)^2(-3i)^3 + 5(1)(-3i)^4 + (-3i)^5$
$(1-3i)^5 = 1 - 15i - 90 + 270i + 405 - 243i$
$(1-3i)^5 = 316 + 12i$

© Houghton Mifflin Company. All rights reserved.

Appendix A

A.1 Exercises

1. $x^2 + y^2 = 16 = 4^2$
$r = 4$

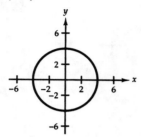

3. $4x^2 + 4y^2 = 40$
$x^2 + y^2 = 10 = \sqrt{10}^2$
$r = \sqrt{10} \approx 3.16$

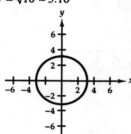

5. Find r using the Distance Formula.
$r = \sqrt{(3-0)^2 + (4-0)^2} = \sqrt{25} = 5$
The equation is
$x^2 + y^2 = 25$

7. Find r using the Distance Formula.
$r = \sqrt{(0-0)^2 + (7-0)^2} = 7$
The equation is
$x^2 + y^2 = 49$

9. $x^2 = y = 4\left(\frac{1}{4}\right)y$

Therefore $c = \frac{1}{4}$. The focus is $F\left(0, \frac{1}{4}\right)$ and the

directrix is $y = -\frac{1}{4}$.

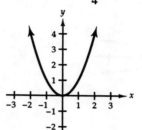

11. $y = -\frac{1}{2}x^2$

$x^2 = -2y = 4(-\frac{1}{2})y$

Therefore $c = -\frac{1}{2}$. The focus is $F\left(0, -\frac{1}{2}\right)$ and

the directrix is $y = \frac{1}{2}$.

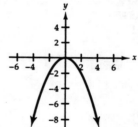

13. $y^2 = -20x = 4(-5)x$
Therefore $c = -5$. The focus is $F(-5, 0)$ and the
directrix is $x = 5$.

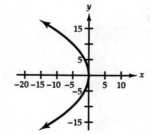

© Houghton Mifflin Company. All rights reserved.

192   *Appendix A*

15. $x = 2y^2$

$y^2 = \frac{1}{2}x = 4\left(\frac{1}{8}\right)x$

Therefore $c = \frac{1}{8}$. The focus is $F\left(\frac{1}{8}, 0\right)$ and the

directrix is $x = -\frac{1}{8}$.

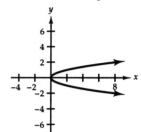

17. Since the focus is $F(0, 4)$, we know $c = 4$. Thus
$x^2 = 4(4)y = 16y$.

19. Since the focus is $F(-6, 0)$, we know $c = -6$. Thus
$y^2 = 4(-6)x = -24x$.

21. Since the directrix is $y = -2$, we know $c = 2$. Thus
$x^2 = 4(2)y = 8y$.

23. A vertical parabola containing $P(6, -3)$ has
equation $x^2 = 4cy$

$6^2 = 4c(-3)$
$4c = -12$
$c = -3$

Thus, the equation is $x^2 = 4(-3)y = -12y$.

25. $\frac{x^2}{2^2} + \frac{y^2}{3^2} = 1$
Vertices: $(0, \pm 3)$
Co-vertices: $(\pm 2, 0)$

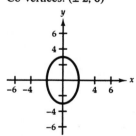

27. $\frac{x^2}{1^2} + \frac{y^2}{6^2} = 1$
Vertices: $(0, \pm 6)$
Co-vertices: $(\pm 1, 0)$

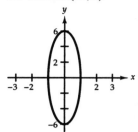

29. $\frac{x^2}{1^2} + \frac{y^2}{5^2} = 1$
Vertices: $(0, \pm 5)$
Co-vertices: $(\pm 1, 0)$

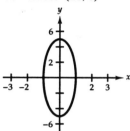

31. $\frac{x^2}{4} + \frac{y^2}{16} = 1$
$\frac{x^2}{2^2} + \frac{y^2}{4^2} = 1$
Vertices: $(0, \pm 4)$
Co-vertices: $(\pm 2, 0)$

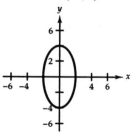

33. If the length of the minor axis is 2, then
$b = \frac{2}{2} = 1$.
Thus
$\frac{x^2}{1^2} + \frac{y^2}{3^2} = 1$
$x^2 + \frac{y^2}{9} = 1$

© Houghton Mifflin Company. All rights reserved.

35. If the foci are at $(0, \pm3)$, then
$$b^2 = a^2 - c^2$$
$$b^2 = 4^2 - 3^2$$
$$b^2 = 7$$
Thus
$$\frac{x^2}{4^2} + \frac{y^2}{7} = 1$$
$$\frac{x^2}{16} + \frac{y^2}{7} = 1$$

37. If the length of the major axis is 10, then
$a = \frac{10}{2} = 5$. If the length of the minor axis is 6,

then $b = \frac{6}{2} = 3$. Thus, $\frac{x^2}{5^2} + \frac{y^2}{3^2} = 1$
$$\frac{x^2}{25} + \frac{y^2}{9} = 1$$

39.
$$\frac{x^2}{a^2} + \frac{y^2}{b^2} = 1$$
$$\frac{0^2}{7^2} + \frac{(-6)^2}{b^2} = 1$$
$$b^2 = 36$$
$$\frac{x^2}{49} + \frac{y^2}{36} = 1$$

41. $\frac{x^2}{3^2} - \frac{y^2}{1^2} = 1$
Vertices: $(\pm3, 0)$
Asymptotes: $y = \pm\frac{1}{3}x$

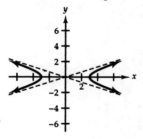

43. $\frac{y^2}{6^2} - \frac{x^2}{6^2} = 1$
Vertices: $(0, \pm6)$
Asymptotes: $y = \pm x$

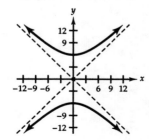

45. $\frac{x^2}{\left(\frac{2}{3}\right)^2} - \frac{y^2}{1^2} = 1$

Vertices: $\left(\pm\frac{2}{3}, 0\right)$

Asymptotes: $y = \pm\frac{3}{2}x$

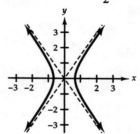

47. $\frac{y^2}{3^2} - \frac{x^2}{2^2} = 1$
Vertices: $(0, \pm3)$
Asymptotes: $y = \pm\frac{3}{2}x$

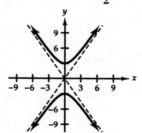

49. If the vertices are at $(\pm4,0)$, $a = 4$. If the foci are at $(\pm5, 0)$, $c = 5$.
$$c^2 = a^2 + b^2$$
$$b^2 = 5^2 - 4^2 = 3^2$$
$$\frac{x^2}{16} - \frac{y^2}{9} = 1$$

51. If the vertices are at $(0, \pm3)$ $a = 3$. If the asymptotes are $y = \pm\frac{3}{4}x$, then
$$y = \frac{3}{4}x$$
$$\frac{a}{b}x = \frac{3}{4}x$$
$$\frac{3}{b}x = \frac{3}{4}x$$
$$b = 4$$
$$\frac{y^2}{3^2} - \frac{x^2}{4^2} = 1$$
$$\frac{y^2}{9} - \frac{x^2}{16} = 1$$

© Houghton Mifflin Company. All rights reserved.

53. If the midpoints of the central rectangle are (±6, 0) and (0, ±4), then

$$\frac{x^2}{6^2} - \frac{y^2}{4^2} = 1$$

$$\frac{x^2}{36} - \frac{y^2}{16} = 1$$

A.2 Exercises

1. $(x+5)^2 + (y-2)^2 = 16 = 4^2$
Center: $C(-5, 2)$
Radius: $r = 4$

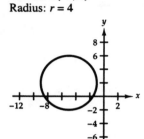

3.
$$x^2 + y^2 + 2x - 4y + 1 = 0$$
$$x^2 + 2x + y^2 - 4y = -1$$
$$x^2 + 2x + 1 + y^2 - 4y + 4 = -1 + 1 + 4$$
$$(x+1)^2 + (y-2)^2 = 4 = 2^2$$
Center: $C(-1, 2)$
Radius: $r = 2$

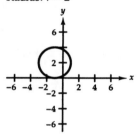

5.
$$x^2 - 8x + y^2 = 0$$
$$x^2 - 8x + 16 + y^2 = 16$$
$$(x-4)^2 + y^2 = 16 = 4^2$$
Center: $C(4,0)$
Radius: $r = 4$

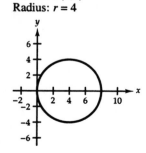

7. $(x-5)^2 + (y-8)^2 = 7^2 = 49$

9.
$$(x-4)^2 + (y+7)^2 = r^2$$
$$(2-4)^2 + (-3+7)^2 = r^2$$
$$4 + 16 = r^2$$
$$r^2 = 20$$
$$(x-4)^2 + (y+7)^2 = 20$$

11. The center is at the midpoint of the line segments

PQ: $M\left(\dfrac{-1+5}{2}, \dfrac{-3+7}{2}\right) = M(2, 2)$

The radius is $\overline{PM}$ or $\overline{MQ}$; use $\overline{PM}$.

$$r = \sqrt{(-1-2)^2 + (-3-2)^2} = \sqrt{9+25} = \sqrt{34}$$

Thus

$$(x-2)^2 + (y-2)^2 = \sqrt{34}^2 = 34$$

13. $(x-3)^2 = 8(y+1) = 4(2)(y+1)$
Thus $c = 2$.
Vertex: $V(3, -1)$
Focus: $F(3, -1 + 2) = F(3, 1)$
Directrix: $y = -1 - 2 = -3$

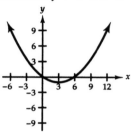

15. $(y-2)^2 = -6(x-1) = 4\left(-\dfrac{3}{2}\right)(x-1)$

Thus $c = -\dfrac{3}{2}$.
Vertex: $V(1, 2)$

Focus: $F\left(1 - \dfrac{3}{2}, 2\right) = F\left(-\dfrac{1}{2}, 2\right)$

Directrix: $x = 1 - \left(-\dfrac{3}{2}\right) = \dfrac{5}{2}$

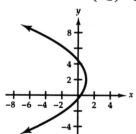

© Houghton Mifflin Company. All rights reserved.

17. $x^2 - 4x - y + 3 = 0$

$$x^2 - 4x = y - 3$$
$$x^2 - 4x + 4 = y - 3 + 4$$
$$(x-2)^2 = y + 1 = 4\left(\frac{1}{4}\right)(y+1)$$

Thus $c = \frac{1}{4}$.

Vertex: $V(2, -1)$

Focus: $F\left(2, -1 + \frac{1}{4}\right) = F\left(2, -\frac{3}{4}\right)$

Directrix: $y = -1 - \frac{1}{4} = -\frac{5}{4}$

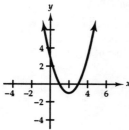

19. $x - y^2 + 6y + 16 = 0$

$$y^2 - 6y = x + 16$$
$$y^2 - 6y + 9 = x + 16 + 9 = x + 25$$
$$(y-3)^2 = 4\left(\frac{1}{4}\right)(x+25)$$

Thus $c = \frac{1}{4}$.

Vertex: $V(-25, 3)$

Focus: $F\left(-25 + \frac{1}{4}, 3\right) = F\left(-\frac{99}{4}, 3\right)$

Directrix: $x = -25 - \frac{1}{4} = -\frac{101}{4}$

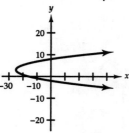

21. If the vertex is at $V(2, 4)$ and the focus at $F(2, 6)$, then $c = 6 - 4 = 2$. This is a vertical parabola.

$$(x-2)^2 = 4(2)(y-4)$$
$$(x-2)^2 = 8(y-4)$$

23. If the vertex is at $V(3, 0)$ and the directrix is $x = 4$, then $c = 3 - 4 = -1$. This is a horizontal parabola.

$$(y-0)^2 = 4(-1)(x-3)$$
$$y^2 = -4(x-3)$$

25. If the focus is at $F(3, -1)$ and the directrix is $x = -1$, then $c = \dfrac{3-(-1)}{2} = 2$ and the vertex is $V(3-2, -1) = V(1, -1)$. This is a horizontal parabola.

$$[y-(-1)]^2 = 4(2)(x-1)$$
$$(y+1)^2 = 8(x-1)$$

27. $\dfrac{(x+2)^2}{3^2} + \dfrac{(y-1)^2}{2^2} = 1$

hus, $a = 3$, $b = 2$, $c = \sqrt{9-4} = \sqrt{5}$

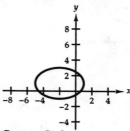

Center: $C(-2, 1)$

Vertices: $A(-2 \pm 3, 1)$ or $A_1(-5, 1)$ $A_2(1, 1)$

Co-vertices: $B(-2, 1 \pm 2)$ or $B_1(-2, -1)$, $B_2(-2, 3)$

Foci: $F(-2 \pm \sqrt{5}, 1)$ or $F_1(-2 - \sqrt{5}, 1)$, $F_2(-2 + \sqrt{5}, 1)$

29. $4(x-3)^2 + 25(y+4)^2 = 100$

$$\frac{(x-3)^2}{5^2} + \frac{(y+4)^2}{2^2} = 1$$

Thus, $a = 5$, $b = 2$, $c = \sqrt{25-4} = \sqrt{21}$

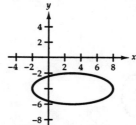

Center: $C(3, -4)$

Vertices: $A(3 \pm 5, -4)$ or $A_1(-2, -4)$, $A_2(8, -4)$

Co-vertices: $B(3, -4 \pm 2)$ or $B_1(3, -6)$, $B_2(3, -2)$

Foci: $F(3 \pm \sqrt{21}, -4)$ or $F_1(3 + \sqrt{21}, -4)$, $F_2(3 - \sqrt{21}, -4)$

© Houghton Mifflin Company. All rights reserved.

31. $x^2 + 49y^2 - 294y + 392 = 0$

$$x^2 + 49(y^2 - 6y) = -392$$

$$x^2 + 49(y^2 - 6y + 9) = -392 + 49 \cdot 9$$

$$x^2 + 49(y-3)^2 = 49$$

$$\frac{x^2}{7^2} + \frac{(y-3)^2}{1^2} = 1$$

Thus, $a = 7$, $b = 1$, $c = \sqrt{49-1} = \sqrt{48} = 4\sqrt{3}$

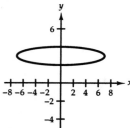

Center: $C(0, 3)$

Vertices: $A(0 \pm 7, 3)$ or $A_1(-7, 3)$, $A_2(7, 3)$

Co-vertices: $B(0, 3 \pm 1)$ or $B_1(0, 2)$, $B_2(0, 4)$

Foci: $F(0 \pm 4\sqrt{3}, 3)$ or $F_1(-4\sqrt{3}, 3)$, $F_2(4\sqrt{3}, 3)$

33. $25x^2 - 200x + 9y^2 - 36y + 420 = 0$

$$25(x^2 - 8x) + 9(y^2 - 4y) = -420$$

$$25(x^2 - 8x + 16) + 9(y^2 - 4y + 4) = -420 + 25 \cdot 16 + 9 \cdot 4$$

$$25(x-4)^2 + 9(y-2)^2 = 16$$

$$\frac{(x-4)^2}{\left(\frac{4}{5}\right)^2} + \frac{(y-2)^2}{\left(\frac{4}{3}\right)^2} = 1$$

Thus, $a = \frac{4}{3}$, $b = \frac{4}{5}$, $c = \sqrt{\frac{16}{9} - \frac{16}{25}} = \frac{16}{15}$.

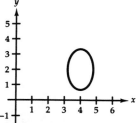

Center: $C(4, 2)$

Vertices: $A\left(4, 2 \pm \frac{4}{3}\right)$ or $A_1\left(4, \frac{10}{3}\right)$, $A_2\left(4, \frac{2}{3}\right)$

Co-vertices: $B\left(4 \pm \frac{4}{5}, 2\right)$ or $B_1\left(\frac{24}{5}, 2\right)$, $B_2\left(\frac{16}{5}, 2\right)$

Foci: $F\left(4, 2 \pm \frac{16}{15}\right)$ or $F_1\left(4, \frac{46}{15}\right)$, $F_2\left(4, \frac{14}{15}\right)$

© Houghton Mifflin Company. All rights reserved.

35. Because the center is at $C(4, -1)$ and a vertex is at $A(4, 4)$, then $a = 4 - (-1) = 5$. Since a co-vertex is at $B(2, -1)$, then $b = 4 - 2 = 2$.
Thus the equation is

$$\frac{(x-4)^2}{2^2} + \frac{[y-(-1)]^2}{5^2} = 1$$

$$\frac{(x-4)^2}{4} + \frac{(y+1)^2}{25} = 1$$

37. Because the vertices are at $A_1(-1, 3)$ and $A_2(5, 3)$,

$a = \dfrac{5-(-1)}{2} = 3$, and the center is at $C(2, 3)$.

Because the minor axis is of length 2, $b = \dfrac{2}{2} = 1$.
Thus the equation is

$$\frac{(x-2)^2}{3^2} + \frac{(y-3)^2}{1^2} = 1$$

$$\frac{(x-2)^2}{9} + (y-3)^2 = 1$$

39. Because the foci are at $F_1(0, 2)$ and $F_2(0, 8)$,

$c = \dfrac{8-2}{2} = 3$, and the center is at $C(0, 5)$. Since

the major axis is of length 12, $a = \dfrac{12}{2} = 6$.

Therefore $b^2 = a^2 - c^2 = 36 - 9 = 27$. Thus the equation is

$$\frac{(x-0)^2}{27} + \frac{(y-5)^2}{6^2} = 1$$

$$\frac{x^2}{27} + \frac{(y-5)^2}{36} = 1$$

41. $\dfrac{(x+2)^2}{3^2} - \dfrac{(y+1)^2}{5^2} = 1$

Thus, $a = 3$, $b = 5$, $c = \sqrt{9+25} = \sqrt{34}$.

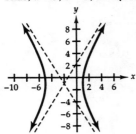

Center: $C(-2, -1)$
Vertices: $A(-2, \pm 3, -1)$ or $A_1(-5, -1)$, $A_2(1, -1)$
Foci: $F(-2 \pm \sqrt{34}, -1)$ or $F_1(-2 + \sqrt{34}, -1)$,
$F_2(-2 - \sqrt{34}, -1)$
Asymptotes: $y + 1 = \pm\dfrac{5}{3}(x+2)$ or $y_1 = \dfrac{5}{3}x + \dfrac{7}{3}$,

$y_2 = -\dfrac{5}{3}x - \dfrac{13}{13}$

43. $4y^2 - 49(x-7)^2 = 196$

$$\frac{y^2}{7^2} - \frac{(x-7)^2}{2^2} = 1$$

Thus, $a = 7$, $b = 2$, $c = \sqrt{49+4} = \sqrt{53}$

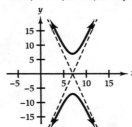

Center: $C(7, 0)$
Vertices: $A(7, 0 \pm 7)$ or $A_1(7, -7)$, $A_2(7, 7)$
Foci: $F(7, 0 \pm \sqrt{53})$ or $F_1(7, -\sqrt{53})$,
$F_2(7, \sqrt{53})$.
Asymptotes: $y = \pm\dfrac{7}{2}(x-7)$ or $y_1 = \dfrac{7}{2}x - \dfrac{49}{2}$,

$y_2 = -\dfrac{7}{2}x + \dfrac{49}{2}$

45.
$$y^2 - x^2 - 2x - 4y + 2 = 0$$
$$y^2 - 4y - (x^2 + 2x) = -2$$
$$y^2 - 4y + 4 - (x^2 + 2x + 1) = -2 + 4 - 1$$
$$(y-2)^2 - (x+1)^2 = 1$$
Thus, $a = 1$, $b = 1$, $c = \sqrt{2}$.

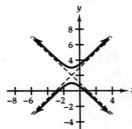

Center: $C(-1, 2)$
Vertices: $A(-1, 2 \pm 1)$ or $A_1(-1, 1)$, $A_2(-1, 3)$
Foci: $F(-1, 2 \pm \sqrt{2})$ or $F_1(-1, 2 - \sqrt{2})$,
$F_2(-1, 2 + \sqrt{2})$
$y_2 = x + 3$

© Houghton Mifflin Company. All rights reserved.

47.
$$x^2 - 16y^2 + 6x - 160y - 407 = 0$$
$$(x^2 + 6x) - 16(y^2 + 10y) = 407$$
$$(x^2 + 6x + 9) - 16(y^2 + 10y + 25) = 407 + 9 - 16 \cdot 25$$
$$(x+3)^2 - 16(y+5)^2 = 16$$
$$\frac{(x+3)^2}{4^2} - \frac{(y+5)^2}{1^2} = 1$$

Thus, $a = 4$, $b = 1$, $c = \sqrt{4^2 + 1^2} = \sqrt{17}$.

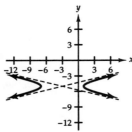

Center: $C(-3, -5)$
Vertices: $A(-3 \pm 4, -5)$ or $A_1(-7, -5)$, $A_2(1, -5)$
Foci: $F(-3 \pm \sqrt{17}, -5)$ or $F_1(-3 - \sqrt{17}, -5)$, $F_2(-3 + \sqrt{17}, -5)$
Asymptotes: $y + 5 = \pm \frac{1}{4}(x+3)$ or $y_1 = -\frac{1}{4}x - \frac{23}{4}$, $y_2 = \frac{1}{4}x - \frac{17}{4}$

49. Because the vertices are at $A_1(0, 2)$ and $A_2(2, 2)$,
$a = \dfrac{2-0}{2} = 1$, and the center is at $C(1, 2)$. Because
the foci are at $F_1(-1, 2)$ and $F_2(3, 2)$,
$c = \dfrac{3-(-1)}{2} = 2$. Therefore
$b^2 = c^2 - a^2 = 4 - 1 = 3$.
Thus the equation is
$$\frac{(x-1)^2}{1^2} - \frac{(y-2)^2}{3} = 1$$
$$(x-1)^2 - \frac{(y-2)^2}{3} = 1$$

51. Because the center is at $C(-3, 1)$ and a vertex is at
$A(-3, 3)$, $a = 3 - 1 = 2$. Because the central
rectangle has width 2, $b = \dfrac{2}{2} = 1$.
Thus the equation is
$$\frac{(y-1)^2}{2^2} - \frac{[x-(-3)]^2}{1^2} = 1$$
$$\frac{(y-1)^2}{4} - (x+3)^2 = 1$$

53. Because the center is at $C(5, 2)$ and a vertex is at
$A(7, 2)$, $a = 7 - 5 = 2$.

Because the slopes of the asymptotes are $\pm \dfrac{1}{2}$,
$$\frac{b}{a} = \frac{1}{2}$$
$$\frac{b}{2} = \frac{1}{2}$$
$$b = 1$$
Thus the equation is
$$\frac{(x-5)^2}{2^2} - \frac{(y-2)^2}{1^2} = 1$$
$$\frac{(x-5)^2}{4} - (y-2)^2 = 1$$

© Houghton Mifflin Company. All rights reserved.